Lecture Notes in Networks and Systems

Volume 1137

AF829499

The series "Lecture Notes in Networks and Systems" publishes the latest developments in Networks and Systems—quickly, informally and with high quality. Original research reported in proceedings and post-proceedings represents the core of LNNS.

Volumes published in LNNS embrace all aspects and subfields of, as well as new challenges in, Networks and Systems.

The series contains proceedings and edited volumes in systems and networks, spanning the areas of Cyber-Physical Systems, Autonomous Systems, Sensor Networks, Control Systems, Energy Systems, Automotive Systems, Biological Systems, Vehicular Networking and Connected Vehicles, Aerospace Systems, Automation, Manufacturing, Smart Grids, Nonlinear Systems, Power Systems, Robotics, Social Systems, Economic Systems and other. Of particular value to both the contributors and the readership are the short publication timeframe and the world-wide distribution and exposure which enable both a wide and rapid dissemination of research output.

The series covers the theory, applications, and perspectives on the state of the art and future developments relevant to systems and networks, decision making, control, complex processes and related areas, as embedded in the fields of interdisciplinary and applied sciences, engineering, computer science, physics, economics, social, and life sciences, as well as the paradigms and methodologies behind them.

Indexed by SCOPUS, EI Compendex, INSPEC, WTI Frankfurt eG, zbMATH, SCImago.

All books published in the series are submitted for consideration in Web of Science.

For proposals from Asia please contact Aninda Bose (aninda.bose@springer.com).

Ferhan Çebi · Ahmet Tezcan Tekin · Adnan Çorum ·
Gizem Deniz Cömert · Gulin Idil S. Bolatan
Editors

Engineering and Technology Management in Challenging Times

Springer

Editors
Ferhan Çebi
Department of Management Engineering
Istanbul Technical University
Istanbul, Türkiye

Ahmet Tezcan Tekin
Department of Management Engineering
Istanbul Technical University
Istanbul, Türkiye

Adnan Çorum
Department of Industrial Engineering
Bahçeşehir University
Istanbul, Türkiye

Gizem Deniz Cömert
Department of Management Engineering
Istanbul Technical University
Istanbul, Türkiye

Gulin Idil S. Bolatan
Department of Industrial Engineering
Alanya Alaaddin Keykubat University
Antalya, Türkiye

ISSN 2367-3370 ISSN 2367-3389 (electronic)
Lecture Notes in Networks and Systems
ISBN 978-3-031-70934-0 ISBN 978-3-031-70935-7 (eBook)
https://doi.org/10.1007/978-3-031-70935-7

This Springer imprint is published by the registered company Springer Nature Switzerland AG
The registered company address is: Gewerbestrasse 11, 6330 Cham, Switzerland

Preface

In today's uncertain environment, engineering and technology management is becoming increasingly crucial. The complexities of the current socio-economic and technical environment pose significant obstacles, requiring sophisticated comprehension and skillful handling of intricacy, unpredictability, and uncertainty. *Engineering and Technology Management in Challenging Times* aims to clarify the various aspects of this concept, providing a scholarly examination of the methods, ideas, and techniques necessary for navigating the turbulent landscapes of the present time.

This volume, tailored to the professional interests of academics, professionals, and influential thinkers from various fields, aims to consolidate and distribute the latest research, empirical findings, and theoretical frameworks related to engineering and technology management in uncertain circumstances. Featuring input from respected academics, professionals, and influential thinkers from various fields, this publication serves as a comprehensive collection of information, skills, and proven methods designed to clarify the difficulties and possibilities in today's environment.

This volume acknowledges the crucial role of engineering and technology management in driving innovation, resilience, and sustainable development. In light of worldwide catastrophes, such as pandemics and climate change, organizations must develop the ability to adapt and have strategic foresight. This book explains the complexities of overcoming uncertainty, focusing on the importance of leadership, organizational culture, and technology innovation in promoting resilience and agility.

Furthermore, this volume surpasses the limits of specific academic disciplines by advocating for an interdisciplinary approach that incorporates knowledge from engineering, management, economics, sociology, and other pertinent topics. Emphasizing technical, social, economic, and environmental interdependence highlights the need for comprehensive and systematic engineering and technology management approaches.

Each chapter in this volume provides a detailed examination of important topics and concepts that are not only academically significant but also highly relevant in practical settings. This collection of contributions provides a thorough and diverse analysis of the challenges and opportunities faced by modern organizations and societies. It covers a wide range of topics, including agile project management methodologies, ethical considerations in technological innovation, risk assessment frameworks, and sustainable business models.

Ultimately, *Engineering and Technology Management in Challenging Times* is evidence of the collaborative effort to understand and tackle the present day's intricacies. We hope the information in these pages will stimulate more investigation, discussion, and action, advancing the conversation on engineering and technology management to new levels of academic inquiry and practical significance.

This book compiles selected and subsequently invited high quality papers from Engineering and Technology Management Summit 2023 (ETMS 2023). After the doubleblind review process, improved versions of 23 papers were included in this book. We would like to thank all the referees who took part in this process and contributed to the selection of highquality papers.

Istanbul, Türkiye Ferhan Çebi
Istanbul, Türkiye Ahmet Tezcan Tekin
Istanbul, Türkiye Adnan Çorum
Istanbul, Türkiye Gizem Deniz Cömert
Antalya, Türkiye Gulin Idil S. Bolatan

Reviewers

Ahmet Ferda Çakmak, Bülent Ecevit University
Akın Menekşe, İstanbul Technical University
Şükrü Alp Baray, İstanbul Cerrahpaşa University
Alper Camcı, Bahçeşehir University
Arzu Karaman Akgül, Yıldız Technical University
Atıl Kurt, Alanya Alaaddin Keykubat University
Aycan Pekpazar, Samsun University
Başar Öztayşi, İstanbul Technical University
Bilge Öztürk, Bayburt University
Birdoğan Baki, Karadeniz Technical University
Büşra Buran, Yıldız Technical University
Ceyda Tanyolaç Bilgiç, İstanbul Cerrahpaşa University
Denizhan Demirkol, Aydın Adnan Menderes University
Dilay Gonidis Çelebi, University of York
Elif Haktanır Aktaş, Bahçeşehir University
Emel Gelmez, Selçuk University
Emrah Ateş, Yalova University
Enis Küçük, Artvin Çoruh University
Eren Kamber, Alanya Alaaddin Keykubat University
Ersin Namlı, İstanbul Cerrahpaşa University
Esra Adıyeke, Bahçeşehir University
Fatih Yiğit, Altınbaş University
Fatma Yaşlı Şen, Eskişehir Technical University
Fuat Kosanoğlu, İstanbul Technical University
Hakkı Okan Yeloğlu, Başkent University
Hasan Kürşat Güleş, Selçuk University
Mehmet Tanyaş, Maltepe University
Merve Cengiz Toklu, Sakarya University
Mete Ömerali, İstanbul Technical University
Ömür Benek, İstanbul University
Ömür Yaşar Saatçioğlu, Dokuz Eylül University

Serhat Çakır, Middle East Technical University
Sıtkı Gözlü, İstanbul Technical University
Tolga Kaya, İstanbul Technical University
Umut Aydın, Bandırma Onyedi Eylül University
Umut Uğurlu, Bahçeşehir University

Contents

A Comparative Analysis of Image Processing Models for Digital Signage Applications

Hakan Bacanak, Enes Öztürk, Emir Alanyalıoğlu, Hakan Erdem, and Basar Oztaysi

Abstract The study investigates the integration of Artificial Intelligence (AI) with digital signage, emphasizing the role of AI-driven image analysis in enhancing its effectiveness. It explores three core areas: demographics analysis, emotion detection, and attention tracking, all crucial for optimizing digital signage impact. Through a comprehensive literature review, we identify emerging trends and advancements in the fusion of image analysis with digital signage. Moreover, we present an overview of available image analysis tools suitable for addressing the challenges of digital signage. To facilitate decision-making, we conduct a formal comparison of these image analysis models, delineating their strengths and weaknesses within the digital signage context. In conclusion, this paper illuminates the potential of AI-powered image analysis in elevating the effectiveness of digital signage. It serves as a practical guide for stakeholders interested in implementing image analysis solutions in this domain and proposes avenues for future research.

Keywords Image processing · Digital signature · Artificial intelligence

1 Introduction

Digital signage refers to the use of digital displays, such as LCD screens, LED boards, projectors, or other electronic devices, to display multimedia content, information, advertisements, or messages in various public or private spaces. These displays can be controlled remotely and updated in real-time, allowing for dynamic and engaging content delivery. Digital signage is commonly found in places like retail stores, airports, hotels, restaurants, corporate offices, educational institutions, and more [1]. Digital signage is important because it provides a versatile and impactful way to

H. Bacanak · E. Öztürk · E. Alanyalıoğlu · H. Erdem
KocSistem Bilgi Ve İletişim Hizmetleri A.Ş, Istanbul, Turkey

B. Oztaysi (✉)
Industrial Engineering Department, Istanbul Technical University, Istanbul, Turkey
e-mail: oztaysib@itu.edu.tr

F. Çebi et al. (eds.), *Engineering and Technology Management in Challenging Times*, Lecture Notes in Networks and Systems 1137,
https://doi.org/10.1007/978-3-031-70935-7_1

communicate information, enhance branding, engage audiences, and drive desired actions, making it a valuable tool for businesses and organizations across various industries.

Artificial Intelligence (AI) can greatly improve digital signage in several ways, enhancing its capabilities, interactivity, and effectiveness. One of the potential developments is to analyze data from cameras and sensors to gather insights about the demographics, behaviors, and preferences of viewers. This data can be used to deliver personalized content, advertisements, and promotions that resonate with specific audience segments. Image analysis which is a subset of computer vision, can be incredibly valuable for audience analytics by analyzing images or video feeds of the audience, AI-powered image analysis techniques can provide valuable insights into demographics, behaviors, emotions, and engagement levels [2].

In the domain of digital signage, image analysis can be used in three main problems. The first one is demographics analysis which mainly focus on the estimation of gender and age of the audience. The second one is emotion detection, by analyzing facial expressions, providing image analysis can provide insights into how viewers are reacting to the content displayed [3]. The last one is attention and engagement tracking which uses image analysis to track eye movement and gaze patterns in order to determine which the engagement of the audience with the displayed content.

The aim of this paper is to provide a literature review on image analysis applications in DS and provide a formal comparison of existing packages which can be used for main image analysis issues of DS. To this end the rest of the paper is as follows. In Sect. 2, a brief literature review on DS related image analysis topics related are provided. In Sect. 3, existing image analysis packages are summarized. In Sect. 4, comparison results of image analysis models are explained. In the final section the conclusions and suggestions on future studies are given.

2 Literature Review

An extended literature provides recent studies on image analysis related with digital signage. One branch of studies focus on gender prediction. In a current study Yildiz et al. [4], the study demonstrated a 93.71% success rate for gender prediction on the VGGFace2 dataset and 85.52% on the Adience dataset, showcasing the potential of AI technologies and deep learning in achieving high performance for tasks like classification and face analysis in image processing and computer vision, ultimately aiming to accurately classify low-resolution images. In another study Hacibeyoglu and Ibrahim [5] focuses on automatic gender classification using CNNs for facial images captured by mobile phones, particularly in the context of online platforms and social media applications. The proposed CNN architecture demonstrates successful gender prediction with 98.87% accuracy on training data and 89.13% accuracy on testing data from the Adience benchmark dataset, emphasizing the significance of network design, activation functions, optimizers, and loss functions in achieving robust image classification performance. In another study Camalan [6] focuses on utilizing biometric data collected from individuals for various applications including transportation, public areas, and security. It develops an image processing-based system using statistical methods, deep learning, and neural networks to predict age range and gender. The system employs RGB and 3D cameras for face and body information, achieving up to 59.1% accuracy for age prediction using CNNs. Integrating 3D body data improves gender and age prediction rates, reaching 99.26% and 99.41% respectively, with separate examination of upper and lower body parts.

An other branch of studies is facial emotion recognition. Facial emotion recognition is the process of detecting human emotions from facial expressions using computer algorithms. There are many works related to facial emotion recognition that have been published in recent years. In a current study, Mehendale [7] propose using FERC model which is based on a two-part convolutional neural network (CNN) that removes the background from the picture and concentrates on the facial feature vector extraction. In this study, the expressional vector (EV) is used to find the five different types of regular facial expression. Supervisory data were obtained from the stored database and the results show that the proposed model provides 96% accuracy. In another study Gautam and Seeja [8] focus on facial emotion recognition using handcrafted features and CNN. In the paper, feature extraction techniques used are HOG and SIFT coupled with the convolutional neural networks. For train and test purposes, datasets entitled CKplus and JAFFE are used. The results show varying accuracy levels between 82 and 98%. Khaireddin and Chen [9] present a comprehensive review of the state-of-the-art performance of facial emotion recognition (FER) on the FER2013 dataset. Different deep learning models and feature extraction methods and provide insights into the challenges and future directions of FER research were compared. The other important papers in the literature are Punuri et al. [10], Chavali et al. [11] and Chowdary et al. [12].

Age prediction is estimating the age of a person from their facial features using machine learning algorithms. In a current study Abirami et al. [13] use Caffenet, a pretrained CNN as a facial feature extractor, for age prediction from real time facial images using CNN. The system gives an accuracy of %63.07 in predicting age correctly. In another study, Akhand et al. [14] focus on age prediction problem by using deep CNNs. Deep CNNs based on transfer learning are the key component of the proposed age estimation method from human facial images. Three datasets namely, UTKFace, CACD, and FGNet are used in the paper and the models ResNet18, ResNet34, ResNet50, Inceptionv3 and DenseNet are compared. In another study, Deng et al. [15] presents a multifeatured learning and fusion network for facial age estimation. They engineered a regression-ranking age-feature estimator to convert the fusion features into the exact age numbers using Age-Net, Race-Net and Gender-Net. They showed experimental results on MORPH2, FG-NET and LAP datasets demonstrated the effectiveness and efficiency of their proposed method on facial age estimation in comparison to previous state-of-the-art methods. The other important papers in the literature are Grupta and Nain [16], Agbo-Ajala et al. [17] and Fang et al. [18].

Another research area is gaze detection which is a technology that involves monitoring and measuring the direction of a person's gaze, or where they are looking. In the context of increasing online exams, the demand for proctors has surged, yet even with individual monitoring, the rate of online examinees resorting to unfair means remains notably higher than offline exams. In a current study Singh et al. [19] to address this, the "Eagle Eye" system was conceived, utilizing AI and Machine Learning for eye gaze detection to enhance monitoring efficiency and combat misconduct in exams. The system's approach involves calibrating the examinee's eye movement and utilizing gathered data to identify fraudulent behavior. In cases of detection, alerts are sent to both examiner and examinee for appropriate actions. This system aims to streamline the process and reduce resource requirements. In another study Sui et al. [20] introduce a novel scanpath prediction method called ScanDMM for 360° images, addressing the time-dependency challenge often overlooked by existing methods. ScanDMM utilizes a Deep Markov Model architecture with a semantics-guided transition function to capture dynamic attentional patterns. The model's state initialization strategy enhances its accuracy by considering viewing starting points. The proposed method outperforms others on four 360° image databases and demonstrates generalizability in applications such as saliency detection and image quality assessment. This approach sheds light on improving these areas by leveraging scanpath prediction models. In another study, Mondal et al. [21] focus on ZeroGaze problem which aims to predict gaze for never-before-searched objects. The authors propose Gazeformer to solve the ZeroGaze problem. The model Gazeformer is compared with other models and the results reveal that proposed Gazeformer model surpasses others by 19–70% in ZeroGaze and standard gaze prediction, and is shown that the model is more than five times faster than the current state-of-the-art solution.

3 Methodology

With the increasing production and sharing of digital image data, the field of image processing has come to the fore. It is an interdisciplinary field in which computers understand, analyze and interpret images. It includes methods with wide application potential from medicine to automotive, security to entertainment.

In the image collection and preparation phase, the acquired images are corrected, resized and, if necessary, subjected to preprocessing such as noise reduction or contrast enhancement. According to [22], "The collection and preparation of images is one of the first and most important stages of the image processing process. At this stage, the acquired images are made suitable for processing. This may include adjusting the resolution, size and color depth of the image, reducing noise and increasing contrast." In the feature extraction phase, features representing important information in the images are extracted. Patterns and structures in the image are revealed using techniques such as edge detection, corner detection, and color histograms. According to [21], "Feature extraction is an important step in image processing. In this stage, features are extracted that represent important information in the image. These features can be used to identify objects in the image, classify them, or perform other specific tasks." In dimension reduction and selection, large and complex feature vectors are often replaced with less dimensional and more meaningful representations. This can help the model learn better while reducing the computational burden.

According to [23], "Dimension reduction and selection is an important step in image processing. In this stage, large and complex feature vectors are often replaced with less dimensional and more meaningful representations. This can help the model learn better while reducing the computational burden." In the learning and classification phase, predefined models are trained using the data. By analyzing the dataset, the model gains the ability to recognize objects, classify them, or perform other specific tasks. According to [22], "Learning and classification is the final stage of image processing. In this stage, predefined models are trained using the data. By analyzing the dataset, the model gains the ability to recognize objects, classify them, or perform other specific tasks." According to [21], "Image processing is used in many different fields. Examples include medical imaging, automotive industry, security, entertainment, manufacturing, production, agriculture, education, and more." According to [23], "Deep learning has revolutionized the field of image processing. Deep neural networks have the capacity to learn complex features and relationships thanks to their layered structure. This has opened up new possibilities for processing more complex tasks and large datasets in image processing."

In conclusion, image processing is an exciting field that enables the conversion of digital image data into meaningful information. With the development of technology, the application areas and methods of image processing are expanding and evolving. Studies in this field enable the development of more intelligent and effective systems in many sectors.

Current studies provide various image/video analysis tools and models for demographics, emotion and attention detection problems. In this study we focus on most commonly used ones.

VGG19: VGG19 is a deep convolutional neural network architecture that was introduced as part of the Visual Geometry Group (VGG) at the University of Oxford. It is a variant of the VGG network and is known for its simplicity and effectiveness in image classification tasks. The VGG19 model was introduced in 2014 and achieved significant success in the ImageNet Large Scale Visual Recognition Challenge (ILSVRC) competition. VGG19 has a depth-rich structure, containing a total of 19 layers (16 convolution layers and 3 fully connected layers). The model has a structure that combines simple 3×3 convolution and 2×2 pooling layers. This simple structure emphasizes learning using more layers rather than more complex and deep networks. VGG19 is often presented as a pre-trained network and can be used in visual tasks such as image classification, object recognition, and so on. VGG19 model can be implemented using popular deep learning libraries such as PyTorch, TensorFlow.

InceptionV3: InceptionV3 is a deep learning model developed by Google and is a member of the Convolutional Neural Network (CNN) family. It was introduced in 2015 and has demonstrated outstanding performance in visual classification competitions such as the ImageNet Large Scale Visual Recognition Challenge (ILSVRC). InceptionV3 is named after its "Inception" modules, where complex and diverse learning layers work in parallel. These modules aim to optimize learning by combining convolution filters of different sizes. This allows the model to learn faster while at the same time achieving lower computational costs. The architecture of the model offers a lightweight and efficient structure compared to both deeper and wider networks. InceptionV3 can be used in a variety of visual tasks, especially in areas such as image classification, object recognition and transfer learning. InceptionV3 can be implemented using popular deep learning libraries such as TensorFlow and Keras.

Convolutional Neural Networks (CNN): CNN are deep learning models specifically designed for visual data analysis and processing. Such networks have achieved great success in visual tasks such as image classification, object recognition, face recognition, and segmentation. The main reason for the success of CNNs is their ability to learn the features of data hierarchically. CNNs consist of convolution and pooling layers to learn the feature maps of the input data. Convolution layers scan using customized filters to detect specific features of the image (e.g., edges, corners, color transitions). Pooling layers reduce the size of feature maps, increasing learning speed and reducing computational burden. CNNs can also contain fully connected layers, which produce the final outputs using the learned features. An important feature is that CNNs can also be used as pre-trained networks. This often makes it easier to achieve better results with less data. Popular deep learning frameworks such as TensorFlow, PyTorch and Keras can be used to implement CNNs.

FER: FER is an open source library for Python. It is capable of face recognition and emotion detection. Fer bundles a model of Keras which is an open-source high-level neural networks API written in Python that allows developers to define and train neural networks with ease. The model includes a convolutional neural network (CNN) architecture with saved weights to a HDF5 file. FER allows to selection of face detector between OpenCV Haar Cascade classifier and MTCNN.

DeepFace: DeepFace is an open source library for Python that includes leading AI model approaches for face recognition. It is a hybrid face recognition framework wrapping state-of-the-art models such as VGG-Face, Google FaceNet, OpenFace, Facebook DeepFace, DeepID, ArcFace, Dlib and SFace. This combined approach allows the selection of different models while trying different tasks such as age, gender, emotion, and race analysis in addition to the face recognition. Moreover, the library allows the selection of different face detectors such as OpenCV (Cascade Classifier), SSD, Dlib, MTCNN, RetinaFace, MediaPipe, YOLOv8 Face and YuNet. Being able to select different models and different face detectors in a single library such as DeepFace, makes the comparisons substantially easier.

4 Comparison of Image Processing Models

In this section we focus on the main digital signage related image analysis problems namely Facial emotion recognition, gender prediction, age prediction. In order to provide a starting reference for further studies, existing packages are used with various existing datasets and the results are compared.

The models incorporated in the study involve VGG19, Inception-v3, deepface (with OpenCV, MTCNN and MEdiaPipe), FER and CNN. In order to provide a comprehensive comparison, possible datasets, FER2013, CK+, CelebA, UTKFaces, and faces are used for comparison purposes.

Table 1 shows the comparison results related with the facial emotion recognition problem. It is observed that the highest accuracy is maintained with VGG19 model in FER2013 dataset, and CNN model maintain the highest accuracy with CNN model.

Table 2 shows the comparison results regarding the gender prediction and age prediction problems. In gender prediction problem, CelebA data set is used and deepface model which use MTCNN provide the highest accuracy level. In the age prediction problem the highest accuracy is provided by CNN model in Faces database.

Table 1 Comparison results regarding the facial emotion recognition problem

Model	Dataset	Loss	Accuracy	Val_loss	Val_acc	Epoch	Label
VGG19	FER2013	0.6269	0.7715	0.9455	0.6821	25	Facial
Inception-v3	FER2013	0.9099	0.6660	1.6050	0.4773	50	Facial
deepface (OpenCV)	FER2013	–	0.56	–	–	–	Facial
deepface (MTCNN)	FER2013	–	0.54	–	–	–	Facial
deepface (MediaPipe)	FER2013	–	0.49	–	–	–	Facial
fer	FER2013	–	0.48	–	–	–	Facial
CNN	CK+	0.4643	0.7803	0.9478	0.6622	50	Facial
deepface (OpenCV)	CK+	–	0.55	–	–	–	Facial
deepface (MTCNN)	CK+	–	0.52	–	–	–	Facial
deepface (MediaPipe)	CK+	–	0.53	–	–	–	Facial
fer	CK+	–	0.51	–	–	–	Facial

Table 2 Comparison results regarding the gender and age prediction problem

Model	Dataset	Loss	Accuracy	Val_loss	Val_acc	Epoch	Label
Inception-v3	CelebA	0.1995	0.9304	0.2269	0.9260	10	Gender
deepface (OpenCV)	CelebA	–	0.94	–	–	–	Gender
deepface (MTCNN)	CelebA	–	0.95	-	–	–	Gender
deepface (MediaPipe)	CelebA	–	0.87	–	–	–	Gender
CNN	UTKFace	0.2866	0.8990	0.2550	0.8880	25	Age-Gender
CNN	Faces	0.2374	0.9067	0.4470	0.8421	10	Age

5 Conclusion

Digital signage has an increasing trend as it provides efficient communication, improve customer experience, and cost efficiency. Using image processing methods, digital signage can be used to provide improved services. To this end, facial emotion recognition, age prediction, and gender prediction problem are among the leading problems. In digital signage domain, effective solution of these problems have a potential to improve service quality, create audience oriented content sharing and content optimization.

In this study, we provide information about already existing packages which can be used for above mentioned problem and related datasets. We also provide comparison results of different models with existing databases. The results show that VGG19 and CNN are the best packages for facial emotion recognition while deepface is the best package for gender prediction. The results also show that for digital signage oriented applications additional models are required to reach desired accuracy level.

In the future studies, a new dataset can be created focusing on Digital Signage problem and the results of the packages can be compared with each other.

References

1. Garaus M, Wagner U, Rainer RC (2021) Emotional targeting using digital signage systems and facial recognition at the point-of-sale. J Bus Res 131:747–762
2. Merad D, Drap P, Lufimpu-Luviya Y, Iguernaissi R, Fertil B (2016) Purchase behavior analysis through gaze and gesture observation. Pattern Recogn Lett 81:21–29
3. Sturari M, Liciotti D, Pierdicca R, Frontoni E, Mancini A, Contigiani M, Zingaretti P, Robust and affordable retail customer profiling by vision and radio beacon sensor fusion. Pattern Recogn Lett 81:30–40
4. Yildiz K, Gunes E, Bas A (2021) CNN-based gender prediction in uncontrolled environments. Düzce Univ J Sci Technol 9(2021):890–898
5. Hacibeyoglu M, Ibrahim MH (2018) Human gender prediction on facial images taken by mobile phone using convolutional neural networks. Int J Intell Syst Appl Eng. ISSN 2147-6799
6. Camalan S (2018) Age and gender prediction from 3d-body and face images. In: Software engineering, Doctor of Philosophy Thesis, Atılım University
7. Mehendale N (2020) Facial emotion recognition using convolutional neural networks (FERC). SN Appl Sci 2:446
8. Gautam C, Seeja KR (2023) Facial emotion recognition using handcrafted features and CNN. Proc Comput Sci 218(2023):1295–1303
9. Khaireddin Y, Chen Z (2021) Facial emotion recognition: state of the art performance on FER2013. Comput Vis Pattern Recogn. https://doi.org/10.48550/arXiv.2105.03588
10. Punuri SB, Kuanar SK, Kolhar M, Mishra TK, Alameen A, Mohapatra H, Mishra SR (2023) Efficient net-XGBoost: an implementation for facial emotion recognition using transfer learning. Mathematics 11(3):776
11. Chavali ST, Kandavalli CT, Sugash TM, Subramani R (2023) Smart facial emotion recognition with gender and age factor estimation. Proc Comput Sci 218:113–123
12. Chowdary MK, Nguyen TN, Hemanth DJ (2021) Deep learning-based facial emotion recognition for human–computer interaction applications. Neural Comput Appl 1–18
13. Abirami B, Subashini TS, Mahavaishnavi V (2020) Gender and age prediction from real time facial images using CNN. Mater Today Proc 33:4708–4712

14. Akhand MAH, Ijaj Sayim M, Roy S, Siddique N (2020) Human age prediction from facial image using transfer learning in deep convolutional neural networks. In Proceedings of international joint conference on computational intelligence: IJCCI 2019. Springer Singapore, pp 217–229
15. Deng Y, Teng S, Fei L, Zhang W, Rida I (2021) A multifeature learning and fusion network for facial age estimation. Sensors 21(13):4597
16. Gupta SK, Nain N (2023) Single attribute and multi attribute facial gender and age estimation. Multim Tools Appl 82(1):1289–1311
17. Agbo-Ajala O, Viriri S, Oloko-Oba M, Ekundayo O, Heymann R (2022) Apparent age prediction from faces: a survey of modern approaches. Front Big Data 5:1025806
18. Fang J, Yuan Y, Lu X, Feng Y (2019) Muti-stage learning for gender and age prediction. Neurocomputing 334:114–124
19. Singh J, Aggarwal R, Tiwari S, Joshi V (2022) Exam proctoring classification using eye gaze detection. IEEE. https://doi.org/10.1109/ICOSEC54921.2022.9951987
20. Sui X, Fang Y, Zhu H, Wang S, Wang Z (2023) ScanDMM: a deep Markov model of scanpath prediction for 360° images. IEEE. https://doi.org/10.1109/CVPR52729.2023.00675
21. Jain AK, Flynn PJ, Dasarathy B (2016) Image processing and analysis. CRC Press
22. Demirhan M (2023) Image processing. Beta Publications
23. Krizhevsky A, Sutskever I, Hinton GE (2017) Deep learning for computer vision. In: IEEE signal processing magazine

A Literature Review of Data Analytics: Tabular and Graphical Analysis

Beyzanur Cayir Ervural

Abstract Data analytics is a hot topic for most disciplines and companies due to its many advantages. Data analytics applications have an effective practice area and an effective instrument for system improvements, resource optimization, and problem solving. In this study, studies on data analytics applications in the literature are analyzed, and statistically qualitative/quantitative changes by application area, country and years are presented in tables. The findings show the importance of the subject and its widespread and effective performance in recent years. This study may provide insights into new application areas for practitioners and researchers working in this field.

Keywords Data analytics · Operations management · Literature reviews

1 Introduction

Data analytics is a branch that involves analyzing datasets to gain insights that will help solve problems in various industries. The purpose of data analytics is to identify, predict, or improve organizational success. It utilizes various disciplines such as computer programing, statistics, and mathematics to provide accurate data analysis. Thanks to data analytics applications in enterprises, resources will be employed effectively and profitability will increase by optimizing the system.

Data analytics approaches are effective problem solving tools that are widely used in different fields [1]. It has a wide application area, particularly in operations research, operations management, energy planning, revenue management, decision making, and optimization. In terms of the used methods, it helps to solve many real world problems such as supply chain and logistics, disaster management, health systems, energy planning, defense industry, and transportation systems.

B. C. Ervural (✉)
Department of Aviation Management, Necmettin Erbakan University, Konya, Turkey
e-mail: bc.ervural@erbakan.edu.tr

F. Çebi et al. (eds.), *Engineering and Technology Management in Challenging Times*, Lecture Notes in Networks and Systems 1137,
https://doi.org/10.1007/978-3-031-70935-7_2

Data analytics applications, also known as business analytics today, are built on three basic pillars. The first domain is basic statistics, the second is database management, and the third is machine learning (ML) and artificial intelligence techniques. As it is seen obviously, data analytics is the intersection of these three basic disciplines. Today, with the increase in system structures based on Industry 4.0, artificial intelligence, and machine learning/deep learning approaches, digitalization has been achieved to a great extent. Using a variety of data analytics approaches, meaningful information is transformed to help businesses achieve their goals. Thus, most companies reach their economic growth targets by increasing profitability and optimizing the efficiency level of the system with the adaptation of data analytics.

Businesses that apply data analytics approaches can compete with an advantageous position in every aspect. Thanks to accurate prediction management based on data analytics methods that work integrated with machine learning and statistical approaches, businesses make very successful demand forecasts and meet customer demands with a low margin of error.

Employing data analytics, enterprises can control exactly what customers are looking for. By analyzing customer data in detail and comprehensively with data analytics approaches, accurate situation determination can be made through customer target group and market research.

Data analytics can provide companies to specify to optimize systems in different ways to grow revenue channels. It aims to detect and eliminate risky/dangerous situations before they become a problem. Using data analytics enables rapid responses to real-time threat detection methods, automated alerts and mitigations. Thus, data fraud can be prevented. One of the main benefits of data analytics is its ability to improve decision-making processes. It supports the improvement of inefficient business processes to increase the efficiency of operations and save time and resources.

Studies on data analytics applications were examined and statistical analyses were performed. The results obtained reveal the prominence of the subject and demonstrate that interest is increasing because the subject has an interdisciplinary field of application. When the keywords data analytics, machine learning, and artificial intelligence are searched in the Web of Science (WoS) database between 2008 and 2023, the obtained graphs show that interest in the subject has increased dramatically in recent years.

2 Data Analytics Techniques

Here, first, the most basic data analytics methods are explained, and then some prominent studies in the literature are mentioned.

2.1 Statistics

Statistics is the most widely used science in many basic fields, and the first thing to be done before many applications is statistical analysis. It is the branch of science that provides the most comprehensive information about the data, and many preliminary examinations and researches such as normal distribution fitness test, skewness, analysis of variance, histograms, sampling, regression, and correlation analyses that examine the relationship between variables can be implemented [2]. Multivariate analysis is an effective tool widely used for business analytics applications because of its fast working principles and satisfactory results. Although data analytics is at the forefront of data analytics applications, it is not sufficient for big data analytics to meet other requirements under heterogeneous information/unstructured datasets [3].

2.2 Machine Learning

Machine learning approaches are artificial intelligence-based algorithms and are solution approaches that gradually increase the accuracy level by imitating the way people learn using high-dimensional data and algorithmic structures and provide successful results.

Machine learning applies many procedures to build mathematical models and generate predictions using historical data or information. Machine learning can be divided into three types on a broad level: supervised learning, unsupervised learning, and reinforcement learning. Support vector machines, deep learning, extreme learning, random forest, artificial neural networks, decision trees are all well-known machine learning techniques for dealing with complicated real-world issues [4].

2.3 Data Mining

Data mining is the algorithmic process of extracting insights from a given dataset [5]. Data mining techniques help search large volumes of raw data to identify meaningful information and useful models. Data mining models are generally developed on the basis of the fields of machine learning and statistics [6]. Thanks to effectively used data mining methods, businesses can develop successful marketing strategies, increase their sales, and reduce their costs in order to survive in market conditions where competition is getting harder. Data mining is an indispensable tool for many disciplines, especially for data analytics and business intelligence applications. It has several application areas such as classification, clustering, regression, and market segmentation [7]. Some challenges associated with data mining problems include the control of massive data, multidimensional and multi-source data structures, and the ever-changing data network.

2.4 *Optimization*

Optimization is a quantitative decision-making method used to find the best possible solution by maximizing or minimizing the objective function given under certain constraints. The major disadvantages of solving problems are memory problems and computational time [8]. Therefore, some metaheuristic optimization algorithms are used to solve complex real-life problems such as genetic algorithms, tabu search, particle swarm, and simulated annealing etc.

Optimization methods are essential for businesses to efficiently manage all of their resources by maximizing system productivity and reducing losses [9]. In this way, businesses reach a definitive solution.

Research shows that data analytics applications have several applications. Data analytics applications can be used in all kinds of problems and optimization studies that may reduce productivity that may be encountered in the business. For example, by examining the factors that cause time loss and labor loss in the enterprise, the data are first analyzed with statistical tools and evaluated graphically and tabularly.

Then, if the problem is an improvement study for cost minimization, a model is created accordingly and solved under the specified constraints. For example, if it is desired to create the correct demand forecasting model for the next 5 years, algorithms are created with the lowest margin of error using machine learning and classical forecasting models. According to the result, the decision of which raw material to buy and which supplier to sign a new agreement with is determined. Accordingly, the investment decision to be made becomes clear.

By analyzing large amounts of data, manufacturers can detect configurations, predict equipment failures, optimize maintenance schedules, and modernize production processes. This allows proactive management decisions, reduces interruption/ disruption, and increases overall operational efficiency. Since optimization supports maximizing the benefits and minimizing the losses and undesirable situations in the system, it is clear from this that the improvement of all processes in general is possible with correctly applied data analytics approaches.

Research in operations management has traditionally focused on models for understanding how companies should operate, mostly at the strategic level. Data Analysis and Operations Research are two intersecting disciplines because, for example, there are many data problems in which Operations Research optimization techniques need to be applied to identify optimal structures in the underlying data.

Today's data-driven systems require the application of operations research using data analytics methods due to the increasing ease of data collection in various fields such as healthcare, supply chain mechanism, revenue management, pricing, maintenance management and scheduling.

The field of machine learning, which lies at the intersection of computer science and statistics, has established approaches for obtaining high-quality predictive models for complex high-dimensional data structures.

3 Literature Review

According to literature reviews, many studies have discussed applications of data analytics to analyze real-world problems such as inventory management, production planning, marketing strategies, forecasting, agriculture planning, transportation optimization, prediction of consumer behavior and pricing policies. In order to understand the customer's purchasing decision, a machine learning algorithm attempts to capture the customer's behavior in a more cost-effective way [10–12].

To examine customer attitudes in shopping centers, the interaction of product type and number and shopping movement are examined, purchasing frequency analysis, and promotional marketing activities on products are evaluated [13]. Some researchers have investigated appropriate electricity price forecasting models using successful ML models [14, 15].

For marketing segmentation activities, researchers evaluate customer behaviours, preferences, needs, physiological/cognitive factors, and other indicators that affect sales policy, such as service time estimation, customer arrival forecasting, and labor planning processes [16, 17].

To analyse crop management, precision of agriculture techniques in harvest supply and improved field efficiency subjects can be used with ML, and decision support systems. ML methods have been applied to increase productivity to estimate soil moisture and nutrient content [18, 19]. The ML algorithms utilized for growing effective irrigation management systems are based on simulation and optimization techniques [20]. ML in irrigation management is used to estimate the evapotranspiration process [21, 22].

Deep neural networks [23] and data mining techniques such as k-mean clustering, k-nearest neighbour, ANN, and SVM [24] are utilized for precise yield prediction. Deep neural networks [23] and data mining techniques such as k-mean clustering, k-nearest neighbour, ANN, and SVM [24] are employed for accurate yield forecasting. The implementation of big data analytics in production planning supports demand forecasting and production planning [25]. ML algorithms help inefficient production planning by reducing setup time and superior demand sensing [26, 27]. ML algorithms are applied for estimating freight [28], inventory management and payment delays [29] estimating product shelf life, dynamic allocation [30, 31], and predicting supply chain risks in fuzzy environment [32]. ML also finds application in developing local food supply chains that provide food security and sustainability in logistics systems [33].

Some recently published studies have been considered in the study as follows: Voorneveld and Groot [34] explored whether spending more on data analytics can help startups thrive and what data analytics capabilities a company should develop to capitalize on the returns from its share of data analytics. The Markov decision process approaches are employed to model progress and spending and to determine the optimal policy. This methodology investigates whether increasing data analytics capabilities in startups can increase their progress. A summary of options for the gaining of data analytics abilities is offered. Willetts and Atkins [35] showed that SMEs' industry positioning tool is useful for gaining competitive advantage by increasing their use of business intelligence and big data analytics. Liu et al. [36] constructed a theoretical model focusing on the mediating role of person-instrument fit and examined the mechanisms by which big data analytics intellectual capability (BDAIC) affects a firm's operational performance. This study applied the analytical approach of structural equation modeling to investigate the impact of BDAIC on an enterprise's operational performance. Naseer et al. [37] discussed that organizations should allow agility in their incident response to quickly respond to various cybersecurity threats, and big data analytics play a key role in supporting agility in the incident response. The proposed framework clarified how big data analytics embedded dynamic proficiencies at the pervasive analytics stage support agility in IR by pervading agile features of flexibility, speed, and learning in incident response. Ma et al. [38] investigated the impact of complex traffic situations on maritime safety, focusing on inland waterway traffic. It presents a big data analytics methodology using data from the Automatic Identification System and historical maritime accident archives. A significant relationship was obtained between the positions of maritime accidents and zones with enlarged ship trajectory density and average speed. These results may provide perceptions for traffic safety management and implies strategies for avoiding maritime accidents. Liu and Yoon [39] launched a theoretical agenda to guide future studies in constructing smart, goal-driven data combinations. Automated data integration solutions need to robustly and flawlessly handle high volumes of different data for various analytical needs or operational arrangements. The proposed original project automates data integration for non-technical data users. Zhou et al. [40] offered a broad multi-stage multi-criteria data analytics method that integrates data envelopment analysis, the best–worst method, and the multi-attributive border approximation area comparison method to offer a robust ranking basis. The proposed method position airports from the best to the worst and assesses the strengths and weaknesses of each of the candidate airports.

As can be seen from the literature research, data analytics methods have been used for many real field problems using statistics, machine learning, and optimization tools as sub-branches of data analytics.

Studies on data analytics applications were reviewed, and statistical analyses are shared in this section. The results obtained reveal the importance of this subject, and show that interest shows an increasing trend since it has an inter-disciplinary field of application.

When the Web of Science (WoS) database is searched for data analytics, machine learning, and artificial intelligence keywords between 2008 and 2023, Fig. 1 shows that interest in the subject has increased in recent years. It is clearly seen in the graphs below that we must conduct studies on this subject as a country. It is seen that especially developed countries have a large share of data analytics applications.

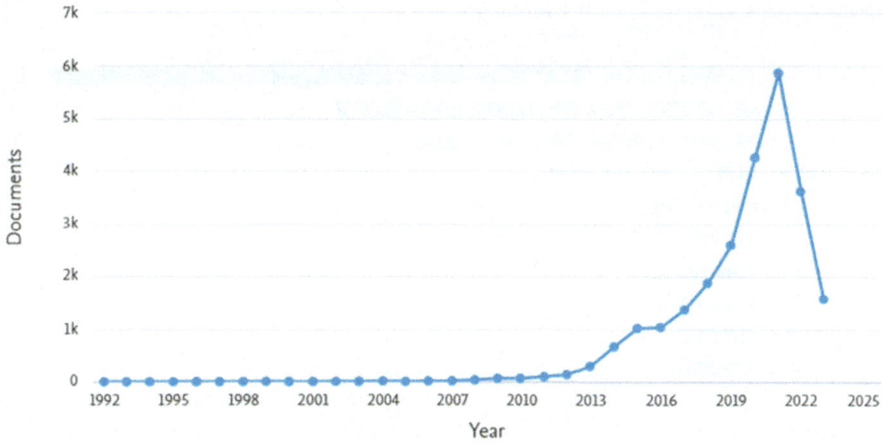

Fig. 1 Distribution of studies based on years

As in every sector, with the outbreak of the pandemic, a decreasing trend was observed in the research and studies conducted on this subject. When the graph is examined, a decrease was observed after 2019 because of the impact of the pandemic. Even though there has been a rapid recovery, the impact continues as scientists are still negatively affected by the negative health, economic, and social effects of the epidemic.

When the publications are evaluated according to the countries, the ranking of Turkey can be easily seen. In Fig. 2, it is seen that most of the studies are conducted by America, China, India, The UK, Australia, Canada, and Germany, respectively. Compared with other countries, Turkey's ranking is slightly behind in this regard. Therefore, it can be said that as a country, we need more research and projects on this subject.

When the distribution of the methods used according to the application areas is examined, it is seen in Fig. 3 that they are mostly used in computer sciences (25.1%), followed by engineering (18.7%), business management (6.4%) and social sciences (6.1%), mathematics (5.4%), medicine (4.5%), decision sciences (4.0%), environmental sciences (3.5%), energy (3.2%) and other disciplines (19.2%).

The distribution of studies in the WoS database according to publications such as books, articles, and book chapters is shown in Fig. 4. When the publications are examined, article studies have the largest share.

Delen, D. (with 34 articles), Kumar, N. (with 31 articles), Chang, V. (with 29 articles), Gunasekaran, A. (with 29 articles), Bibri, S.E, (with 27 articles), Sarker, I.H. (with 27 articles), and Akter, S. (with 25 articles) are the most productive researchers in this area. The frequencies of data analytics application articles by the authors are shown in Fig. 5. Particularly, researchers from the U.S.A, China, India, U.K, Australia, and Canada have been studying this field.

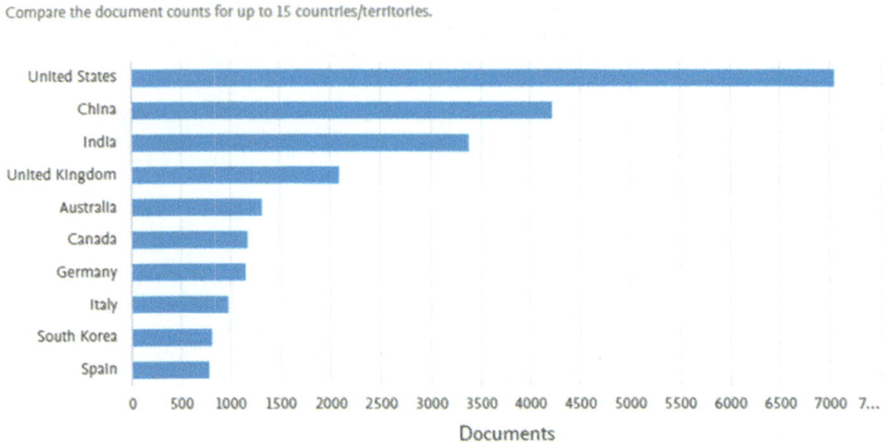

Fig. 2 Distribution of studies by country

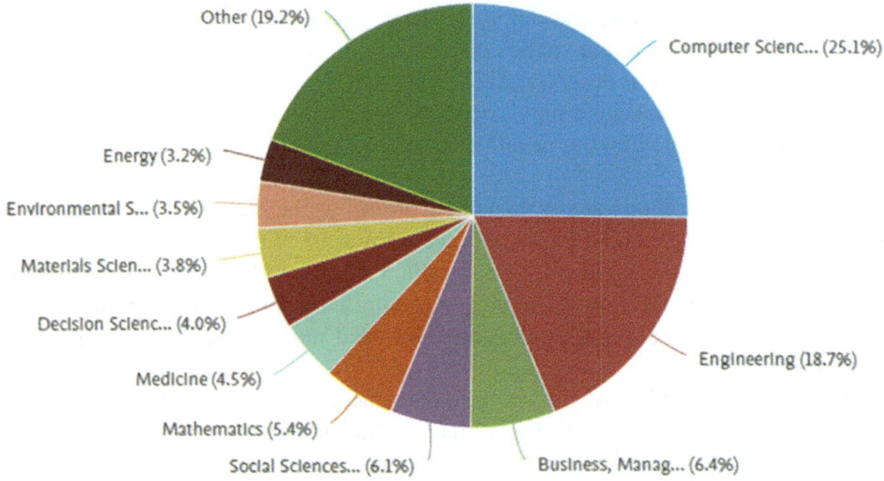

Fig. 3 Distribution of studies by subject area

4 Future of Data Analytics Methods

It is certain that there will always be a need for big data analytics applications with industry 4.0 implementations in the digitalization and big data eras. Considering the studies/researches on data analytics, it can be predicted that these tools will be employed in many ways in the future, as they have a very common and effective area of use due to their successful and practical results. It solves the most basic problems of many enterprises by evaluating it from a multidimensional perspective

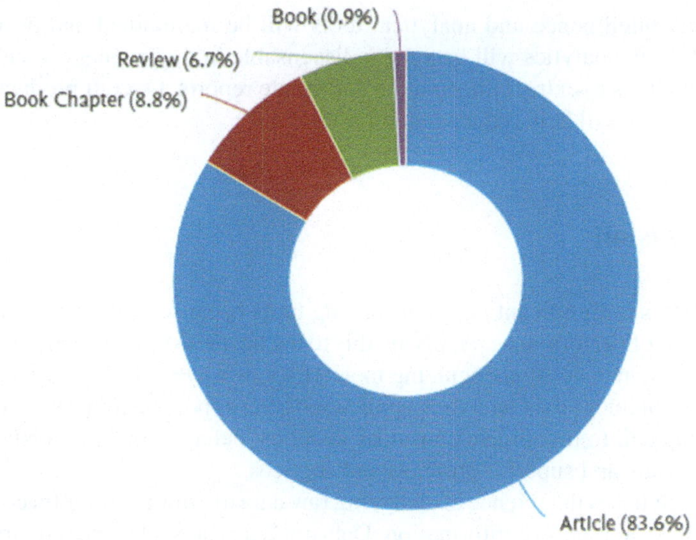

Fig. 4 Distribution of studies by document type

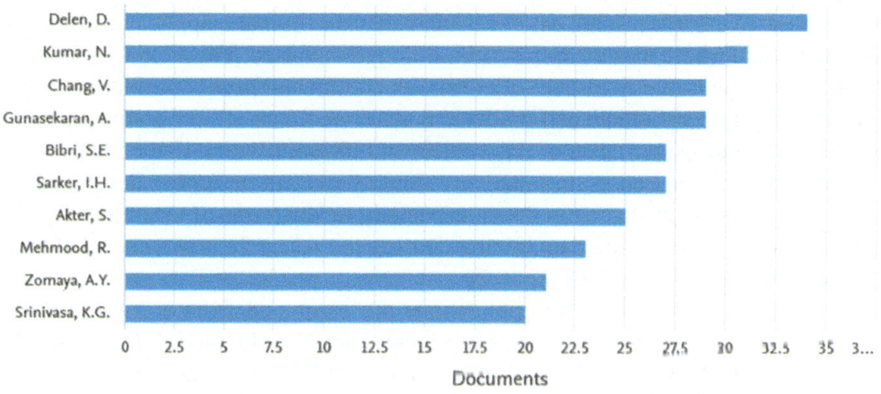

Fig. 5 Documents by the authors

and because of its flexible solution possibilities, it is indispensable for such field problems.

When the application field is investigated from literature reviews, most of the operations management issues are closely related to the overall optimization of the system flow and constrained resources; therefore, the topics are supported by data analytics. For instance, forecasting, inventory management, revenue management and marketing, transportation, supply chain management, risk analysis, healthcare systems, and workforce-shift problems.

Business intelligence and analytical tools will be maintained and developed in the future. Data analytics will become indispensable in the business world and will allow individuals to extract information and create reports. Overall productivity will increase, which will help reduce human limitations.

5 Conclusion

Data analytics is significant decision-making tools because of the development and optimization of performance. Applying this to the business model means that companies can help cut costs by recognizing more efficient ways of doing business. A firm may also implement data analytics applications to support better professional decision making and to investigate consumer fashions, behaviours, and satisfaction that can lead to new and superior products and services.

Data analytics is the science of analyzing raw data to draw meaningful conclusions about massive amounts of information. Data analytics aids a business to optimize its performance, operate more efficiently, maximize revenues, or make more strategic decisions. There are various methods for data analytics, such as descriptive analytics, diagnostic analytics, predictive analytics, and prescriptive analytics.

Because of its successful, practical, and interdisciplinary application, there is almost no field where data analytics methodology is not used. This field will continue to be an indispensable field due to its serious and meaningful contributions to the sector. Many undergraduate and graduate courses and master's programs are being opened due to the importance of this subject and its contribution to the industry.

The studies discussed in this study are limited and can be further developed for future research and evaluated in detail, especially in key terms of machine learning, optimization, statistics, and data mining. During the research, it was seen that there is a need for data-driven studies.

In this study, data analytics applications are examined from a wide range of perspectives. Literature research was conducted, and the obtained statistical results were shared with the readers. The importance of the subject and the interest of researchers in this subject show that these tools can solve many more complicated problems. In this study, we have explained several big data-related analytics methods. In fact, we have underlined the significance of statistics, optimization models, data mining, and machine learning for supporting big data analytics. We consider these discoveries valuable to scholars and practitioners concerned with how big data analytics can be utilized in real-life problems.

The research provided a comprehensive review of recently developed data analytics methods. For further research, scholars advise comparing the performances of big data analytics approaches against existing problem sets in the literature in terms of solution quality, application area, and execution time.

References

1. Sarker IH (2021) Machine learning: algorithms, real-world applications and research directions. SN Comput Sci 2(3):1–21. https://doi.org/10.1007/S42979-021-00592-X/FIGURES/11
2. Huang WA, Chaovalitwongse S (2015) Computational optimization and statistical methods for big data analytics: applications in neuroimaging. INFORMS Tutorial Oper Res 2015:71–88
3. Chen CLP, Zhang CY, Data-intensive applications, challenges, techniques and technologies: a survey on big data. Inf Sci 275:314–347
4. Sun F, Huang G, Wu QMJ, Song S, Wunsch II DC, Efficient and rapid machine learning algorithms for big data and dynamic varying systems. IEEE Trans Syst Man Cybern Syst 47(10):2625–2626
5. Choi TM, Chan HK, Yue X (2017a) Recent development in big data analytics for business operations and risk management. IEEE Trans Cybern 47(1):81–92
6. Choi TM, Gao J, Lambert JH, Ng CK, Wang J (2017b) Optimization and control for systems in the big-data era: theory and applications. Springer, New York
7. Fan S, Li X, Zhao JL (2017) Collaboration process pattern approach to improving teamwork performance: a data mining based methodology. INFORMS J Comput 29(3):438–456
8. Potvin JY (2009) Evolutionary algorithms for vehicle routing. INFORMS J Comput 21(4):518–548
9. Huang S, Chaovalitwongse WA (2015) Computational optimization and statistical methods for big data analytics: applications in neuroimaging. INFORMS Tutorial Oper Res 71–88
10. Peker S, Kocyigit A, Eren PE (2017) A hybrid approach for predicting customers' individual purchase behavior. Kybernetes 46(10):1614–1631. https://doi.org/10.1108/K-05-2017-0164/FULL/PDF
11. Martínez A, Schmuck C, Pereverzyev S, Pirker C, Haltmeier M (2020) A machine learning framework for customer purchase prediction in the non-contractual setting. Eur J Oper Res 281(3):588–596. https://doi.org/10.1016/J.EJOR.2018.04.034
12. Arasu BS, Seelan BJB, Thamaraiselvan N (2020) A machine learning-based approach to enhancing social media marketing. Comput Electr Eng 86:106723. https://doi.org/10.1016/J.COMPELECENG.2020.106723
13. Paolanti M, Pietrini R, Mancini A, Frontoni E, Zingaretti P (2020) Deep understanding of shopper behaviours and interactions using RGB-D vision. Mach Vis Appl 31(7–8). https://doi.org/10.1007/S00138-020-01118-W
14. Yang W, Sun S, Hao Y, Wang S (2022) A novel machine learning-based electricity price forecasting model based on optimal model selection strategy. Energy 238:121989. https://doi.org/10.1016/J.ENERGY.2021.121989
15. Sueyoshi T, Tadiparthi GR (2005) A wholesale power trading simulator with learning capabilities. IEEE Trans Power Syst 20(3):1330–1340. https://doi.org/10.1109/TPWRS.2005.851948
16. Jagabathula S, Subramanian L, Venkataraman A (2017) A model-based embedding technique for segmenting customers. SSRN Electron J. https://doi.org/10.2139/SSRN.2696161
17. Smirnov D, Huchzermeier A (2020) Analytics for labor planning in systems with load-dependent service times. Eur J Oper Res 287(2):668–681. https://doi.org/10.1016/J.EJOR.2020.04.036
18. Morellos A et al (2016) Machine learning based prediction of soil total nitrogen, organic carbon and moisture content by using VIS-NIR spectroscopy. Biosyst Eng 152:104–116. https://doi.org/10.1016/J.BIOSYSTEMSENG.2016.04.018
19. Prasad R, Deo RC, Li Y, Maraseni T (2018) Soil moisture forecasting by a hybrid machine learning technique: ELM integrated with ensemble empirical mode decomposition. Geoderma 330:136–161. https://doi.org/10.1016/J.GEODERMA.2018.05.035
20. Safavi HR, Esmikhani M (2013) Conjunctive use of surface water and groundwater: application of support vector machines (SVMs) and genetic algorithms. Water Resour Manag 27(7):2623–2644. https://doi.org/10.1007/S11269-013-0307-2/FIGURES/12

21. Saggi MK, Jain S (2019) Reference evapotranspiration estimation and modeling of the Punjab Northern India using deep learning. Comput Electron Agric 156:387–398. https://doi.org/10.1016/J.COMPAG.2018.11.031
22. Torres AF, Walker WR, McKee M (2011) Forecasting daily potential evapotranspiration using machine learning and limited climatic data. Agric Water Manag 98(4):553–562. https://doi.org/10.1016/J.AGWAT.2010.10.012
23. Mehra M, Saxena S, Sankaranarayanan S, Tom RJ, Veeramanikandan M (2018) IoT based hydroponics system using Deep Neural Networks. Comput Electron Agric 155:473–486. https://doi.org/10.1016/J.COMPAG.2018.10.015
24. Mucherino A, Papajorgji P, Pardalos PM (2009) A survey of data mining techniques applied to agriculture. Oper Res 9(2):121–140. https://doi.org/10.1007/S12351-009-0054-6/TABLES/1
25. Feng Q, Shanthikumar JG (2018) How research in production and operations management may evolve in the era of big data. Prod Oper Manag 27(9):1670–1684. https://doi.org/10.1111/POMS.12836
26. Lorena AC et al (2011) Comparing machine learning classifiers in potential distribution modelling. Expert Syst Appl 38(5):5268–5275. https://doi.org/10.1016/J.ESWA.2010.10.031
27. Mehdizadeh E, Niaki STA, Hemati M (2018) A bi-objective aggregate production planning problem with learning effect and machine deterioration: modeling and solution. Comput Oper Res 91:21–36. https://doi.org/10.1016/J.COR.2017.11.001
28. Krisztin T (2018) Semi-parametric spatial autoregressive models in freight generation modeling. Transp Res Part E Logist Transp Rev 114:121–143. https://doi.org/10.1016/J.TRE.2018.03.003
29. Shukla M, Jharkharia S (2013) Agri-fresh produce supply chain management: a state-of-the-art literature review. Int J Oper Prod Manag 33(2):114–158. https://doi.org/10.1108/01443571311295608/FULL/XML
30. Lin CWR, Chen HYS (2010) Dynamic allocation of uncertain supply for the perishable commodity supply chain 41(13):3119–3138. https://doi.org/10.1080/00207540310000088147
31. Piramuthu S, Zhou W (2013) RFID and perishable inventory management with shelf-space and freshness dependent demand. Int J Prod Econ 144(2):635–640. https://doi.org/10.1016/J.IJPE.2013.04.035
32. Luangkesorn KL, Klein G, Bidanda B (2016) Analysis of production systems with potential for severe disruptions. Int J Prod Econ 171:478–486. https://doi.org/10.1016/J.IJPE.2015.09.014
33. Saetta SA, Caldarelli V, Tiacci L, Lerche N, Geldermann J (2015) A logistic network to harmonise the development of local food system with safety and sustainability. Int J Integr Supply Manag 9(4):307–328. https://doi.org/10.1504/IJISM.2015.070530
34. Voorneveld M, Groot M (2024) Optimal investment strategy on data analytics capabilities of startups via Markov decision analysis. Decis Anal J 100438
35. Willetts M, Atkins AS (2024) Evaluation of a software positioning tool to support SMEs in adoption of big data analytics. J Electron Sci Technol 22(1):100229
36. Liu Y, Qiao H, Wang J, Jiang Y (2024) Influencing mechanism of the intellectual capability of big data analytics on the operational performance of enterprises. Heliyon 10(3)
37. Naseer A, Naseer H, Ahmad A, Maynard SB, Siddiqui AM (2023) Moving towards agile cybersecurity incident response: a case study exploring the enabling role of big data analytics-embedded dynamic capabilities. Comput Secur 135:103525
38. Ma Q, Tang H, Liu C, Zhang M, Zhang D, Liu Z, Zhang L (2024) A big data analytics method for the evaluation of maritime traffic safety using automatic identification system data. Ocean Coast Manage 251:107077
39. Liu D, Yoon VY (2024) Developing a goal-driven data integration framework for effective data analytics. Decis Support Syst 114197
40. Zhou Y, Kundu T, Goh M, Chakraborty S, Bai X (2023) A multi-stage multi-criteria data analytics approach to rank commercial service airports. J Air Transp Manage 111:102410

A Mobile Application Design of Cancer Follow-up Care for Breast Cancer Survivors

Hossein Pourghaemi and Hatice Camgöz Akdağ

Abstract This work designs a mobile application to assist breast cancer survivors in managing their follow-up care. Breast cancer survivors encounter the challenge of tracking their follow-up appointments, which provide medical and psychological support post-treatment. While existing digital solutions document and track follow-up care, they primarily target professionals and care providers. To the best of our knowledge, no digital solution currently caters to patient use. Our primary objective is to facilitate follow-up care and motivate breast cancer patients to continue their treatment. We design a mobile application that generates a personalized follow-up care plan based on the latest literature. Patients can schedule follow-up appointments, document and share complaints, and track lifestyle habits and interventions. Push notifications remind patients of upcoming appointments, ensuring adherence to the follow-up plan. Furthermore, user data collected in the long term serve as reliable information for forecasting optimal follow-up care plans for future breast cancer survivors. Considering the application's target audience of non-professionals with limited medical knowledge, we employ simple questions instead of medical terms. The application features user-friendly elements like checkboxes and drop-down menus, offering a simple and intuitive patient experience. We conducted two semi-structured independent interviews with medical oncologists to evaluate usability and gather suggestions. The interviews encompassed an overview of the application, its benefits for breast cancer survivors, and sought feedback on user information.

Keywords Breast cancer · Follow-up care · Cancer survivorship

---

H. Pourghaemi · H. Camgöz Akdağ (✉)
Istanbul Technical University, Istanbul, Turkey
e-mail: camgozakdag@itu.edu.tr

H. Pourghaemi
e-mail: hossein.pourghaemi@gmail.com

F. Çebi et al. (eds.), *Engineering and Technology Management in Challenging Times*, Lecture Notes in Networks and Systems 1137,
https://doi.org/10.1007/978-3-031-70935-7_3

1 Introduction

Breast cancer is initiated when breast tissue cells undergo uncontrolled growth; strictly speaking, it is the result of a change within the DNA of breast tissue cells. Breast cancer is the most common cancer in women, but it can also be rarely experienced by men. Specific signs or symptoms are usually absent in the early stages. However, sometimes, during self-examination, a person may detect a tumor in the breast. It is worth noting that not all breast tumors are life-threatening or cancerous, but some benign breast masses elevate the risk of breast cancer in women [1]. To reduce the risk of cancer, it is recommended that any lump (mass of tissue) or change in the appearance of the breast be assessed by a medical professional.

The breast consists of different parts, namely: Lobules, Ducts, Nipple, and Areola. Most common starting point for breast cancer is the ducts, known as Ductal cancer. Some start in the Lobules, which is termed lobular or invasive cancer. Other types of breast cancer, such as Phyllodes or Angiosarcoma, are less common [2]; these cancers commonly occur in younger women who are in their 40s or 50s [3]. A limited number of cancers originate in other breast tissues, which are called Sarcomas or Lymphomas; these are not classified as breast cancer [3].

At times, cancer is transported through the bloodstream or lymphatic system to other parts of the body, causing damage to organs such as the lungs, liver, and brain; it is referred to as metastasized breast cancer. Breast cancer may reoccur in another part of the body months or years after initial diagnosis and treatment [4].

It is essential for breast cancer survivors to maintain contact with care providers for an extended period; hence, all cancer survivors should be furnished with a follow-up care plan. Since the possibility of breast cancer returning spans up to 10 years or even more for some cancer types, follow-up should continue for at least 5 years [5].

In this work, with the intention of enhancing follow-up care for breast cancer survivors, a mobile application has been designed for tracking the follow-up care plan. For this, the latest follow-up plans in the literature, as well as patient-specific information, are utilized to generate a follow-up care plan. Based on the main pain points of breast cancer survivors extracted from the literature, four different application features are introduced in this work; (1) two different sources are used to prepare a patient-specific follow-up care plan (follow-up care routine recommendation based on the state of the art literature, and recommendations provided by the care providers of the patient), (2) a complaint management feature is designed in order to assist with documenting and sharing symptoms and side effects, (3) a feature to help the patient keep track of their lifestyle habits and interventions is provided, and (4) a notification system is designed to automatically remind the patient about their next appointments. To validate and evaluate the application features, two independent semi-structured interviews were conducted with medical oncologists, and their suggestions and recommendations were utilized to improve the application. Furthermore, data gathered from this application will be a reliable source that can be used in forecasting the best possible customized follow-up care plan in the future.

2 Literature Review

2.1 *Breast Cancer*

A change within the DNA of breast tissue cells results in breast cancer, which is the most common cancer in women. Although there is usually no obvious reason behind it, research has shown that genetic causes, smoking, alcohol, and other lifestyle habits can lead to cancer [6]. Other factors that increase the risk of breast cancer in women include pregnancies, multiple pregnancies and miscarriages, use of birth control pills, obesity, red meat consumption, and caffeine consumption [7].

Breast cancer molecular targets have been categorized in two main groups, namely Estrogen Receptor Alpha (ER) and epidermal growth factor 2 (ERBB2); another related receptor is Progesterone Receptor (PR), which is considered an enabler of ER [8, 9]. If at least 1% of breast cancer tumors have ER or PR, the cancer is categorized as HR+ [10]. Breast cancer is divided into three major subtypes: hormone receptor positive/ERBB2 negative (HR+/ERBB2−), ERBB2 positive (ERBB2+), and triple negative. The most common breast cancer subtype is HR+/ERBB2−, which comprises 70% of patients, and ERBB2+comprises 15–20% of patients [9]; triple-negative cancer, which lacks ER, PR, or ERBB2 target, comprises 15% of breast cancer [11, 12]. Systemic therapy for breast cancer is determined by these subtypes [13].

Although the majority of people with breast cancer will not die from the disease, comorbidities (such as obesity, hypertension, and diabetes mellitus) will likely have an impact on disease-free survival and, eventually, overall survival [14]. Diet, physical activity, and weight all play a role in breast cancer patient survival. A total-health approach to enhancing long-term survival in breast cancer involves a focus on healthy nutrition, an active lifestyle, and a balanced weight [15]. Mehra et al. investigate the relationship between BMI and cancer recurrence in breast cancer survivors and suggest that keeping BMI between 18.5 and 35 kg/m^2 increases the chance of survival [16].

2.2 *Follow-up Care*

Breast cancer follow-up care is usually administered to patients by their cancer care team after treatment. Patients who have successfully completed treatment and do not have advanced breast cancer are often provided with psychological support and are given the opportunity to voice their concerns to doctors. Conversely, patients suffering from advanced breast cancer may require ongoing treatment beyond the primary course. During follow-up care appointments, any problems and complaints a patient may have need to be shared. Additionally, depending on the treatment, patients may experience side effects either immediately after therapy or in the long term. Patients are required to attend all follow-up sessions and provide the results of necessary tests to their care providers.

Several works have studied the post-treatment problems experienced by breast cancer survivors. A significant biographical change occurs after a cancer diagnosis [17, 18]. The STEP project focused on the physical aspects of treatment for 1748 breast cancer patients and specified post-cancer needs [19, 20]. It revealed that the most frequently reported symptoms are fatigue (66.6%), weight changes (57%), weakness (61.6%), and sleep disruption (60%).

Vandendorpe et al. examined the complaints of cancer survivors and the connection between post-cancer symptoms and the fear of cancer recurrence [21]. Even though a 5-year survival can be indicative of successful treatment for most types of cancer, there exists a possibility of recurrence for more than 10 years after treatment [22]. Additionally, for some patients, the annual risk of recurrence is higher than others [23]. Ten-year survival rates for breast cancer exceed 70% in most European regions, with 89% survival for local and 62% for regional disease [24]. The lifetime risk of breast cancer recurrence in patients with specific mutations (e.g., BRCA1) ranges from 65 to 90%; there is also a 25% to 31% possibility of contralateral breast cancer within 10 years [25]. Therefore, another significant objective of follow-up care is the identification of contralateral breast cancer.

Another main objective of follow-up care is the detection of curable in-breast and loco-regional recurrences in early stages [1]. Bock et al. conducted twelve studies involving a total of 5,045 patients, in which 378 isolated locoregional recurrences were identified [26]. For asymptomatic patients, 40% of locoregional recurrences are identified during routine visits or tests, with 47% of those diagnosed after mastectomy, and 36% after breast-conserving therapy [26]. Furthermore, recurrence of locoregional breast cancer is associated with a significant increase in the risk of metastasis [27]. Senkus et al. [28] summarize the goals of follow-up care as follows:

- Identifying new local recurrences or contralateral breast cancer.
- Assessing and providing treatment for treatment-related complications.
- Encouraging motivation in patients to continue Endocrine Therapy (ET).
- Assisting treated patients in returning to normal life through psychological support and information provision.

2.3 Routine Follow-up Examinations

During the follow-up appointment, patient's medical history needs to be reviewed, any treatment-related side effects must be identified, the clinical examination should be conducted, and patient should be asked to provide test results. For instance, young individuals, particularly those with dense breast tissue and hereditary or family predispositions, may require an MRI of the breast in addition to regular mammography and breast ultrasonography [29, 30].

Table 1 AGO grades of recommendation [1]

Symbol	Explanation
++	This examination or therapeutic intervention is extremely advantageous to patients, and it should be carried out without reservation
+	This examination or therapeutic intervention has a little benefit for patients, but it can be done
±	This examination or therapeutic intervention has not been proved to help patients and should only be used in exceptional circumstances. A broad advice is not possible based on existing understanding
–	This examination or therapeutic intervention may be harmful to patients, and it may not be carried out
–	In any event, examination or therapeutic intervention is clearly harmful to patients and should be avoided

The independent Gynecological Oncology group AGO [1] investigated and ranked the grade of recommendation for such experiments and tests. The grade of recommendation is denoted using +, ++, ±, −, −− signs (Please refer to Table 1 for the meaning behind each symbol). Based on AGO's work, routine follow-up experiments and their respective grades of recommendation can be listed as follows: history (++), physical examination (++), self-examination (+), mammography (++), breast sonography (++), MRI of the breast (±), pelvic examination (++), DXA-scan (+), routine biochemistry (−), blood tests for monitoring severe toxicities (+), liver ultrasound (−), bone scan (−), chest X-ray (−), CT-scan of the chest, CT-scan of the abdomen, CT-scan of the pelvis (−), identification of isolated or circulating tumor cells (−), positron emission tomography scan (−), and whole-body MRI (−).

Furthermore, follow-up should not be viewed solely in terms of the physical aspects; patients' anxiety levels typically increase once therapy is completed, as their interaction with the treatment team diminishes [29]. For example, in the months following the completion of adjuvant chemotherapy and radiotherapy, depression and extreme exhaustion are commonly experienced [31].

2.4 Follow-up Care Routine Recommendation

Routine visits every 3–4 months for the first 2 years, every 6 months for years 3–5, and yearly afterward, balancing patient demands and follow-up expenditures [28]. Follow-up care routine recommendations for patients with invasive and non-invasive breast cancer can be outlined as follows [1, 29, 32, 33]:

- In the first two years, regular follow-up visits are advised every 3–4 months (every 6 months for low-risk and Ductal carcinoma in situ (DCIS) individuals), every 6–8 months from years 3 to 5, and yearly after that. History, physical examinations, and counseling are recommended in these sessions.

- A monthly self-examination is recommended.
- Annual bilateral (after BCT) and/or contralateral (after mastectomy) mammography is advised, with ultrasonography and breast MRI if needed.
- Imaging, modalities, and biochemistry should be considered only in case of complaints, clinical findings, or the possibility of cancer return.
- Echocardiography is recommended 6–12 months and 5 years after the patient completes the cardiotoxic therapy.

2.5 *Present Solutions for Follow-up Care*

There are a few advanced digital solutions for documenting and tracking follow-up care of breast cancer survivors, such as ODSeasyNet, Celsius37, and Onkys; these software systems are referred to as Tumor Documentation Systems (TDS) [34–36]. such systems are designed for use by professional care providers who utilize this system to document and manage all treatment and after-treatment information. To the best of our knowledge, there is no digital solution that can be used by patients at this time. Two main follow-up methods are used by patients, (1) keeping track of the follow-up plan manually, and (2) using the follow-up form provided by the American Society of Clinical Oncology (ASCO) [37].

Manuel Follow-up: The main follow-up method used by breast cancer survivors is the manual keeping of required follow-up records and information. In this method, after the cancer treatment, a follow-up plan is provided by the care provider team, primarily based on state-of-the-art routine follow-up recommendations. In this method, patients are usually responsible for keeping track of their follow-up plan. As the follow-up period is long and can be complicated, this method is typically challenging and error prone.

ASCO Follow-up Care Form: Another method, provided by the ASCO, offers a form to organize information needed for follow-up care of breast cancer patients. This form consists of three main parts: general information, treatment summary, and follow-up care plan. The ASCO Follow-up Care Form provides essential information all in one form; filling and updating this form helps the patient to keep track of her follow-up care plan and share this form with her care provider when needed. Patients can print a copy or fill out and update it as a digital file.

Even though using ASCO form is a much easier and more effective follow-up method, just like manual follow-up, it has some disadvantages. In both methods, the patient needs to employ a digital or analog reminder in order not to miss appointments. Furthermore, even if the patient manages to follow the sessions in the short term, it would be very difficult to keep track of the different versions of documents in the long term. As software development and cloud storage technologies are now more advanced and available, it is essential to transition the follow-up care process to a digital environment.

3 Methodology

After an extensive literature review, the main resources for follow-up breast cancer care were identified. Some resources, like AGO [1], are not suitable to be used directly by patients, as these guidelines are prepared for specialists. ASCO follow-up care is a resource that is mainly used by patients and is utilized as one of the resources in this work. As these resources are considered the most trusted references used at the time, this work was heavily reliant on these resources. This work is prepared in three main steps: in the first step, all necessary information needed in the follow-up care of patients with breast cancer was gathered; in the second step, features of the application were designed based on the main pain points of breast cancer patients; finally, in the last step, two independent interviews were conducted with experienced breast cancer oncologists and surgeons to evaluate the application.

3.1 Application Features

Breast cancer survivors experience three main pain points, as below:

- As the follow-up plan can be as long as 10 years or even more in some cases, patients struggle with keeping track of their follow-up plan.
- Most breast cancer survivors -even those with a less severe cancer- experience symptoms and side effects. Documenting and sharing such complaints with care providers is one of the most important aims of follow-up care.
- Interventions affecting lifestyle risks and comorbidity need to be tracked by the patient to minimize the negative impact on disease outcomes.

Application features are designed based on these pain points. So, four main features based on those pain points were designed in this application. In the following paragraphs, each feature is explained in detail.

Customized Follow-up Plan. A patient specific follow-up care plan was prepared using two different sources: first, follow-up care routine recommendations based on state-of-the-art literature [ASCO-ACS recommendations 2016, NCCN 2021, ESMO 2019, and S3-guidelines 2017], and second, recommendations provided by the patient's care providers. In the case of the first source, the surgery date of the patient is required to calculate the exact date of each follow-up session a simple algorithm is used to calculate all subsequent follow-ups based on the surgery date and the recommended follow-up routine in the literature. For example, if the surgery date of the patient is 01/01/2022, the algorithm sets a monthly self-examination on 02/01/2022, 03/01/2020, and so on. In the case of the second source, when the care provider recommends a new appointment for the patient, the patient can manually create a new follow-up appointment in the application.

Complaint Management. To assist breast cancer survivors in documenting and sharing symptoms and side effects, a complaint management feature has been designed. This feature allows patients to select from common symptoms and side effects, which are prepared based on the side effects of Tamoxifen, Aromatase Inhibitors, GnRH agonist, and other common post-treatment symptoms. The list includes endometrial cancer, bone loss, major blood clots, vision issues, joint/muscle aches, vaginal dryness, hair thinning, fatigue, menopausal symptoms, numbness, pain, and depression [ASCO]. Furthermore, patients can manually input notes about any other possible symptoms and side effects. To enhance the complaint management feature, the patient is enabled to take pictures related to her complaint as well. Finally, to facilitate the sharing task, the patient is enabled to share a copy of a saved complaint with her doctor through a messaging platform or email.

Lifestyle Guide. The aim of this feature is to help the patient keep track of her lifestyle habits and interventions. For this, the patient needs to enter some information, including height, weight, average nightly fasting hours, amount of alcohol consumption per week, smoking status, and physical activities per week. The evaluation of these intervention parameters altogether and coming up with a unique value that describes the overall impact of these interventions is not within the scope of this work. To the best of our knowledge, there is no study that investigates the effect of these parameters all together on the improvement of overall survival. Hence, all information asked from the patient in this feature is evaluated individually, and the patient is informed if any of her interventions are not going right. For this, we use interventions recommended by the Ago guideline.

Notification System. A notification system has been developed to remind the patient about her next appointments. Such notifications help the patient to be prepared for the next session; for example, patients will be notified to take a blood test before an appointment. These notifications are set based on the auto-generated and manual follow-up plan.

3.2 *Interview*

Two semi-structured and independent interviews were conducted with medical oncologists and breast cancer surgeons to evaluate the usefulness of the application and to elicit suggestions and improvement ideas. Interviewees had an average field experience of 10 years, and each interview lasted approximately 30 min. The interviews were conducted in two parts: first, a brief explanation about the application, its users, and possible benefits for the breast cancer survivors was provided. Then, interviewees were asked to focus on two specific aspects of the application. Firstly, interviewees were asked to carefully review all information asked from the patients in the sign-up process, evaluate the importance of this information in follow-up care, and specify if any of those are non-required information. Secondly, they were asked to navigate the application, examine the features and provide their feedback regarding the utility of the application and possible improvements. Based on the feedback gathered from the oncologists, (1) optional and required fields were specified considering the importance of such information in the quality of the follow-up care, and the quality of the data which will be gathered in the long term, and (2) questions were simplified to be more understandable by patients.

3.3 Application Design

Using mobile applications for managing the personal tasks has gain a great popularity in recent years. Using mobile phone to set reminders and check daily activities is more convenient than using personal computers, hence a mobile application prototype is designed in this work. To design the mobile application prototype, Uizard web-based software is used as a digital prototyping tool [38]. Uizard provides essential tools including different shapes, text boxes, icons, buttons, and forms. Furthermore, it 18 allows to upload images and and other graphical elements from personal computers. Forms elements such as input stepper, input field, drop-down menu, switch on/off, and checkboxes enhance the usability of the application and make the patient interface more pleasant. As for the breast cancer follow-up care we need to gather much information from the user, in this work form elements are heavily used to make the data gathering entering more comfortable for the user.

4 Application Design

This application comprises various pages: a sign-up page, two pages for collecting the patient's diagnosis and treatment history, two pages for gathering additional information about ongoing treatment and care providers, a home page, two pages for scheduling appointments and monitoring recommended follow-up care, a page for recording and sharing complaints, and finally, a page to assist the patient with lifestyle habits.

4.1 Sign-up Page

The sign-up page is the initial page that the patient encounters upon installing the application. Here, the patient provides basic information required to create an account, including their name, date of birth, gender, and education. Additionally, the patient is prompted to provide an email address and password for future logins.

4.2 Diagnosis and Treatment History Pages

Following sign-up, two consecutive pages are dedicated to capturing details about the patient's diagnosis and treatment history. On these pages, the patient is prompted to furnish essential information pertaining to their cancer diagnosis and treatment. This information can be shared with care providers to facilitate a more comprehensive follow-up plan. In this page, patient is asked to provide a diagnosis date and surgery date (which will be used to schedule the routine follow-up care plan). A drop-down list is utilized to assist the patient in selecting among possible options for cancer location (left, right, both) and receptor type (ERBB2-, ERBB2+, and triple-negative). Additionally, the patient is inquired about any family history of cancer, with options including mother, sister, aunt, daughter, and others.

On the treatment history page, patient is required to respond to three questions and select from provided options: type of surgery (mastectomy, and conservation therapy), results of the genetic test (BRCA1, BRCA2), and type of lymph removal surgery (Axillary Dissection, and Sentinel Biopsy). Furthermore, the patient is also asked to indicate if have undergone radiotherapy, chemotherapy [before surgery (Neoadjuvant), or after surgery (Adjuvant)], and hormonal therapy.

4.3 Ongoing Treatment Pages

The ongoing treatment section comprises two pages titled "Complete Your Profile 1" and "2". The first page pertains to the medications that the patient is currently taking. Patient can select from options including Tamoxifen, Aromatase Inhibitors (anastrozole, exemestane, and letrozole), and GnRH agonist (Zoladex, Lupron). Although the planned duration in months is also requested, it is not a mandatory field. Additionally, patients can click on the information icon next to each medicine to view potential side effects. The second page prompts the patient to provide the names of their primary care provider, surgeon, radiation oncologist, and medical oncologist. It's important to note that these fields are optional. Another optional question on this page pertains to the agents used during chemotherapy; the patient can select from options including 5-Fluorouracil, Carboplatin, Cyclophosphamide, Docetaxel, Doxorubicin, Epirubicin, Methotrexate, Paclitaxel, Pertuzumab, and Trastuzumab.

4.4 Home Page

Important follow-up care notifications can be viewed here, and patients can navigate to other pages (appointment, complaint, lifestyle, and setting page), receive important notifications, and access lifestyle recommendations. Additionally, the user can see how complete her profile is and is encouraged to navigate to the profile page to enter further information about her treatment. Lifestyle recommendations presented on this page are based on the recommendations provided by AGO [1].

4.5 Appointment Setting and Recommended Follow-Up Care Pages

An appointment or test reminder can be easily set on this page. All essential options for setting an appointment or reminder are provided on this page, including test type, specialization, doctor name, location, and appointment date. Provided options for test type are prepared based on AGO guidelines and include Mammography, breast Sonography, MRI of breast, DXA-scan, biochemistry, blood tests, liver Ultrasound, bone scan, chest X-ray, CT-scan of the chest, CT-scan of the abdomen, CT-scan of the pelvis, Positron Emission Tomography scan, and whole-body MRI.

4.6 Complaint Page

This page helps the patient record complaints and share them. At the top of this page, all previous complaints are listed, allowing the patient to navigate and view each one in detail. To record a complaint, the patient can select from common side effects. Enabling the patient to take personal notes and photos helps to document complaint accurately and share them.

4.7 Interventions Guide Page

On this page, essential information to evaluate interventions and lifestyle habits is gathered from the patient by asking simple questions. Questions include weight and height, amount of overnight fasting hours, smoking status, alcohol consumption, and amount of exercise per week. After filling these fields and pressing the calculate button, each answer is evaluated individually, and the patient is informed if any of them are not in a healthy state.

5 Conclusion

A mobile application for follow-up care of breast cancer survivors was designed in this work. This allows patients who have finished their primary treatment to keep track of their long-term follow-up care treatment. Hence, more comprehensive tools and systems are needed by care providers; this mobile application is designed to be used by patients themselves.

This mobile application utilizes state-of-the-art resources to generate an auto-generated routine follow-up care plan; the patient is also given a simple feature to add other related appointments to the auto-generated list. The patient can easily document and share symptoms and side effects with her care provider. It also provides an easy-to-use feature to assist the user in keeping track of her lifestyle habits and interventions, including BMI, alcohol consumption, overnight fasting, and other lifestyle habits. Mobile notifications, considered the most effective notifying tool, are employed to notify the patient about the next upcoming appointment and test dates.

To validate and evaluate the application features, two independent semi-structured interviews were conducted with medical oncologists, and their suggestions and recommendations were utilized to improve the application.

References

1. AGO breast cancer follow-up guidelines. www.agoonline.de/fileadmin/agoonline/downloads/_leitlinien/kommission_mamma/2022/englisch/Einzeldateien/AGO_2022E_16_Breast_Cancer_Follow-up.pdf. Accessed 6 Feb 2023
2. www.verywellhealth.com. Accessed 1 Apr 2023
3. Hartmann LC, Sellers TA, Frost MH, Lingle WL, Degnim AC, Ghosh K, Vierkant RA, Maloney SD, Pankratz VS, Hillman DW, Suman VJ, Johnson J, Blake C, Tlsty T, Vachon CM, Melton LJ, Visscher DW (2005) Benign breast disease and the risk of breast cancer. N Engl J Med 353(3):229–237
4. Chen YH, Tu YL, Chen HK, Shih SL (2022) Invasive lobular carcinoma enclosed by a benign phyllodes tumor: a case report. World J Surg Oncol 91:106804
5. Nath DAR, Thomas DM, Mathews DR (2021) Synchronous malignant phyllodes tumour and invasive ductal carcinoma in contralateral breasts—"A rare co-existence"
6. Pérez-Solis MA, Maya-Nuñez G, Casas-González P, Olivares A, Aguilar-Rojas A (2016) Effects of the lifestyle habits in breast cancer transcriptional regulation. Cancer Cell Int 16(1):7
7. Onsory K, Ranapoor S (2011) Cancer breast and the effect of environmental factors involved
8. Joshi H, Press MF (2018) Molecular oncology of breast cancer. In: Bland KI, Copeland EM, Klimberg VS, Gradishar WJ (eds) The breast. Elsevier, Philadelphia, PA
9. Waks AG, Winer EP (2018) Breast cancer treatment: a review. JAMA 321(3):288–300
10. Hammond ME, Hayes DF, et al (2016) American Society of Clinical Oncology/College of American Pathologists guideline recommendations for immunohistochemical testing of estrogen and progesterone receptors in breast cancer. J Clin Oncol
11. Denkert C, Liedtke C, et al (2017) Molecular alterations in triple-negative breast cancer - the road to new treatment strategies. The Lancet
12. Gluz O, Nitz UA, Christgen M, Kates RE, Shak S, Clemens M, Kraemer S, Aktas B, Kuemmel S, Reimer T, Kusche M, Heyl V, Lorenz-Salehi F, Just M, Hofmann D, Degenhardt T, Liedtke C, Svedman C, Wuerstlein R, Kreipe HH, Harbeck N (2016) West German study group phase III plan B trial: first prospective outcome data for the 21-gene recurrence score assay and concordance of prognostic markers by central and local pathology assessment. J Clin Oncol 34(20):2341–2349
13. Fisher B, Anderson S, Bryant J, Margolese RG, Deutsch M, Fisher ER, Jeong JH, Wolmark N (2002) Twenty-year follow-up of a randomized trial comparing total mastectomy, lumpectomy,

and lumpectomy plus irradiation for the treatment of invasive breast cancer. N Engl J Med 347(16):1233–1241
14. Bodai BI, Tuso P (2015) Breast cancer survivorship: a comprehensive review of long-term medical issues and lifestyle recommendations. Permanente J 19(2):48–79
15. Aiello Bowles EJ, Boudreau DM, Chubak J, Yu O, Fujii M, Chestnut J, Buist DS (2012) Patient-reported discontinuation of endocrine therapy and related adverse effects among women with early-stage breast cancer. JNCI Monographs 8(6):e149–e157
16. Mehra K, Berkowitz A, Sanft T (2017) Diet, physical activity, and body weight in cancer survivorship. Cancer 101(6):1151–1165
17. Smith T, Stein KD, Mehta CC, Kaw C, Kepner JL, Buskirk T, Stafford J, Baker F (2007) The rationale, design, and implementation of the American Cancer Society's studies of cancer survivors
18. van de Poll-Franse LV, Horevoorts N, van Eenbergen M, Denollet J, Roukema JA, Aaronson NK, Vingerhoets A, Coebergh JW, de Vries J, Essink-Bot ML, Mols F (2011) The patient reported outcomes following initial treatment and long term evaluation of survivorship registry: scope, rationale and design of an infrastructure for the study of physical and psychosocial outcomes in cancer survivorship cohorts. Eur J Cancer 47(14):2188–2194
19. Molassiotis A, Yates P, Li Q, So W, Pongthavornkamol K, Pittayapan P, Komatsu H, Thandar M, Yi M, Titus Chacko S, Lopez V, Butcon J, Wyld D, Chan R, Doolan M, Litam ME, Onofre R, Lluch C, Nacion R, Ombao ML, Soe ZW, Myint T, Ang E, Arao H, Yagasaki K, Ravindran V, Rhenius RV, Lucas A, Kujur LP, Princy A, Chow Choi K, Ping Choy Y, Pui Lee Y, Yip Shiu C, Xu Y (2017) Mapping unmet supportive care needs, quality-of-life perceptions and current symptoms in cancer survivors across the Asia-Pacific region: results from the international STEP study. Ann Oncol 28(10):2552–2558
20. Chan RJ, Yates P, Li Q, Komatsu H, Lopez V, Thandar M, Chacko ST, So WK, Pongthavornkamol K, Yi M, Pittayapan P (2017) Oncology practitioners' perspectives and practice patterns of post-treatment cancer survivorship care in the Asia-Pacific region: results from the STEP study. BMC Cancer
21. Vandendorpe B, Drouet Y, Ramiandrisoa F, Guilbert P, Costa B, Servagi-Vernat S (2021) Psychological and physical impact in women treated for breast cancer: need for multidisciplinary surveillance and care provision. Cancer/Radiothérapie 25(4):330–339
22. Francisci S, Capocaccia R, Grande E, Santaquilani M, Simonetti A, Allemani C, Gatta G, Sant M, Zigon G, Bray F, Janssen-Heijnen M (2009) EUROCARE working group. The cure of cancer: a European perspective. Eur J Cancer 45(6):1067–1079
23. Park S, Koo JS, Kim MS, Park HS, Lee JS, Lee JS, Kim SI, Park BW (2012) Characteristics and outcomes according to molecular subtypes of breast cancer as classified by a panel of four biomarkers using immunohistochemistry. World J Surg 21(1):50–57
24. Allemani C, Minicozzi P, Berrino F, Bastiaannet E, Gavin A, Galceran J, Ameijide A, Siesling S, Mangone L, Ardanaz E, Hédelin G, Mateos A, Micheli A, Sant M (2013) EUROCARE working group. Predictions of survival up to 10 years after diagnosis for European women with breast cancer in 2000–2002. Eur J Cancer 132(10):2404–2412
25. Risch HA, McLaughlin JR, Cole DEC, Rosen B, Bradley L, Fan I, Tang J, Li S, Zhang S, Shaw PA, Narod SA (2006) Population BRCA1 and BRCA2 mutation frequencies and cancer penetrances: a kin-cohort study in Ontario, Canada. JNCI J Natl Cancer Inst 98(23):1694–1706
26. De Bock G, Bonnema J, van Der Hage J, Kievit J, van de Velde C (2004) Effectiveness of routine visits and routine tests in detecting isolated locoregional recurrences after treatment for early-stage invasive breast cancer: a meta-analysis and systematic review. Ann Surg Oncol 22(19):4010–4018
27. Fisher B, Anderson S, Fisher ER, Redmond C, Wickerham DL, Wolmark N, Mamounas EP, Deutsch M, Margolese R (1991) Significance of ipsilateral breast tumour recurrence after lumpectomy 338(8763):327–331
28. Senkus E, Kyriakides S, Ohno S, Penault-Llorca F, Poortmans P, Rutgers E, Zackrisson S, Cardoso F (2015) ESMO clinical practice guidelines for diagnosis, treatment and follow-up. Ann Oncol 26:v8–v30

29. Cardoso F, Kyriakides S, Ohno S, Penault-Llorca F, Poortmans P, Rubio I, Zackrisson S, Senkus E (2019) Early breast cancer: ESMO clinical practice guidelines for diagnosis, treatment and follow-up. Ann Oncol 30(8):1194–1220
30. Committee Opinion No. 601: Tamoxifen and uterine cancer. Obstet Gynecol 123(6):1394–1397 (2014)
31. Bower JE (2014) Cancer-related fatigue: mechanisms, risk factors, and treatments. J Natl Cancer Inst Monogr 11(10):597–609
32. Oncology ASoC. ASCO cancer treatment and survivorship care plans. [Online]. www.asco.org/sites/new-www.asco.org/files/content-files/practice-and-guidelines/documents/breast-cancer-treatment-summary-and-survivorship/-care-plan.docx. Accessed 2023
33. Gradishar WJ, Moran MS, Abraham J, Aft R, Agnese D, Allison KH, Blair SL, Burstein HJ, Dang C, Elias AD, Giordano SH, Goetz MP, Goldstein LJ, Hurvitz SA, Isakoff SJ, Jankowitz RC, Javid SH, Krishnamurthy J, Leitch M, Lyons J, Matro J, Mayer IA, Mortimer J, O'Regan RM, Patel SA, Pierce LJ, Rugo HS, Sitapati A, Smith KL, Smith ML, Soliman H, Stringer-Reasor EM, Telli ML, Ward JH, Wisinski KB, Young JS, Burns JL, Kumar R (2021) NCCN guidelines® insights: breast cancer, Version 4.2021: featured updates to the NCCN guidelines. J Natl Compr Cancer Netw 19(5):484–493. https://jnccn.org/view/journals/jnccn/19/5/article-p484.xml
34. www.asthenis.de. Accessed 15 May 2022
35. www.celsius37.com. Accessed 15 May 2022
36. www.tevaris.de. Accessed 16 May 2023
37. www.asco.org/sites/new-www.asco.org/files/content-files/practice-and-guidelines/documents/breast-cancer-treatment-summary-and-survivorship/-care-plan.docx. Accessed 2023
38. www.uizard.io. Accessed 2 July 2023

A Model Proposal for Selecting the Most Appropriate Radiotherapy Devices for Hospitals

Gonca Turna and Ahmet Beşkese

Abstract Choosing the appropriate radiotherapy device is an immensely important decision for hospitals, particularly in the field of oncology, where these devices come in a variety of technological forms, each with unique attributes, functionalities, and considerations. This study aims to select the most suitable radiotherapy device for healthcare institutions. To address the complexity and uncertainty inherent in this decision-making process, hesitant fuzzy AHP is employed, allowing for the incorporation of hesitant preferences provided by decision-makers. The proposed hierarchical decision model comprises five main criteria: Technical characteristics (radiation leakage limit, dose rate, field size, flattening filter, multileaf collimator, IGART and respiratory motion management system), training, installation, and utilization (training of user/s, ease of use, software/hardware upgrade availability and delivery time), warranty and maintenance (maintenance tasks, ease of maintenance design, service contract type, spare parts availability post- warranty, uptime guarantee), economic factors (cost efficiency, payment terms), and brand's features (reliability, relationships, corporate social responsibility). Through a real-world case study, the use of this model and methodology is illustrated. Following a thorough evaluation involving specialists with diverse expertise, it was determined that the paramount factor in the selection process of a radiotherapy device is its technical characteristics.

Keywords Medical device selection · Hospital · Radiotherapy device · Multi-criteria decision making · Hesitant fuzzy AHP

G. Turna (✉)
Graduate School, Bahçeşehir University, Istanbul, Turkey
e-mail: turnagonca@gmail.com; gonca.turna@bahcesehir.edu.tr

A. Beşkese
Department of Industrial Engineering, Bahçeşehir University, Istanbul, Turkey
e-mail: ahmet.beskese@bau.edu.tr

F. Çebi et al. (eds.), *Engineering and Technology Management in Challenging Times*, Lecture Notes in Networks and Systems 1137,
https://doi.org/10.1007/978-3-031-70935-7_4

1 Introduction

Radiotherapy constitutes a fundamental treatment approach within the realm of oncology. It encompasses the utilization of elevated-energy radiation to pinpoint and eliminate cancer cells and tumors, either independently or in conjunction with surgical procedures or chemotherapy [1].

A radiotherapy device refers to the specialized equipment used to deliver radiation to the targeted area of the body. These devices play a pivotal role in delivering accurate radiation doses precisely to cancerous cells while limiting exposure to neighboring healthy tissues [2].

The selection of an appropriate radiotherapy device is a critical decision for hospitals, as it directly impacts the quality and efficacy of patient care in the field of oncology. Each hospital has unique requirements and constraints that must be considered when choosing a radiotherapy device. Therefore, the decision-making process for radiotherapy device selection requires a comprehensive and systematic approach that can handle uncertainties, subjectivity, and imprecision inherent within the process of making choices.

Multi-Criteria Decision Making (MCDM) encompasses various tools aiding managers in complex decision-making by simultaneously considering multiple qualitative and quantitative criteria, often with inherent conflicts, relying on expert opinions to determine criteria weights and assess alternatives [3, 4]. Because of its very nature, radiotherapy device selection problem constitutes a perfect application area for MCDM.

Among various MCDM methods, the Analytic Hierarchy Process (AHP) stands out as particularly appropriate for selecting radiotherapy devices, thanks to its numerous advantages tailored to the complexities inherent in decision-making processes of this nature. AHP provides a clear and structured frame for organizing the decision-making process [5]. It enables decision-makers to decompose the intricate challenge of selecting devices into a hierarchical framework of criteria and sub-criteria, simplifying its management and comprehension the interrelationships between different factors [6].

The fuzzy AHP method can be regarded as an advanced analytical approach that has developed from the conventional AHP [7]. Fuzzy AHP is frequently selected due to its capability to manage uncertainty and imprecision, and vagueness in decision-making processes [8]. Fuzzy AHP's ability to tackle ambiguity and lack of precision in the decision-making process, along with its capability to incorporate expert opinions and handle qualitative linguistic expressions, makes it a valuable method for radiotherapy device selection. It empowers decision-makers to arrive at better-informed choices and well-rounded decisions, considering a broad spectrum of variables and uncertainties [9], ultimately leading to improved decision outcomes in the field of radiotherapy.

In this context, the destination is the hesitant fuzzy set theory which is an extension of fuzzy set theory. It enables decision-makers to articulate their uncertainty or indecision or ambiguity in making pairwise comparisons [10]. By using hesitant fuzzy sets, decision-makers can represent their degree of preference or hesitation when comparing the significance of criteria and options providing a more accurate and flexible decision-making framework [11].

Considering these, this study aims to delve into the intricate process of selecting the most suitable radiotherapy device, recognizing its central role in cancer management and healthcare, and present an original decision model for the problem at hand. To accomplish this aim, it utilizes the AHP methodology and incorporates hesitant fuzzy sets into the decision-making procedure. With its theoretical foundation, original decision model, and illustrative application, it is expected to contribute to the advancement of radiotherapy device selection practices in hospitals.

The rest of this paper is structured as follows: Sect. 2 provides a review of literature on multi-criteria decision making in the selection of medical devices. Section 3 outlines the methodology focusing on HFAHP. Section 4 presents the hierarchical decision model and the brief definitions of criteria. Section 5 provides a real-world case application and examines the findings. Section 6 discusses the sensitivity analysis results. Lastly, Sect. 7 outlines the findings, discusses limitations, and provides perspectives on potential future research paths.

2 Literature Review

The Analytic Hierarchy Process (AHP) is a method that involves comparing elements in pairs, relying on expert assessments to establish priority scales [12]. Fuzzy AHP extends the traditional AHP to address uncertainties and imprecisions in decision-making [7]. It effectively addresses uncertainty and offers increased robustness and flexibility in decision-making [13]. Unlike classical sets, fuzzy sets allow elements to have membership degrees ranging from 0 to 1 [14]. In Fuzzy AHP, linguistic terms used in pairwise comparisons are converted into fuzzy numbers, reflecting the degree of membership in a specific set [7]. Hesitant Fuzzy AHP further extends the traditional and Fuzzy AHP to handle decision-making situations in which decision makers exhibit hesitation or have difficulty expressing precise judgments [15]. There are several studies and research articles that have employed AHP, fuzzy AHP, and hesitant fuzzy AHP techniques in the context of medical device selection. Park et al. [16] developed a structured evaluation framework for the purpose of choosing the most suitable medical devices. This framework encompasses three key components: technology evaluation, economic evaluation, and usability evaluation. Kritchanchai et al. [17] employed AHP to make decisions concerning sustainable medical devices. Their focus was on criteria such as price, quality, service, reliability, appearance, and end-of-life attributes. Shbool et al. [18] proposed a hybrid MCDM integrating AHP with the Elimination Et Choice Translating Reality (ELECTRE) method. The model comprised 10 criteria such as imaging, size, ease of use, durability, connectivity, storage price, number of probes, warranty and years of service, and through inputs from professional physicians, seven ultrasound machines were evaluated. The calculations in the proposed model were illustrated with a case study examining the optimal selection of an ultrasound machine for a gynecology clinic in the Kingdom of Jordan. Abdel-Basset et al. [19] presented an innovative method for evaluating the selection of smart medical devices (SMDs) within a group decision-making (GDM) context, especially in uncertain decision environments. They offered a methodology combining neutrosophics with bipolar numbers and Technique for Order Preference by Similarity to Ideal Solution (TOPSIS) under group decision making. They used seven criteria such as safety, cost, flexibility, quality, ease of use, maintenance requirements, and service for selecting smart medical devices for analyzing sugar levels in diabetic patients. Otay [20] applied a fuzzy integrated AHP and TOPSIS methodology to choose surgery robots. The main criteria considered were cost, reliability, adaptability, serviceability, ease of use, and man–machine interface. In the assessment of suppliers providing medical examination gloves for hospitals, fuzzy AHP and fuzzy TOPSIS methods were used. The criteria in this evaluation included service quality, cost, and risk [21]. Öztürk and Tozan [22] presented a decision support framework designed to assist in choosing between high-flux and low-flux dialyzers for treating chronic kidney failure patients undergoing dialysis. They introduced eight main criteria, including cost, membrane material, medical evaluation, technical infrastructure, expertise, clearance, ultrafiltration coefficient, and toxin removal mechanism. There appears to be limited literature on the application of hesitant fuzzy AHP in medical device selection. Büyüközkan and Mukul [23] conducted a study assessing smart health technologies using hesitant fuzzy linguistic MCDM techniques, including HFL AHP. Their evaluation focused on technology-based, human-based, system-based, and organization-based factors, each with specific contributions to the decision-making process.

3 Methodology

Hesitant Fuzzy AHP adheres to the procedure outlined in Buckley's method [24], which comprises the following steps:

The procedure starts with Step 1 where the primary and subsidiary factors (main and sub-criteria) exerting an impact on the decision-making process are delineated, and subsequently, a hierarchical model depicting the decision problem is constructed.

Step 2 entails the creation of pairwise comparison matrices for criteria, sub-criteria, and alternatives. During these comparisons, experts articulate their evaluations using linguistic terms in Table 1.

In Step 3, the linguistic terms gathered from Step 2 are transformed into triangular fuzzy numbers based on the scale delineated in Table 1.

Step 4 incorporates the use of the fuzzy envelope approach, as introduced by Liu and Rodríguez [25], to amalgamate the assessments provided by the experts. Parameters "a" and "d" for the trapezoidal fuzzy membership function, which determine are computed using the OWA operation as follows:

Table 1 Linguistic scales for hesitant fuzzy AHP

Rank	Linguistic term	Abbreviation	Triangular fuzzy number
10	Absolutely high importance	AHI	7, 9, 9
9	Very high importance	VHI	5, 7, 9
8	Essentially high importance	ESHI	3, 5, 7
7	Weakly high importance	WHI	1, 3, 5
6	Equally high importance	EHI	1, 1, 3
5	Exactly equal	EE	1, 1, 1
4	Equally low importance	ELI	0.33, 1, 1
3	Weakly low importance	WLI	0.2, 0.33, 1
2	Essentially low importance	ESLI	0.14, 0.2, 0.33
1	Very low importance	VLI	0.11, 0.14, 0.2
0	Absolutely low importance	ALI	0.11, 0.11, 0.14

$$a = min\left\{a_L^i, a_M^i, a_M^{i+1}, \ldots, a_{M,}^j a_R^j\right\} = a_L^i \tag{1}$$

$$d = max\left\{a_L^i, a_M^i, a_M^{i+1}, \ldots, a_{M,}^j a_R^j\right\} = a_R^j \tag{2}$$

$$b = \{a_M^i, \ \ if \ \ i+1 = \\ j \ \ OWA_{w^2\left(a_m^j,\ldots,a_m^{(i+j)/2}\right),\ if\ \ i+j\ is\ even} \ \ OWA_{w^2\left(a_m^j,\ldots,a_m^{(i+j+1)/2}\right),\ if\ \ i+j\ is\ odd}\Big\} \tag{3}$$

$$c = \{a_M^{i+1}, \ \ if \ \ i+1 = \\ j \ \ OWA_{w^1\left(a_{m,}^j a_{m,}^{j-1}\ldots,a_m^{(i+j)/2}\right),\ if\ \ i+j\ is\ even} \ \ OWA_{w^1\left(a_{m,}^j a_{m,}^{j-1}\ldots,a_m^{(i+j+1)/2}\right),\ if\ \ i+j\ is\ odd}\Big\} \tag{4}$$

where;

$$w_1^1 = \alpha_2, w_2^1 = \alpha_2(1-\alpha_2), \ldots, w_n^1 = \alpha_2(1-\alpha_2)^{n-2} \tag{5}$$

$$w_1^2 == \alpha_1^{n-1}, w_2^2 = (1-\alpha_1)\alpha_1^{n-2}, \ldots, w_n^2 = 1-\alpha_1 \tag{6}$$

$$\alpha_1 = \frac{g-(j-i)}{g-1} \tag{7}$$

$$\alpha_2 = \frac{(j-i)-1}{g-1} \tag{8}$$

Also, the reciprocal values in each matrix are filled in to end up with the fully completed individual decision matrices. The reciprocal value of a trapezoidal fuzzy number, $\tilde{a}_{ij} = (a_{ij_l}, a_{ij_{m1}}, a_{ij_{m2}}, a_{ij_u})$ can be defined using Eq. 9.

$$\tilde{a}_{ji} = \left(\frac{1}{a_{ij_u}}, \frac{1}{a_{ij_{m2}}}, \frac{1}{a_{ij_{m1}}}, \frac{1}{a_{ij_l}} \right) \tag{9}$$

In Step 5, the consistency of each fuzzy pairwise comparison matrix is assessed. To evaluate consistency, the fuzzy pairwise comparison values are transformed from fuzzy values to crisp values using the graded mean integration method. This conversion involves transforming a trapezoidal fuzzy number $\tilde{a}_{ij} = (a_{ij_l}, a_{ij_{m1}}, a_{ij_{m2}}, a_{ij_u})$ into a crisp number using Eq. 10.

$$\mathrm{A} = \frac{l + 2m_1 + 2m_2 + u}{6} \tag{10}$$

Equations 11 and 12 can be applied to get the consistency ratio (CR).

$$CI = \frac{\lambda_{max} - n}{n - 1} \tag{11}$$

$$CR = \frac{CI}{RI} \tag{12}$$

When discussing the Consistency Index (CI), denotes the principal eigenvector of the matrix, n corresponds to the number of criteria, and RI represents the random index, which varies depending on the specific n. If the calculated CR value is less than 0.1, it can be deemed consistent.

After confirming consistency in each individual matrix, the next step is to aggregate these matrices using geometric means to create a consolidated or aggregate matrix as in Eq. 13.

$$\tilde{C} = \left[1 \cdots \tilde{c}_{xy} \vdots \ddots \vdots \tilde{c}_{yx} \cdots 1 \right] \tag{13}$$

such that

$$\tilde{c}_{ij} = \sqrt[K]{\prod_{k=1}^{n} \tilde{a}_{ij}^{k}} \tag{14}$$

Here, k represents the individual expert, and K is the total number of these experts.

In Step 7, the fuzzy geometric mean for each row ($\tilde{r}_i$) of the aggregate pairwise comparison matrix is estimated using Eq. 15.

$$\tilde{r}_i = (\tilde{c}_{i1} \otimes \tilde{c}_{i2} \cdots \otimes \tilde{c}_{in})^{1/n} \tag{15}$$

In Step 8, the fuzzy weight ($\tilde{w}_i$) of each criterion, sub-criterion, and alternative is calculated using Eq. 16.

$$\tilde{w}_i = \tilde{r}_i \otimes (\tilde{r}_1 \tilde{r}_2 \ldots \tilde{r}_n)^{-1} \tag{16}$$

In Step 9, the trapezoidal fuzzy numbers obtained are converted into crisp values using Eq. 10.

In Step 10, the defuzzified weights are normalized to calculate the local weights for the criteria. The overall weights for the sub-criteria are determined by multiplying their individual local weights by the weights of the corresponding main criteria.

In Step 11, the final score for each alternative is computed using Eq. 17.

$$S_i = \sum_{j=1}^{n} w_j s_{ij}, \forall i \tag{17}$$

4 Model

To establish the criteria employed in this research, an exhaustive examination of pertinent literature was the initial step. This process encompassed a thorough review of publications issued by the World Health Organization, focusing on radiotherapy equipment. Subsequently, these criteria were formulated based on the crucial aspects elucidated in these publications. To ensure their relevance and appropriateness, the criteria underwent rigorous assessment through consultations with experts. Since this study is related to radiotherapy device selection and requires special competence, all the participating experts have extensive experience in the field, spanning several years. The experts involved in the study consist of radiation oncology doctors, physicists working in the radiation oncology department and biomedical engineers working in the hospital's purchasing department. The finalization of the criteria list transpired following comprehensive discussions with these experts. While determining both main and sub-criteria, those elements that, while vital in device selection, did not exhibit significant variations among devices and brands were excluded from consideration. Subsequently, the criteria list was fine-tuned and solidified.

The hierarchical decision model proposed in this study (see Fig. 1) encompasses five primary criteria for radiotherapy device selection, namely "technical characteristics", "training, installation, and utilization", "warranty and maintenance", "economic factors", and "device brand specifications", and a total of 21 sub-criteria as can be defined below:

4.1 *Technical Characteristics (C1)*

The **radiation leakage limit (TC1)**.within a radiotherapy device denotes the highest acceptable amount of radiation that may escape beyond the treatment area during device operation [26]. This criterion holds significant importance as it safeguards both patients and healthcare personnel against undue radiation exposure. The specific limit of radiation leakage is contingent upon factors such as the radiation type

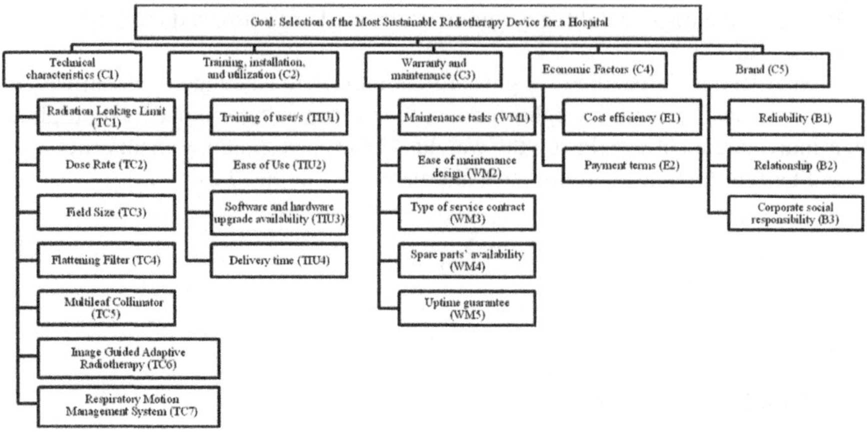

Fig. 1 Hierarchical decision model for most suitable radiotherapy device selection

and its energy, the particular design and protective measures incorporated into the radiotherapy devices, as well as the intended clinical purpose [27].

The **dose rate (TC2)** in radiotherapy equipment signifies the speed at which radiation is delivered to the patient during treatment [28]. It plays a crucial role in treatment planning, impacting both treatment duration and total radiation dosage. Selection of an optimal dose rate is determined by tumor characteristics, disease stage, and healthy tissue radiation tolerance.

Field size (TC3) is pivotal in radiotherapy planning, directly impacting tumor coverage and healthy tissue exposure [29]. Precise matching of field size to the target area is crucial to ensure the tumor receives the intended radiation dose while minimizing exposure to nearby healthy organs and tissues. Field size is influenced by the design and capabilities of the specific radiotherapy equipment in use, with differences in collimation systems and beam delivery mechanisms across device types leading to variations in available field sizes.

The **flattening filter (TC4)**, a crucial element in a linear accelerator, plays a key role in modifying the energy distribution of the radiation beam exiting the treatment head [30]. This flattened beam ensures consistent and accurate delivery of radiation treatment, facilitating precise dosage administration to the target area and appropriate dosing to surrounding healthy tissues. Various radiotherapy devices may employ different flattening filter types or alternative methods to achieve beam flattening.

Multileaf collimators (MLCs) (TC5) are vital components in contemporary linear accelerators, serving to adjust the intensity of radiation beams during radiotherapy [31]. They are crucial for delivering precise doses of radiation to the target area while reducing exposure to nearby healthy tissues. MLC design and features can differ across radiotherapy devices and manufacturers, including variations in the number, width, speed, penumbra of leaves, and integration with treatment planning systems [32].

IGART (TC6), a sophisticated radiotherapy approach, employs real-time imaging to guide and adjust radiation treatment based on changes in patient anatomy and tumor position during therapy [33]. The primary goal is to improve treatment precision by accommodating daily anatomical variations, ensuring accurate tumor targeting while sparing critical organs and healthy tissues. IGART systems vary across radiotherapy devices, with differences in imaging modalities, integration with treatment delivery, adaptive planning software, automation, and workflow [34].

Respiratory motion management (**TC7**) involves strategies and technologies aimed at addressing internal organ motion induced by patient breathing [35]. Such motion can compromise imaging and radiation treatment accuracy, particularly in organs like the lungs and liver [35]. The objective is to mitigate breathing-induced movement's impact on image quality and treatment precision. Variations exist in respiratory motion management systems across different radiotherapy devices.

4.2 Training, Installation, and Utilization (C2)

Training of users (**TIU1**) plays a crucial role in the safe and effective utilization of advanced medical equipment, like radiotherapy devices [36]. Training options may differ across various brands of radiotherapy devices.

The user-friendliness or **ease of use** (**TIU2**) of a radiotherapy device pertains to its simplicity and ease of operation for healthcare providers. A device that is easy to use can greatly enhance the efficiency, safety, and effectiveness of radiation treatments [16].

Software/hardware upgrade availability (**TIU3**) is vital for maintaining equipment relevance with the latest technological advancements, enhancing treatment capabilities, and meeting evolving regulatory requirements. Access to timely upgrades significantly prolongs the lifespan and enhances the value of the radiotherapy device.

The delivery time (**TIU4**) of a radiotherapy device may fluctuate based on multiple factors, such as the device type, manufacturer, delivery location (country or region), and equipment complexity.

4.3 Warranty and Maintenance (C3) [36]

Maintenance tasks (**WM1**) for a radiotherapy device are crucial for maintaining its peak performance, precision, and safety. Regular upkeep aids in identifying and addressing potential issues early on, preventing them from developing into critical problems, and ensures the device's compliance with regulatory standards.

Ease of maintenance (**WM2**) design in a radiotherapy device encompasses deliberate manufacturer decisions aimed at streamlining and simplifying maintenance and servicing procedures.

The type of service contract (WM3) selection relies on the particular requirements and financial resources of the healthcare institution.

Spare parts' availability (WM4) is very important for healthcare facilities. They should assess manufacturer support policies and explore service agreements or in-house inventory solutions to ensure uninterrupted operations and minimize downtime post-warranty.

Uptime guarantee (WM5) represents a pledge from a manufacturer, supplier, or service provider regarding the minimum duration during which a specific equipment or system will remain accessible and operational without experiencing significant downtime.

4.4 Economic Factors (C4)

Cost efficiency (E1) pertains to achieving desired outcomes while minimizing expenses and resource usage [16]. Each radiotherapy device may vary in features, capabilities, and cost structures, impacting overall cost efficiency. Payment terms for medical equipment denote agreed-upon conditions and schedules for vendor payments.

Payment terms (E2) are typically outlined in the purchasing agreement between the healthcare facility and the equipment provider [37]. These terms can be negotiated and may vary depending on factors like order size, the relationship between parties, and equipment complexity.

4.5 Brand (C5)

Reliability (B1) entails a brand's capacity to consistently provide reliable radiotherapy equipment and services [38]. It encompasses various factors contributing to the brand's reputation for producing trustworthy, safe, and effective medical devices.

Relationship (B2) Rezaei and Ortt [39] denotes the connections and interactions between a company and a healthcare facility. Cultivating robust and positive relationships is crucial for fostering successful and expanding business relations. This also encompasses business eagerness, which reflects the company's motivation to engage with target customers effectively, catering to their needs, wants, and preferences. Additionally, flexibility in business models enables adaptation to evolving market conditions and customer demands.

Corporate Social Responsibility (B3) pertains to a company's dedication to conducting its operations ethically and in a socially responsible manner, while also contributing to the betterment of society and the environment [40]. It is a voluntary initiative taken by businesses to go beyond their legal obligations and actively address social, environmental, and ethical issues.

5 Application and Findings

The criteria and sub-criteria in Fig. 1 served as the foundation for the selection of five LINAC device alternatives which are the choice of a collaborative effort with the participation of experts. These five devices were identified as Viewray MRdien, Electa Versa HD, Accuracy Radixact, Varian Truebeam and Electa Unity. For ethical reasons, the devices were kept anonymous, and they are denoted as A1, A2, A3, A4, and A5 randomly.

Given the specialized nature of this study, the participating experts possess significant expertise in the field. These experts collectively bring years of experience to the research. The expert panel comprises a radiation oncology physician, a medical physicist affiliated with the radiation oncology department, and a biomedical engineer responsible for procurement within the hospital.

Data were gathered from the participating experts through the utilization of a comparative questionnaire. This questionnaire facilitated the experts in making pairwise comparisons among criteria, sub-criteria, and alternative options. The questionnaire commenced with evaluations of the main criteria, followed by assessments of sub-criteria in relation to one another. Ultimately, each alternative was appraised based on the sub-criteria.

Considering the assessments provided by three experts regarding the primary criteria, it becomes evident that the technical characteristics criterion holds the highest degree of importance. Subsequently, the training, installation, and utilization criterion holds the second most significant position. The third in order of significance is the economic factors, which is reaffirmed by the consensus among our experts. On the other hand, the brand criterion is deemed the least important, with the warranty and maintenance criterion ranking just above it in terms of significance. Table 2 provides an overview of the importance weights calculated for all main and sub-criteria, and crisp global weights of alternatives against sub-criteria.

In technical characteristics, notably, the IGART system and dose rate criteria hold the highest and equal weight, indicating their critical significance. Respiratory motion management follows closely in importance, while the multileaf collimator criterion comes next in line. Conversely, the dose rate criterion holds relatively lower importance, with flattening filter and radiation leakage limit being the least prioritized. In the Training, Installation, and Utilization criteria, the training of users receives the highest weight, underlining its paramount importance. Ease of use and the availability of software/hardware upgrades follow closely in significance, while delivery time is deemed the least important consideration. Within the warranty and maintenance criteria, post-warranty spare parts availability is of utmost importance, ensuring quick device restoration in case of failure. Ease of maintenance design is ranked as the least prioritized factor, while the uptime guarantee holds the second-highest position. It's anticipated that the type of service contract and maintenance task will have diminishing influence over time. Moving to the economic criteria, payment terms carry more importance than cost efficiency, indicating its higher significance.

Table 2 Global weights of radiotherapy device options in terms of sub criteria

Main criteria (Weights)	Sub-criteria	Global weights of sub-criteria	Crisp global weights of alternatives				
			A1	A2	A3	A4	A5
C1 (0.491)	TC1	0.035	0.008	0.008	0.007	0.006	0.006
	TC2	0.050	0.012	0.011	0.010	0.009	0.009
	TC3	0.105	0.062	0.019	0.012	0.006	0.006
	TC4	0.036	0.008	0.008	0.007	0.007	0.006
	TC5	0.060	0.020	0.019	0.008	0.007	0.007
	TC6	0.105	0.004	0.014	0.014	0.035	0.037
	TC7	0.101	0.004	0.013	0.023	0.028	0.033
C2 (0.219)	TIU1	0.077	0.018	0.016	0.015	0.014	0.013
	TIU2	0.057	0.003	0.014	0.026	0.008	0.006
	TIU3	0.053	0.002	0.015	0.015	0.008	0.013
	TIU4	0.031	0.005	0.010	0.011	0.003	0.002
C3 (0.087)	W1	0.014	0.003	0.003	0.003	0.003	0.002
	W2	0.013	0.002	0.004	0.005	0.001	0.001
	W3	0.016	0.002	0.003	0.002	0.003	0.006
	W4	0.026	0.003	0.009	0.009	0.003	0.002
	W5	0.019	0.001	0.003	0.004	0.005	0.006
C4 (0.129)	E1	0.056	0.010	0.009	0.011	0.012	0.013
	E2	0.074	0.012	0.016	0.018	0.014	0.014
C5 (0.074)	B1	0.043	0.010	0.009	0.009	0.008	0.007
	B2	0.015	0.003	0.003	0.003	0.003	0.003
	B3	0.015	0.002	0.004	0.002	0.003	0.005

In terms of brand features, reliability ranks the highest, followed by relationship and company social responsibility, in decreasing order of importance.

Table 3 outlines the compliance hierarchy among radiotherapy devices, with A3 emerging as the most suitable choice, closely followed by A2 and A5. This comprehensive analysis provides valuable guidance for the informed selection of the optimal radiotherapy device, considering a multitude of criteria.

Table 3 Final scores of the alternatives

Alternatives	Weights
A1	0.195
A2	0.209
A3	0.214
A4	0.184
A5	0.198

6 Sensitivity Analysis

The objective of this analysis is to assess how alterations in the assigned weights to these criteria might influence the ultimate ranking of the options. This is achieved by individually adjusting the weights of the criteria within the range of 0–1, with increments of 0.1. Simultaneously, the weights of the remaining criteria are kept constant in proportion to their original values.

In Fig. 2a, a discernible trend is evident, where altering the weight of the technical characteristic criterion initially doesn't affect the preference for alternative A3. However, beyond a weight of 0.60, A3's ranking diminishes, and A1 temporarily becomes favored, reshuffling the overall order. At the 0.6 threshold, the sequence changes to A1, A5, A4, A2, and finally A3. This shift aligns with expectations, highlighting the interplay between criterion importance and alternative attributes, showcasing the decision-making process's sensitivity to varying weights. Figure 2b reveals sensitivity to changes in the training, installation, and utilization criterion weight. Below 0.1, A5 is preferred; above 0.2, A3 emerges as the clear choice, with the gap between A3 and other alternatives increasing, enhancing decision clarity. In Fig. 2c and d, altering weights for warranty and maintenance, and economic factors lead to minor changes in alternative order, while A3 remains the consistent choice. Stability in A3's preference underscores the model's robustness, holding true regardless of adjustments to these criterion weights. Figure 2e demonstrates a pattern were modifying the brand features criterion's weight impacts A3's preference. Lowering the weight maintains A3's position but increasing it towards 0.3 displaces A3 in favor of A2.

7 Conclusion

This study aimed to select the optimal Linear Accelerator (LINAC) radiotherapy device for hospital-based radiotherapy treatment. The research commenced with a comprehensive literature review to establish criteria influencing device selection, supplemented by expert consultations. Following this, refined criteria were determined, leading to the identification of 5 primary and 21 sub-criteria. Main criteria encompassed technical specifications, training, installation, use, warranty and maintenance, economic considerations, and brand attributes. Five LINAC device alternatives were assessed anonymously using the Hesitant Fuzzy Analytical Hierarchy Process method. Experts from radiation oncology and hospital procurement then ranked criteria and sub-criteria via pairwise comparisons. Technical characteristics were found crucial due to their impact on treatment outcomes, patient safety, and care quality. Healthcare professionals prioritize these attributes to ensure optimal patient treatment. The second most important main criterion is found to be training, installation, and utilization, reflecting experts' consensus on its significance in ensuring comprehensive device understanding and effective usage. Economic factors secured

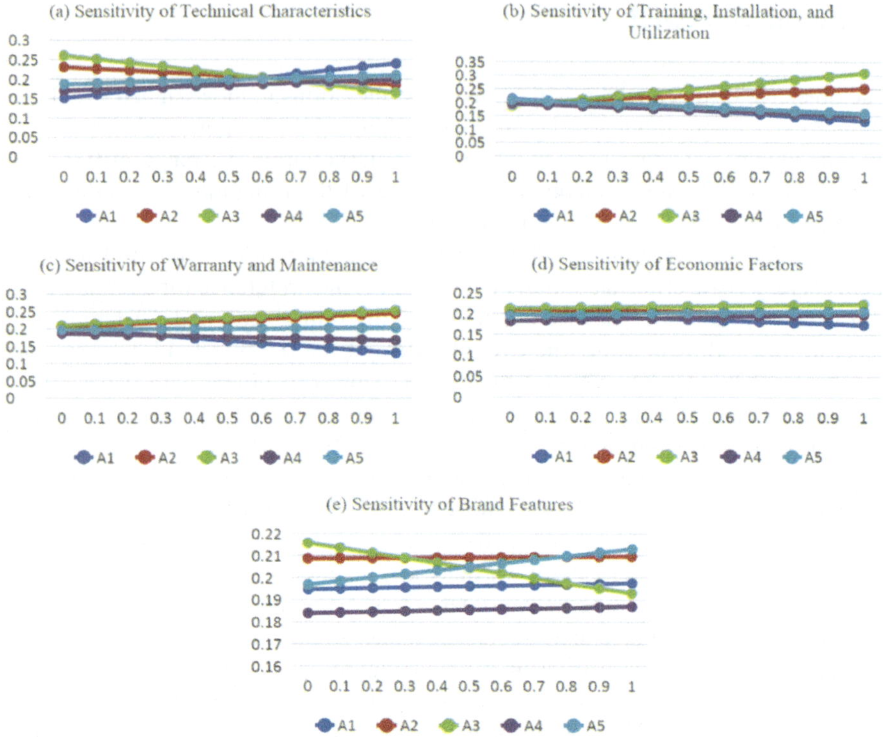

Fig. 2 Sensitivity analysis of main criteria with respect to different weights

the third position, important once technical specifications and usage prerequisites are met. Following this, warranty and maintenance gained significance, impacting long-term economic efficiency. The least significant criterion relates to vendor attributes. Limited number of reputable vendors in the medical device market reduced this criterion's impact. Companies' industry reputation diminished its importance. Together with these, sub-criteria evaluations provided valuable insights, too. IGART and field size were found vital in technical attributes, enhancing treatment accuracy and outcomes. Training, installation, utilization emphasized user training's role, ease of use, and upgrades. Economic factors favored payment terms over cost efficiency. In the warranty and maintenance sub-criteria, post-warranty spare parts availability holds the highest importance, emphasizing rapid device restoration in case of failure. Brand attributes highlighted reliability's paramountcy. Overall, the analysis provides insights into device selection's complex process. It integrates technical precision, training, efficiency, economics, and brand influence. Sub-criteria weights offer a framework for informed decisions balancing care, efficiency, finance, and sustainability. Comprehensive evaluation led to the conclusion that A3 was the most suitable radiotherapy device. Despite not excelling in all individual criteria, it emerged as optimal considering the comprehensive multi-criteria framework. It was followed by A2 and A5, while A4 ranked the lowest behind A1.

The study had some limitations that could impact its scope or findings. First of all, to illustrate the practical applicability of the method proposed, an application was conducted with staff from a private hospital in Istanbul. However, it is important to note that different hospitals may yield varied results due to their unique needs and circumstances. Hence, every hospital needs to select their own decision makers and use their evaluations to end up with the most appropriate importance weights for the criteria at hand tailored for their specific situation.

This research focused on LINAC devices, which are commonly used for radiotherapy treatment and are highly favored among such devices. While some parameters overlap with previous studies involving different devices, it's beneficial to tailor the selection criteria to accommodate the unique characteristics of new devices.

Furthermore, despite the suitability of the hesitant fuzzy AHP method for medical device selection, there is a scarcity of studies in this area. Therefore, it is suggested that future research explore the application of this method in medical device selection to enhance decision-making processes in healthcare settings.

References

1. Baskar R, Lee KA, Yeo R, Yeoh KW (2012) Cancer and radiation therapy: current advances and future directions. Int J Med Sci 9(3):193–201
2. Abshire D, Lang MK (2018) The evolution of radiation therapy in treating cancer. Semin Oncol Nurs 34(2):151–157
3. Kumar A, Garg P, Pant S, Ram M, Kumar A (2022) Multi-criteria decision-making techniques for complex decision-making problems. Math Eng Sci Aerosp (MESA) 13(2):791–803
4. Acar C, Haktanir E, Temur GT, Beskese A (2024) Sustainable stationary hydrogen storage application selection with interval-valued intuitionistic fuzzy AHP. Int J Hydrogen Energy 49(D):619–634
5. Taherdoost H, Madanchian M (2023) Multi-criteria decision making (MCDM) methods and concepts. Encyclopedia 3(1):77–87
6. Canco I, Kruja D, Iancu T (2021) AHP, a reliable method for quality decision making: a case study in business. Sustainability 13(24):13932
7. Liu Y, Eckert CM, Earl C (2020) A review of fuzzy AHP methods for decision-making with subjective judgements. Expert Syst Appl 161:113738
8. Aliyev R, Temizkan H, Aliyev R (2020) Fuzzy analytic hierarchy process-based multi-criteria decision making for universities ranking. Symmetry 12(8):1351
9. Nieto-Morote A, Ruz-Vila F (2011) A fuzzy AHP multi-criteria decision-making approach applied to combined cooling, heating, and power production systems. Int J Inf Technol Decis Mak 10(3):497–517
10. Torra V (2010) Hesitant fuzzy sets. Int J Intell Syst 25(6):529–539
11. Wang JQ, Wang J, Chen QH, Zhang HY, Chen XH (2014) An outranking approach for multi-criteria decision-making with hesitant fuzzy linguistic term sets. Inf Sci 280:338–351
12. FSM Russo R, Camanho R (2015) Criteria in AHP: a systematic review of literature. Proc Comput Sci 55:1123–1132
13. Božić V (2023) Fuzzy approach to risk management: enhancing decision-making under uncertainty

14. Onay A, Karamasa C, Sarac B (2016) Application of fuzzy AHP in selection of accounting elective courses in undergraduate and graduate level. J Account Finance Auditing Stud 2(4):20–42
15. Ranjbar M, Effati S (2023) Group decision making in the analytic hierarchy process by hesitant fuzzy numbers. Sci Rep 13(1):21864
16. Park J, Kim E, Shin K (2019) Developing an evaluation framework for selecting optimal medical devices. J Open Innov Technol Market Complex 5(3):64
17. Kritchanchai D, Wahab SN, Tan A, Mak T (2022) An analytical hierarchy process-based decision making for sustainable medical devices development. LogForum 18(4):481–493
18. Shbool MA, Arabeyyat OS, Al-Bazi A, AlAlaween WH (2021) An integrated multi-criteria decision-making framework for a medical device selection in the healthcare industry. Cogent Eng 8(1):1968741
19. Abdel-Basset M, Manogaran G, Gamal A, Smarandache F (2019) A group decision making framework based on neutrosophic TOPSIS approach for smart medical device selection. J Med Syst 43(2):1–13
20. Otay I (2022) Intuitive fuzzy multi-expert & multi-criteria decision-making methodology: an application in healthcare industry. J Fac Eng Archit Gazi Univ 37(2):1047–1062
21. Goh M, Shuya Z, De Souza R (2018) Operational framework for healthcare supplier selection under a fuzzy multi-criteria environment. In: Proceedings of the 23rd international symposium on logistics (ISL 2018), Bali, Indonesia, pp 129–136
22. Öztürk N, Tozan H (2015) A fuzzy-based decision support model for selecting the best dialyser flux in hemodialysis. J Healthc Eng 6:303–324
23. Büyüközkan G, Mukul E (2020) Evaluation of smart health technologies with hesitant fuzzy linguistic MCDM methods. J Intell Fuzzy Syst 39(5):6363–6375
24. Buckley JJ (1985) Fuzzy hierarchical analysis. Fuzzy Sets Syst 17(3):233–247
25. Liu H, Rodríguez RM (2014) A fuzzy envelope for hesitant fuzzy linguistic term set and its application to multicriteria decision making. Inf Sci 258:220–238
26. Balog J, Lucas D, DeSouza C, Crilly R (2005) Helical tomotherapy radiation leakage and shielding considerations. Med Phys 32(3):710–719
27. Beeksma B, Lehmann J (2019) Excessive applicator radiation leakage for a common therapeutic kilovoltage system. Br J Radiol 92(1094):20180743
28. Dale RG (1996) Dose-rate effects in targeted radiotherapy. Phys Med Biol 41(10):1871
29. Méndez I, Casar B (2021) A novel approach for the definition of small-field sizes using the concept of superellipse. Radiat Phys Chem 189:109775
30. Xiao Y, Kry SF, Popple R, Yorke E, Papanikolaou N, Stathakis S, Xia P, Huq S, Bayouth J, Galvin J, Yin F-F (2015) Flattening filter-free accelerators: a report from the AAPM therapy emerging technology assessment work group. J Appl Clin Med Phys 16(3):12–29
31. Fiveash JB, Murshed H, Duan J, Hyatt M, Caranto J, Bonner JA, Popple RA (2002) Effect of multileaf collimator leaf width on physical dose distributions in the treatment of CNS and head and neck neoplasms with intensity modulated radiation therapy. Med Phys 29(6):1116–1119
32. Boyer A (2001) AAPM radiation therapy committee: basic applications of multileaf collimators. American Association of Physicists in Medicine, Alexandria, VA
33. Goyal S, Kataria T (2014) Image guidance in radiation therapy: techniques and applications. Radiol Res Pract 1–10
34. Gupta T, Narayan CA (2012) Image-guided radiation therapy: physician's perspectives. J Med Phys/Assoc Med Phys India 37(4):174
35. Chi A, Nguyen NP, Komaki R (2014) The potential role of respiratory motion management and image guidance in the reduction of severe toxicities following stereotactic ablative radiation therapy for patients with centrally located early-stage non-small cell lung cancer or lung metastases. Front Oncol 4:151
36. World Health Organization: Technical specifications of radiotherapy equipment for cancer treatment (2021)
37. Bahadori M, Hosseini SM, Teymourzadeh E, Ravangard R, Raadabadi M, Alimohammadzadeh K (2020) A supplier selection model for hospitals using a combination of artificial neural network and fuzzy VIKOR. Int J Healthc Manag 13(4):286–294

38. Stević Ž, Pamučar D, Puška A, Chatterjee P (2020) Sustainable supplier selection in healthcare industries using a new MCDM method: measurement of alternatives and ranking according to compromise solution (MARCOS). Comput Ind Eng 140:106231
39. Rezaei J, Ortt R (2013) Multi-criteria supplier segmentation using a fuzzy preference relations based AHP. Eur J Oper Res 225(1):75–84
40. Govindan K, Shankar M, Kannan D (2018) Supplier selection based on corporate social responsibility practices. Int J Prod Econ 200:353–379

A Multi-objective Optimization Model for Waste-to-Energy Facility Location in Sustainable Municipal Solid Waste Management

Busra Cicekalan, Aycan Pekpazar, and Ferhan Çebi

Abstract Efficient municipal solid waste (MSW) management has a vital role in securing economic, environmental, and social sustainability. Waste-to-Energy (WtE) is a significant method that converts waste into electricity or heat energy, offering a crucial approach to reducing MSW. In this study, a new multi-objective mixed-integer programming model was developed in order to determine the location of the WtE facility with three objectives. The three objectives of the mathematical model were as follows: (1) minimizing total costs, including investment cost of incineration facility, transportation cost, and operating costs; (2) maximizing energy generation; and (3) minimizing greenhouse gas emission. The proposed model incorporates constraints related to the monetary equivalence of objectives, facility location, capacity constraints, mass balances, and flow balances across municipalities, transfer stations, recycling centers, and landfills. The proposed model was applied to a practical scenario in Istanbul, Türkiye, demonstrating its real-world applicability. Based on the model result of the Multi Obj, the optimal location of the WtE facility was found to be E3 and A2 for the European and Asian sides, respectively. The model results revealed that the optimal location for the WtE facility is close to the recycling facility and landfill, primarily to minimize transportation costs between waste source sites and these facility ies. stations, recycling centers, and incineration facilities. We applied this model to a case study in Istanbul, Türkiye. The findings from our study

B. Cicekalan (✉)
Department of Environmental Engineering, Faculty of Civil Engineering, Istanbul Technical University, Maslak, Istanbul 34469, Turkey
e-mail: cicekalan@itu.edu.tr

A. Pekpazar
Department of Industrial Engineering, Faculty of Engineering, Samsun University, 55420 Samsun, Turkey
e-mail: aycan.pekpazar@samsun.edu.tr

F. Çebi
Department of Management Engineering, Faculty of Management Engineering, Istanbul Technical University, Maslak, Istanbul 34469, Turkey
e-mail: cebife@itu.edu.tr

F. Çebi et al. (eds.), *Engineering and Technology Management in Challenging Times*, Lecture Notes in Networks and Systems 1137,
https://doi.org/10.1007/978-3-031-70935-7_5

have significant implications for initiating new MSW management projects with a focus on sustainability.

Keywords Mathematical model · Municipal solid waste (MSW) management · Waste-to-energy (WtE)

1 Introduction

Nowadays, municipal solid waste (MSW) management regarding its harmful effects, like acid compounds, carbon dioxide, and sludge waste, on the environment has become prominent. MSW is a type of waste generated in urban areas that municipalities collect and dispose of. Furthermore, electricity and heat can be generated from MSW, which can be used for the distribution of electricity, heating, cooling, and industrial processes [1]. The energy can be produced throughout the waste in direct (e.g., incineration) and indirect approaches (e.g., combustion, pyrolysis, and biogas). Energy recovery with incineration, a waste-to-energy (WtE), produces energy from the waste with direct combustion. Energy generation from waste helps governments achieve sustainable energy goals [2].

Energy recovery from waste through incineration is one of the sustainable solutions to address energy demand. Therefore, several studies have been conducted on developing mathematical models to solve facility location problems for achieving efficient solid waste management. The majority of the studies addressed the hazardous waste management [3–7]. However, there are a few studies on the location problem of the WtE process for MSW management [8–11]. Habibi et al. [8] developed a multi-objective mathematical model for the MSW management system to minimize the total cost and greenhouse gas emissions. Ayvaz-Cavdaroglu et al. [9] developed mathematical models to compare waste management approaches, including material recovery, incineration, composting, landfill, and anaerobic digestion. In the study of Hrabec et al. [10], a mathematical model was developed to determine the location of the WtE facility in the Czech Republic to minimize investment and transportation costs. Ooi et al. [11] proposed a multi-objective optimization model integrated with the MSW-emissions trading scheme (ETS) framework to achieve environmental and economic benefits.

The mathematical models to design MSW management systems in the literature were mainly focused on two objective functions, including the environmental and economic, and these studies excluded the influence of energy generation. Hence, this study aims to fill the gap in the literature by proposing a multi-objective mixed-integer programming model to determine the locations of Waste-to-Energy (WtE) facilities. The model encompasses three primary objectives: (1) minimizing total costs, which include the investment cost of opening an incineration facility, transportation costs, and operating costs; (2) maximizing energy generation from MSW; and (3) minimizing greenhouse gas (GHG) emissions. These objectives represent a comprehensive approach addressing environmental, economic, and energy generation concerns.

The proposed mathematical model was applied in a real-life case to determine the locations of incineration facilities in Istanbul, Türkiye. The proposed model was first solved separately with GAMS/CPLEX for each objective, and subsequently, it was solved using the monetary equivalent of each objective simultaneously.

The subsequent sections of this paper are structured as follows: In Sect. 2, the literature review regarding mathematical models for MSW management systems is provided. Section 3 includes the formulation of the proposed mathematical model, and a large-scale implementation of the proposed mathematical model in Türkiye is given in Sect. 4. In Sect. 5, the application of the proposed methodology and the results are presented. Lastly, in Sect. 6, conclusions and future research areas are provided.

2 Literature Review

There are several studies on developing mathematical models to solve facility location–routing problems related to providing efficient solid waste management. The majority of the studies addressed the hazardous waste problem in the literature [3–7, 12–16]. Emek and Kara [4] focused on the hazardous waste management problem and proposed a cost-based mathematical model in which the satisfaction of air pollution standards. The Gaussian Plume equation in measuring the air pollution concentrations at population centers was used in the model [4]. Aydemir-Karadag [6] developed a mixed-integer mathematical model to solve the hazardous waste location-routing problem by including different hazardous waste types, waste management principles, electricity generation, and waste-to-technology compatibilities. Yu et al. [7] developed a stochastic bi-objective mixed-integer linear program, and the model was applied to a real-life case study of hazardous waste management in Wuhan, China to demonstrate its applicability. A multi-objective stochastic mixed-integer nonlinear programming model for hazardous waste management was formulated by Rabbani et al. [15]. Saeidi-Mobarakeh et al. [16] developed a robust optimization model for decision-making in hazardous waste management on three main activities which are collection, treatment, and disposal of hazardous waste.

On the other hand, the literature has provided mathematical models for location evaluation and selection problems for MSW management [8, 17–24]. Son and Louati [17] suggested a vehicle routing model for MSW collection by including multiple transfer stations. Habibi et al. [8] proposed a multi-objective mathematical model for the MSW management system, which aims to minimize the total cost and greenhouse gas emissions. Yadav et al. [18] developed an algorithm based on interval analysis to solve facility location problems by dealing with the uncertainties in the MSW system. They performed a hypothetical application rather than a real-life application. Moreover, Darmian et al. [19] developed a multi-objective location-districting model for the collection of MSW in Iran. They reported that MSW collection centers might be located to minimize environmental pollutants, and the cost of MSW collection might be reduced by 20–30% [19].

Three mathematical models in order to compare different MSW technologies including material recovery, incineration, composting, landfill, and anaerobic digestion, which are used to treat different types of waste components (e.g., paper, organic, plastics, metals, glass, and others) were developed by Ayvaz-Cavdaroglu et al. [9]. The proposed methodology was applied to existing facilities in Istanbul, Türkiye [9]. Similarly, Yousefloo and Babazadeh [20] introduced a multi-objective mixed-integer linear programming model aimed at the optimization of total costs and risk objectives within MSW management technologies. The proposed model was applied to the MSW management network in Iran to verify the model [20].

Hrabec et al. [10] applied a mixed integer programming model to find out the best location for the WtE facility to minimize total cost including investment and transportation costs in the Czech Republic. Ilbahar et al. [21] developed a fuzzy linear programming model to determine the most suitable locations for the WtE facility in the Central Anatolia Region of Türkiye. In the study of Hu et al. [22] a robust two-stage model was developed to identify the best location of the WtE facility by considering economic and environmental aspects. A two-stage stochastic mixed-integer linear programming model was proposed by Pluskal et al. [23]. The results obtained from the model provided insights into the economics of operation and identified important factors in the sustainability of the MSW system. However, as highlighted above, the proposed mathematical models regarding MSW management were mainly developed by considering economic and environmental (greenhouse gas emissions) aspects excluding energy generation from MSW. Hence, the study will contribute to the literature by using a multi-objective mixed-integer programming model for the location selection problem of the WtE facility aimed at minimizing the total cost, minimizing greenhouse gas emission, and maximizing energy generation. Furthermore, this study will indicate its suitability for real-world decision-making problems.

3 Mathematical Model

The network construction for the proposed model in the study is depicted in Fig. 1. The transportation network can be considered a key factor for MSW management. In the study, the transportation network consisted of generation nodes (Municipalities), transportation stations, existing recycling facilities, possible incineration facility locations, and disposal facilities. If the distance between a municipality or recycling center and a potential incineration facility exceeds 50 km, the waste first goes to a transfer station before reaching the incineration facility. However, if the distance is less than 50 km, the waste is directly sent to the incineration facility.

The sets, parameters, decision variables, and constraints required for the mathematical model were given as follows:

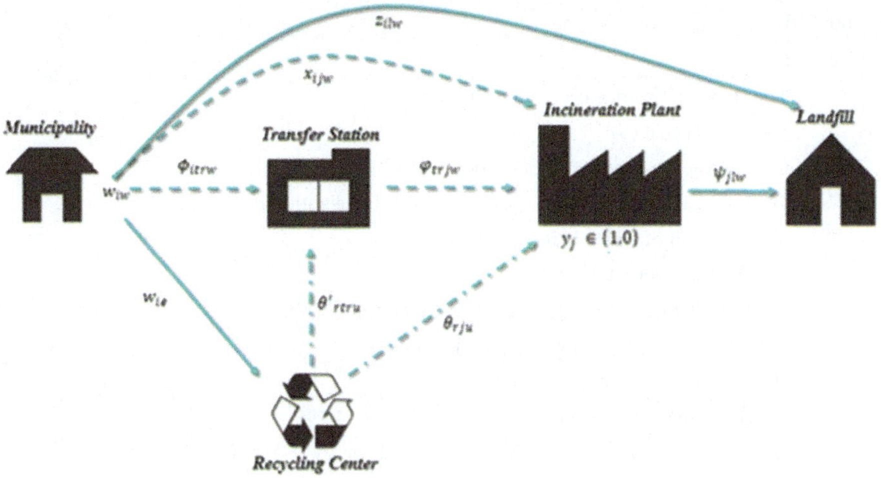

Fig. 1 The proposal model for MSW management

Sets	
P: Population center	$P = \{1, \ldots, p\}$
J: Potential incineration centers	$J = \{1, \ldots, j\}$
M: Municipalities	$M = \{1, \ldots, i\}$
T: Transfer station	$T = \{1, \ldots, t\}$
R: Recycling centers	$R = \{1, \ldots, r\}$
L: Landfill	$L = \{1, \ldots, l\}$

Parameters	
mw_i	Amount of municipal solid waste at municipality $i \in M$
rw_r	Amount of recyclable waste at municipality $i \in M$
α_r	Reduction rate for waste type municipal solid waste at the recycling center $r \in R$
β_r	Reduction rate for recyclable waste at recycling center $r \in R$
γ_i	Reduction rate for waste type incombustible-municipal solid waste at municipality $i \in M$
θ_j	Ratio of residual waste from incineration center
g_a	GHG emissions associated with flue gas treatment by incineration facility (kg CO_2 eq/ton)
g	GHG emissions associated with transportation node (kg CO_2 eq/km)
Cap_j	Capacity of incineration at site j
Cap_j^{min}	Minimum capacity required for incineration at the site j
TC	Unit transportation cost

(continued)

(continued)

Parameters	
d_{ij}	The shortest distance between i and j (km)
IC	Installation cost ($)
GC	Cost of flue gas treatment for air pollutants ($)
C_j	Operation and maintenance (O&M) costs of incineration facility $j \in J$ ($)
ϖ_T	Total amount of waste at incineration
η_e	Electrical conservation (Efficiency) (%)
ω	Conversion rate of total energy to electrical energy (KWh/ton)
ε_{elj}	Electric energy at incineration facility $j \in J$
ε_{hj}	Heat energy at incineration facility $j \in J$
CT	Carbon tax ($/kg CO_2 eq)
rv	Revenue from the sale of 1 kWh of electrical energy ($/KWh)
ɜ	$\begin{cases} 1 \; d_{ij} \geq 50\,km \; i \in M \\ 0 \; otherwise \end{cases}$
æ	$\begin{cases} 1 \; d_{rj} \geq 50\,km \; r \in R \\ 0 \; otherwise \end{cases}$
Decision Variables	
y_j	$\begin{cases} 1 \; The\ incineration\ is\ opened\ at\ jth\ candidate\ site\ n \in J, \\ 0 \; Otherwise \end{cases}$
x_{ij}	Amount of municipal solid waste that goes from municipality i to incineration j
w_{ir}	Amount of municipal solid waste that goes from municipality to recycling center r
z_{il}	Amount of incombustible-municipal solid waste that goes from municipality i to landfill l
ϕ_{it}	Amount of MSW that goes from municipality i to transfer station t
φ_{tj}	Amount of MSW that goes from transfer station t to incineration j
θ_{rj}	Amount of unrecyclable waste that goes from recycling center r to incineration j
θ'_{rt}	Amount of unrecyclable waste that goes from recycling center r to transfer station t
ψ_{jl}	Amount of waste residue transported from incineration j to landfill l

3.1 Mathematical Model

$$Obj1.Min\,\text{Cost} = \sum_{i,t} TC * d_{it} * \phi_{it} + \sum_{t,j} TC * d_{tj} * \varphi_{tj} + \sum_{i,j} TC * d_{ij} * x_{ij} + \sum_{i,r} TC * d_{ir} * w_{ir} + \sum_{r,t} TC * d_{rt} * \theta'_{rt} + \sum_{r,j} TC * d_{rj} * \theta_{rj}$$

$$+\sum_{j,l} TC * d_{jl} * \psi_{jl} + \sum_{i,l} TC * d_{il} * z_{il} + \sum_{i,j} IC * y_j$$
$$+\sum_{i,t,j} C_J * (\varphi_{tj} + x_{ij} + \theta_{rj}) * y_j + \sum_{i,j} GC * y_j \tag{3.1}$$

$$Obj2.\,Min.GHG = \sum_j (\varphi_{tj} + x_{ij} + \theta_{rj}) * g_a$$
$$+ 2g\left[\sum_{i,t} d_{it} + \sum_{i,j} d_{ij} + \sum_{i,r} d_{ir} + \sum_{r,t} d_{rt} \right.$$
$$\left. + \sum_{r,j} d_{rj} + \sum_{j,l} d_{jl} + \sum_{i,l} d_{il}\right] \tag{3.2}$$

$$Obj3.Max.EN = \sum_j \varpi_T * \omega * y_j \tag{3.3}$$

and

$$Multi\,Obj.Max = rv * Obj3 - CT * Obj2 - Obj1 \tag{3.4}$$

s.t.

$$\sum_i x_{ij} + \sum_t \varphi_{tj} + \sum_r \theta_{rj} \le y_j * Cap_j \quad \forall j \in J \tag{3.5}$$

$$\sum_i x_{ij} + \sum_t \varphi_{tj} + \sum_r \theta_{rj} \ge y_j * Cap_j^{min} \quad \forall j \in J \tag{3.6}$$

$$mw_i = \sum_t \phi_{it} * 3 + \sum_j x_{ij} * 3 + \sum_r w_{ir} + \sum_l z_{il} \quad \forall i \in M \tag{3.7}$$

$$\sum_j \varphi_{tj} = \sum_r \theta'_{rt} + \sum_i \phi_{it} \quad \forall t \in T \tag{3.8}$$

$$\sum_i w_{ir} = \sum_t \theta'_{rt} + \sum_j \theta_{rj} \quad \forall r \in R \tag{3.9}$$

$$rw_r = \alpha_r * \sum_i w_{ir} \quad \forall r \in R \tag{3.10}$$

$$\sum_j \theta_{rj} + \sum_t \theta'_{rt} = 3 * (1 - \beta_r) * rw_r \quad \forall r \in R \tag{3.11}$$

$$\sum_l z_{il} = \gamma_i * mw_i \quad \forall i \in M \tag{3.12}$$

$$\sum_{l} \psi_{jl} = \theta_j * \left(\sum_{r} \theta_{rj} + \sum_{i} x_{ij} + \sum_{t} \varphi_{tj} \right) \quad \forall l \in L \tag{3.13}$$

$$\left[\varpi_T = \sum_{r,j} \theta_{rj} + \sum_{i,j} x_{ij} + \sum_{t,j} \varphi_{tj} - \sum_{j,l} \psi_{jl} \right] \tag{3.14}$$

$$\sum_{j} \varepsilon_{elj} = \omega * \eta_e * \sum_{j} \varpi_T * y_j \tag{3.15}$$

$$\sum_{j} \varepsilon_{hj} = (1 - \eta_e) * \omega * \sum_{j} \varpi_T * y_j \tag{3.16}$$

$$y_j \in \{1, 0\} \quad \forall j \in J \tag{3.17}$$

$$x_{ij}, z_{il}, \phi_{it}, \varphi_{tj}, \theta_{rj}, \theta'_{rt}, \psi_{jl} \geq 0 \tag{3.18}$$

The first objective (3.1) endeavors to minimize the overall cost, including investment, transportation, and operating costs, as well as the gas treatment cost of opening an incineration facility. The second objective (3.2) aims to minimize the carbon dioxide emissions from operating an incineration facility and the transportation between facilities. The third objective (3.3) is to maximize the energy generation from an operational incineration facility. Constraint (3.4) consolidates the monetary representation of the three objective functions. Constraints (3.5) ensure that the total flow into an incineration facility cannot exceed its specified capacity. Constraints (3.6) ensure the decided open incineration center meets the minimum capacity requirements.

Constraint (3.7) illustrates the mass balance constraints at each Municipality. Constraints (3.8) ensure that the total amount of waste the transfer station receives from municipalities and recycling centers equals the total amount sent from the transfer station to the incineration center. Constraint (3.9) ensures that the total amount of waste received by the recycling center from municipalities equals the total amount of waste sent from the recycling center to the incineration facility and the transfer station. Constraint (3.10) calculates the amount of recyclable waste generated from the total municipal solid waste received based on the reduction rate at the recycling center. A conversion factor is used for each recycling center to find the amount of recyclable waste sent to the recycling center. In addition, Constraint (3.11) is the mass balance constraint for the recycling centers, including a conversion factor to determine the amount of remaining waste after the recycling process for each recyclable waste type and each recycling center. The unrecyclable amount is sent to the incineration facility. Constraints (3.12) and (3.13) represent the waste sent to the landfill from municipalities and incineration centers. Constraints (3.14), (3.15), and (3.16) indicate the total amount of electric and heat energy produced from the waste at the incineration facilities. Constraints (3.17) and (3.18) represent the non-negativity constraints and the conditions defining the binary variables, respectively.

4 Application in Türkiye

The model was applied to Istanbul, Türkiye's most populated city, with a population of 15 million [25]. Istanbul has three landfills, three recycling facilities, and eight transfer stations. Based on the MSW management in Istanbul, all waste is sent to eight transfer stations within the city. The waste is then transferred to a recycling and composting facility. The recyclable waste is stored in a recycling facility, while the remaining waste is sent to a composting facility. All the residual and non-recyclable materials are sent to landfills. In the proposed model, ten possible locations of the incineration facility were determined; four of them are on the European side, and the remaining six possible locations are on the Asian side (Fig. 2). The ideal location for the incineration facility should be near industrial zones, downtown areas, existing landfills, or recycling facilities. Additionally, a primary consideration for establishing the incineration facility is to achieve the lowest total cost and optimal energy distribution while adhering to air pollution standards. Furthermore, the facility should not be near green fields, protected or sensitive areas, or water resources.

In the model, the amount of the MSW in Istanbul was projected based on the studies conducted by Kor et al. [26] and Ozturk [27]. It was assumed that 65% of the MSW was recyclable waste and 9.7% was non-combustible. The remaining MSW was defined as combustible waste. The objective functions and all parameters required for the problem formulation are presented in Table 1.

Fig. 2 Possible location for incineration facility

Table 1 Objective functions and the other parameters required for the problem

Functions/Parameter	Value	Adapted from
Unit transportation cost	2.17 $/km.ton	Koushki et al. [28]
Installation cost	120 10^6 $	Ozturk [27]
Operating cost	40–80 $/ton	Ozturk [27]
GHG emissions	1189 CO_2 kg.eq/ton	Korkut [29]
GHG emissions associated with transportation node	0.69 CO_2 kg.eq/km	Yılmaz-Balaman [30]
The conversion rate of total energy to electrical energy	~500 kWh/ton	Ozturk [27]
Carbon fee	50–100 $/$CO_2$ton	OECD [31]
Revenue from the sale of 1 kWh of electrical energy	0.02 $/kWh	[32]

5 Results and Discussion

Obj1, Obj2, and Obj3 were individually solved with GAMS/CPLEX for both Asian and European sides. Based on the model results (Table 2), the optimal location of the incineration facility for the Asian side was determined to be A1 and A2, whereas it was found to be E3 and E2 for the European side. The electricity and heat energy percentages of the total energy were approximately 30% and 70% for each side, respectively. In general, the heat energy might be used to meet the facility's own energy needs, whereas electricity can be supplied to the national grid for sale and distribution.

According to the formulation given Multi Obj. (see Eq. 3.4), the model was solved by using the monetary equivalent of each objective simultaneously (Table 3). The result of the model presented that the optimal WtE facility was E3 and A2 for the European and Asian sides, respectively. E3 is close to the existing Incineration

Table 2 The model results of Obj1, Obj2, and Obj3

Objectives	Side	Location	Metric	Value
Obj1	European	E3	Unit total cost ($/ton)	341.6
	Asian	A2		356.8
Obj2	European	E3	GHG emission (kgCO_2eq/year)	7.42×10^8
	Asian	A1		6.07×10^8
Obj3	European	E2	Total energy (MWh/year) Electricity (MWh/year) Heat (MWh/year)	1.181730 367,262 (31%) 814,468 (69%)
	Asian	A1		1.031060 282,208 (28%) 748,852 (72%)

Table 3 The model results of Multi obj

Metric	European side	Asian side
Selected site	E3	A2
Unit total cost ($/ton)	360	340
GHG emission (kgCO_2eq/year)	7.42×10^8	6.08×10^8
Total energy (MWh/year)	1,161,931	904,015
• Electricity	326,588 (28%)	262,164 (29%)
• Heat	835,343 (72%)	641,850 (71%)

Facility, which meets the energy demand of almost 1.4 million people. The model results suggested that the optimal location for the WtE facility is close to the recycling facility and landfill, primarily to minimize transportation costs between waste source sites and these facilities. In the study, the WtE facility's unit cost was around 360 $/ton and 340 $/ton for E3 and A2, respectively. The unit total cost of the WtE facility was reported in the range of 300 to 900 $/ton in the literature [33]. Moreover, in the study, GHG emissions from energy generated through waste incineration were 590 g CO_2/kWh for E2 and 672 g CO_2/kWh for A2. This finding was in line with the results of UKWIN [34], in which GHG emission was about 540–937 g CO_2/kWh of waste incinerated in the United Kingdom.

6 Conclusion

In this study, we explored the waste-to-energy (WtE) facility location problem and formulated a multi-objective mixed-integer programming model. This model aimed to maximize energy generation from municipal solid waste (MSW), minimize the total cost, and reduce GHG emissions. The model incorporated constraints that specify the capacity and location of the WtE facilities, manage waste flow, determine the need for transfer stations, handle the operations of recycling facilities, and guide the energy recovery processes. The proposed model offered a significant contribution to the establishment of effective and eco-friendly waste management systems by balancing waste flow and transformation between municipalities, transfer stations, recycling centers, and WtE facilities. We applied this model to a case study in Istanbul, Türkiye. Based on the model result of the Multi Obj, the optimal WtE facility location was found to be E3 and A2 for the European and Asian sides, respectively. The model results revealed that the optimal location for the WtE facility is close to the recycling facility and landfill, primarily to minimize transportation costs between waste source sites and these facilities. It was found that the electricity and heat energy percentages of the total energy were approximately 30% and 70% for each side, respectively. In general, the heat energy might be used to meet the facility's own energy needs, whereas electricity can be supplied to the national grid for sale and distribution. In the study, the unit cost of the WtE facility was almost 360 $/ton and

340 $/ton for E3 and A2, respectively. The findings from our study have significant implications for initiating new waste management projects with a focus on sustainability. However, there are some limitations and potential future research. While our study successfully employed an exact algorithm to solve the problem optimally using GAMS/CPLEX, the case study we solved is a small-scale problem. For future studies, running the model on larger samples and conducting economic analyses would be beneficial, highlighting potential areas for improvement and broader applicability. Furthermore, future research focusing on developing heuristic algorithms, such as genetic algorithms, may be beneficial for efficiently addressing large-scale problems.

References

1. EPA (2019) Combined heat and power (CHP) partnership. U.S. environmental protection agency
2. Zhang Z, Mao L, Nithya K, Loh KC, Dai Y, He Y, Tong YW (2019) Optimizing mixing strategy to improve the performance of an anaerobic digestion waste-to-energy system for energy recovery from food waste. Appl Energy 249:28–36
3. Wyman MW, Kuby M (1995) Proactive optimization of toxic waste transportation, location, and technology. Locat Sci 3(3):167–185
4. Emek E, Kara BY (2007) Hazardous waste management problem: the case for incineration. J Comput Oper Res 34:1424–1441
5. Erkut E, Karagiannidis A, Perkoulidis G, Tjandra SAA (2008) Multicriteria facility location model for municipal solid waste management in north Greece. Eur J Oper Res 187(3):1402–1421
6. Aydemir-Karadag A (2018) A profit-oriented mathematical model for hazardous waste locating-routing problem. J Clean Prod 202:213–225
7. Yu H, Sun X, Solvang WD, Laporte G, Lee CKM (2020) A stochastic network design problem for hazardous waste management. J Clean Prod 277:123566
8. Habibi F, Asadi E, Sadjadi SJ, Barzinpour F (2017) A multi-objective robust optimization model for site-selection and capacity allocation of municipal solid waste facilities: a case study in Tehran. J Clean Prod 166:816–834
9. Ayvaz-Cavdaroglu N, Coban A, Firtina-Ertis I (2019) Municipal solid waste management via mathematical modeling: a case study in İstanbul, Turkey. J Environ Manage 244:362–369
10. Hrabec D, Šomplák R, Nevrlý V, Viktorin A, Pluháček M, Popela P (2020) Sustainable waste-to-energy facility location: influence of demand on energy sales. Energy 207:118257
11. Ooi JK, Hoy ZX, Hossain MU, Zhang Z, Khan M, Woon KS (2023) An integrated multi-objective optimization framework for municipal solid waste management and emissions trading scheme. Clean Technol Environ Policy 1–15
12. Tavares G, Zsigraiová Z, Semiao V (2011) Multi-criteria GIS-based siting of an incineration plant for municipal solid waste. Waste Manage 31(9–10):1960–1972
13. Tirkolaee EB, Abbasian P, Weber GW (2020) Sustainable fuzzy multi-trip location-routing problem for medical waste management during the COVID-19 outbreak. Sci Total Environ 756:143607
14. Boyer O, Hong TS, Pedram A, MohdYusuff RB, Zulkifli N (2013) A mathematical model for the industrial hazardous waste location-routing problem. J Appl Math 7:1–10
15. Rabbani M, Heidari R, Yazdanparast R (2019) A stochastic multi-period industrial hazardous waste location-routing problem: integrating NSGA-II and Monte Carlo simulation. Eur J Oper Res 272(3):945–961

16. Saeidi-Mobarakeh Z, Tavakkoli-Moghaddam R, Navabakhsh M, Amoozad-Khalili H (2020) A bi-level and robust optimization-based framework for a hazardous waste management problem: a real-world application. J Clean Prod 252:119830
17. Son LH, Louati A (2016) Modeling municipal solid waste collection: a generalized vehicle routing model with multiple transfer stations, gather sites and inhomogeneous vehicles in time windows. Waste Manage 52:34–49
18. Yadav V, Bhurjee AK, Karmakara S, Dikshit AK (2017) A facility location model for municipal solid waste management system under uncertain environment. Sci Total Environ 603–604:760–771
19. Darmian SM, Moazzeni S, Hvattum LM (2020) Multi-objective sustainable location-districting for the collection of municipal solid waste: two case studies. Comput Ind Eng 150:106965
20. Yousefloo A, Babazadeh R (2020) Designing an integrated municipal solid waste management network: a case study. J Clean Prod 244:118824
21. Ilbahar E, Kahraman C, Cebi S (2021) Location selection for waste-to-energy plants by using fuzzy linear programming. Energy 234:121189
22. Hu C, Liu X, Lu J (2017) A bi-objective two-stage robust location model for waste-to-energy facilities under uncertainty. Decis Support Syst 99:37–50
23. Pluskal J, Šomplák R, Hrabec D, Nevrlý V, Hvattum LM (2022) Optimal location and operation of waste-to-energy plants when future waste composition is uncertain. Oper Res Int J 22(5):5765–5790
24. Nevrlý V, Somplak R, Putna O, Pavlas M (2019) Location of mixed municipal waste treatment facilities: cost of reducing greenhouse gas emissions. J Clean Prod 239:118003
25. TUIK: Address based population registration system results in 2022, 2023 (in Turkish) (2020)
26. Kor N, Ozturk I, Tuyluoglu S, Ozabali Sabuncugil A, Tezer H, Allar AD, Cavus E, Karisan H, Akca A (2006) MoEF solid waste master plan I. Phase project, Final report, MIMKO Inc. together with ITU Project Team (in Turkish)
27. Ozturk İ (2015) Solid waste management and EU-compliant applications. ISTAC Technical Books Series Publishing, Istanbul (in Turkish)
28. Koushki PA, Al-Duaij U, Al-Ghimlas W (2014) Collection and transportation cost of household solid waste in Kuwait. Waste Manage 24(9):957–964
29. Korkut NE, Yaman C, Kucukaga Y, Jaunich MK, Demir I (2018) Greenhouse gas contribution of municipal solid waste collection: a case study in the city of Istanbul, Turkey. Waste Manage Res 36(2):1–9
30. Yılmaz-Balaman Ş, Matopoulos A, Wright DG, Scott J (2018) Integrated optimization of sustainable supply chains and transportation networks for multi technology bio-based production: a decision support system based on fuzzy ε-constraint method. J Clean Prod 172:2594–2617
31. OECD (2018) Effective carbon rates 2018 pricing carbon emissions through taxes and emissions trading
32. https://enerjienstitusu.org/
33. IJgosse J (2019) Waste incineration and informal livelihoods: a technical guide on waste-to-energy initiatives. WIEGO technical brief no. 11. WIEGO, Manchester, UK
34. UKWIN (2018) Evaluation of the climate change impacts of waste incineration in the United Kingdom. UKWIN climate change report 2018

A Novel Model for Assessment of Industry 4.0 Maturity in Defense Industry

Ahmet Can Ataman, Bilge Varol, and Gül Tekin Temur

Abstract Technological developments have led to radical changes in all areas of life. The defense industry is also one of the sectors that experiences these radical changes. Defense qualifications, which have a great proposition on the part of countries' independence and political powers, require large investments for countries that do not want to have an outsourced defense system. Developed countries have made great strides in this respect, but the changing circumstances have increased their tendency towards defense systems, where technological developments are closely tracked and applied. Within the scope of this research, firstly, the effects of digital transformation on the defense industry are examined. Then, an Industry 4.0 maturity model including five main criteria (strategy, management and organization, human and R&D culture, product and technology, operation) is proposed with a novel multi-criteria decision making (MCDM) approach called hesitant fuzzy analytic hierarchy process (HFAHP). This approach, in which fuzzy logic is used, is chosen to handle the high uncertainties in the structure of the problem. Later, a real-life case study application is conducted in a pioneer company in Turkey. The results of this study show that strategy has the highest importance followed by product and technology, operation, human and R&D culture and management and organization.

Keywords Defense industry · Industry 4.0 · Digital transformation · Maturity assessment · Multi criteria decision making · Hesitant fuzzy analytic hierarchy process

---

A. C. Ataman
Netas Telecommunication Inc., Istanbul 34912, Turkey

B. Varol (✉)
Faculty of Engineering, Turkish-German University, Istanbul 34820, Turkey
e-mail: bilge.varol@tau.edu.tr

G. T. Temur
Faculty of Engineering and Natural Sciences, Bahcesehir University, Istanbul 34353, Turkey
e-mail: gul.temur@bau.edu.tr

F. Çebi et al. (eds.), *Engineering and Technology Management in Challenging Times*,
Lecture Notes in Networks and Systems 1137,
https://doi.org/10.1007/978-3-031-70935-7_6

1 Introduction

The 1st Industrial Revolution began with the water and steam-powered mechanical production facilities in the production lines in the late eighteenth century. In parallel with the increasing production capacities, the workforce has also increased. At the beginning of the twentieth century, the introduction of mass production based on the division of labor and electrical energy has followed and the foundations of the 2nd Industrial Revolution were laid. In 1969, the foundations of the 3rd Industrial Revolution were laid, which succeeded in carrying the manufacturing automation to the next level with the first programmable logic controller (PLC) where PLCs remain the best option for industrial automation [1] and developed this rapid development of electronics and information technologies for many years. Today, the value-producing companies have reached the 4th Industrial Revolution, namely Industry 4.0. To provide competitive power, Industry 4.0 aims to use both hardware and software structures that can share real-time information, technology activities and products, services, and workforce for the same purpose in the most appropriate way to reduce costs and increase efficiency.

Industry 4.0 is creating new business models and new ways of manufacturing, renewing the industry for the so-called digital transformation [2]. Increasing technological developments have made this technological transformation, namely, digital transformation compulsory in the defense industry. With this transformation, it has been possible to develop more efficient and well-equipped products and systems. As a result of changing conditions, land wars remain in the second or even third plan, and technology-oriented air and sea wars come as a reflection of the digital transformation in the defense industry. Today, countries are ranked not as much on the number of soldiers, but according to their hardware and software power that refers to their technological power.

Based on the above information, in today's highly competitive environment, defense industry products should be produced in a working environment that incorporates Industry 4.0 digital transformation into all its processes. Therefore, the main goal of a company should be to implement the transformation processes within the framework of digital maturity parameters. Depending on these developments, a technology-centered structure called defense industry 4.0 has emerged. This makes the hand of countries selling defense systems much stronger.

Maturity is a measure for checking the ability of an organization and maturity models are the methodologies for measuring how much these abilities are applied in these organizations. Maturity models can be applied for various abilities; however, Industry 4.0 maturity models are one of the most common types of maturity models in the literature. Organizations that do not keep up with Industry 4.0 developments will not be able to survive in the future, and therefore companies need Industry 4.0 maturity models to determine how they can develop these abilities and check where they are.

Maturity models propose a measurement system, but it is also required to develop a metric for determination of integration level [3]. In addition, selecting the main criteria and sub-criteria occupies an important role in the evaluation phase. After deciding the criteria and scoring the options with some experts, a selected methodology can be applied to the model, and then some evaluations are done with the data handled by experts. Maturity models can be used as a roadmap for companies or organizations to ability of integrating Industry 4.0 in themselves.

Considering this information, this study aims to propose an Industry 4.0 maturity model in order to guide the experts from the defense industry to improve their transition to Industry 4.0 or to create a roadmap for how they will transition. In the proposed maturity model for the defense industry, 5 main criteria and 17 sub-criteria are defined in a hierarchy. After defining the criteria, the weights of the main criteria and sub-criteria are calculated, and the prioritization of these criteria is carried out using the HFAHP method. The proposed model is applied to a company in Turkey. Therefore, a real-life case study is presented and carried out. Finally, the priorities of the criteria were determined and presented to the experts as preliminary information and a route.

2 Literature Review

In this section, a brief literature survey on industry 4.0 maturity models, defense industry and HFAHP is presented, and each subject is reviewed independently.

Maturity models have been widely employed as frameworks to identify the capabilities and improvement needs of the companies toward successful Industry 4.0 implementation. The studies that handled Industry 4.0 readiness issues and maturity models can be summarized as: Basl and Doucek [4] reviewed the macro (i.e., country-level) and micro (i.e., enterprise level and more specific) Industry 4.0 readiness models. They presented a seven-level metamodel of readiness indexes and maturity models which gradually comes down from society-level view of Industry 4.0 to sub-dimension within an individual dimension of an area in an enterprise (e.g., information technologies). Their metamodel covers country-wide, as well as enterprise maturity dimensions. They argued that sector-specific models are not adequate, and IT dimension, cross-sectional aspects and security (e.g., cyber-security) has not been sufficiently elaborated in the current models. Mogos et al. [5] explored the factors that affect the readiness and attitude of the companies from Oil and Gas industry (OG), as well as various Manufacturing Industries (MI) in Norway, with an emphasis on training and continuing professional development aspect that transformation for Industry 4.0 necessitates. Their survey revealed notable differences between OG and MI companies with regards to Industry 4.0, which supported different industry lines should be assessed with different approaches. Moreover, the authors found the Industry 4.0 knowledge among the surveyed companies superficial. It can be implied that generic I4MMs, particularly when applied to a range of companies regardless of their industry lines, might not yield reliable or well-informed responses. The researchers should focus not only on the structure of the maturity models, also the application and evaluation methods to ensure the quality of the data gleaned from the companies. Castelo-Branco et al. [6] assessed Industry 4.0 readiness of manufacturing industry in European Union (EU). They used the data collected via Eurostat's questionnaire applied across EU countries in the manufacturing companies with a workforce of 10 or more employees. The data indicated the presence of the necessary elements for Industry 4.0 implementation across two factors; Industry 4.0 Infrastructure (i.e., interconnectivity, interoperability, virtualization), and Big Data Maturity (i.e., information transparency). The examples of these elements are the mobile internet connection for business software applications (i.e., interconnectivity), the link between the companies and their suppliers and/or customers (i.e., interoperability), and the analysis capability for the big data coming from the company's smart devices (i.e., information transparency). They found through a cluster analysis that Netherlands and Finland are the leaders of Industry 4.0 readiness. Castelo-Branco et al. [6] applied their study in manufacturing sector but did not differentiate among the various manufacturing fields. However, the readiness can well vary among different manufacturing fields. Frank et al. [7] developed a model to assess the maturity of manufacturing companies in the adoption of Industry 4.0 front-end technologies (i.e., Smart manufacturing, Smart Working, Smart Supply Chain and Smart Product) and base technologies (i.e., Internet of things, Cloud, Big Data, Analytics). They applied this framework via a survey among 92 companies building machinery and equipment across various industrial sectors in Brazil. They clustered the participant companies based on their Smart Manufacturing adoption level as; low, moderate, and advanced adopters. They found that digital platforms for linking companies with suppliers and customers, virtual reality, augmented reality were adopted at a low level. According to their findings, the company size affected Industry 4.0 maturity level in that large scale companies can invest in and implement advanced Industry 4.0 technologies more effectively compared to small companies. Frank et al. [7] highlighted that the characteristics of the companies can affect the outcomes of framework applications. Kohnová et al. [8] evaluated and compared the Industry 4.0 readiness among a large set of companies in Slovakia, Czech Republic, Austria, Germany and Switzerland, across seven dimensions (e.g., internal system of education, knowledge management and development, availability of technologies and knowledge to develop new products/services). Among the evaluated countries, Germany was found to be the main leader.

Apart from these studies, numerous studies offered I4MMs at different scales. Büyük et al. [9] proposed a digital maturity assessment model for evaluating weaknesses and strengths in Smart Agriculture. A list of necessities for sector-environment-human dimensions were figured out from sustainable perspective. They utilized one of recent decision-making approach named Best Worst Method (BWM) to reveal the importance of the effective parameters of maturity, and they were used in simple weighted average that helps to form the scores of the companies that address their maturity level. Horvat et al. [10] presented an I4MM0 model for the manufacturing companies particularly in the emerging economies. Their model included five dimensions; technology, management and strategy, employees and communication, organization of production and logistics, and interfirm cooperation. They assessed the maturity degree of the company across four stages, ranging from the lowest technical and structural capabilities to the highest level of technical capacity and qualifications. They applied their model by collecting data using on-site interviews and a survey among manufacturing companies in Kazakhstan, the outcomes of which are readiness levels of individual companies, as well as an overall maturity stage for the whole industry. The model of Horvat et al. [10] covers the whole manufacturing sector, without addressing the particularities of various manufacturing fields. PwC surveyed the senior executives from more than 2000 companies globally about Industry 4.0 issues [11]. The survey results highlighted that Industry 4.0 related digitalization of the processes were expected to bring additional revenue, gains in efficiency, and significant cost reductions [11]. PwC [11] offered the PwC Maturity Model as a tool that can help companies evaluate their level and guide them toward understanding the aspects they need to improve. PwC Maturity Model had seven dimensions; Digital business and customer access, Digitization of product and service offerings, Digitization and integration of vertical and horizontal value chains, Data and Analytics as core capability, Agile IT architecture, Compliance, security legal and tax, and Organization, employee and Digital Culture. Based on their maturity level across these dimensions, PwC model grouped the companies into four; Digital novice, Vertical integrator, Horizontal collaborator, and Digital champion. Rockwell Automation [12] developed The Connected Enterprise Maturity Model for implementation of smart manufacturing using novel technologies such as Internet of Things (IoT). Their 5-Stage model incorporated both assessment (i.e., Stage-1) and implementation aspects (i.e., Secure and upgraded network and controls, Defined and organized working data capital, Analytics, Collaboration, Stages 2–5 respectively). The assessment sought to understand the ability of the company to advance to the advanced technologies and included four elements: information infrastructure, controls and devices, information networks, and security policies. More than just an assessment tool, Rockwell's model offered an action plan toward adopting smart technologies. Yagiz Akdil et al. [13] reviewed the I4MMs, proposed a new model and applied it in the retail sector. Their model included three dimensions: Smart Products and Services, Smart Business Processes, and Strategy and Organization, and 13 associated fields. The Smart Business Processes dimension further had three subdimensions: The model evaluated the maturity of a company in four stages; Level 0: Absence, Level: 1 Existence, Level: 2 Survival, and Level: 3 Maturity. In the assessment, the model graded the maturity level by survey questions rated across these levels 0–3 and averaged under dimensions. The ranges of average grade values identified the maturity level. Lichtblau et al. [14] developed IMPULS model as a self-check tool for companies to assess their Industry 4.0 maturity across six dimensions; Strategy and Organization, Smart Factory, Smart Operations, Smart Products, Data-Driven Services and Employees, which included 18 assessment items in total. Based on the assessment outcomes, the IMPULS model grouped the companies into six levels; Outsider, Beginner, Intermediate, Experienced, Expert, and Top Performer. They applied IMPULS model to German companies and offered insights into the maturity levels of the manufacturing companies, as well as action items for improvement towards Industry 4.0 implementation. Temur et al. [15] applied IMPULS model to assess the Industry 4.0 maturity levels of three Turkish companies from construction, textile and wire production sectors [3, 15]. They additionally assessed the perception and awareness of the companies in terms of operational and socio-economic factors regarding Industry 4.0. They argued that fuzzy methods can be applied to reduce the effect of subjectivity in the maturity assessment. Schumacher et al. [16] expanded the maturity model developed by Schumacher et al. [17] into a 10-step realization model for the manufacturing companies, which encompassed knowledge-sharing, maturity assessment, analysis of maturity gaps in the company, and development of the realization paths. The knowledge-sharing aimed to align company members with the experts on the understanding of Industry 4.0 concept so that the data gleaned during the assessment would reflect informed responses. The maturity assessment was performed via application of a questionnaire which included importance rating as well, so that a weighting factor can be applied. The realization paths were shaped by the current state of the company and the gaps detected during the maturity assessment. This extended maturity model had eight dimensions; Technology, Products, Customers and Partners, Value Creation Processes, Data and Information, Corporate Standards, Employees, and Strategy and Leadership, including 65 items in total across the dimensions. They validated the content and the structure of their model in three plants of an Austrian manufacturing company.

Most maturity studies focus on the model development, selecting dimensions and sub-factors that would allow for effective assessment of the companies. However, to glean reliable data and to offer realistic, actionable insights into the maturity stages of the companies, as well as potential improvements, evaluation method is an important component maturity studies. Colli et al. [18] argued that maturity assessments should be performed in dialogue with company members through multiple data collection efforts, and the suggestions to improve the company toward higher maturity stages should be context specific. They adapted the Acatech maturity model developed by Schuh et al. [19] and tested the model via three case studies in manufacturing companies. The dimensions in their model were: Governance, Technology, Connectivity, Value Creation and Competences. They identified the maturity level of the companies across six stages; None, Basic, Transparent, Aware, Autonomous and Integrated. They applied a multiple-choice survey, and then conducted a workshop with company representatives to discuss the issues emerged from the survey results. The workshop allowed the researchers to have an in-depth understanding of company-specific aspects of the maturity dimensions, which enabled them to build and offer sound improvement suggestions that would address the company.

Most I4MMs assume equal weights for the maturity dimension and assessment items. However, depending on the industry or company characteristics, each dimension or item can assume different importance levels. A few studies apply weights to the I4MM items. Schumacher et al. [17] developed A I4MM for the industrial companies through understanding of Industry 4.0 concept, semi-structured interviews with researchers and practitioners, and literature survey. Their model included nine dimensions; Strategy, Leadership, Customers, Products, Operations, Culture, People, Governance and Technology, each of which had different numbers of maturity items. They collected assessment data using a questionnaire with two types of questions. The questions related to the maturity assessed the level of the company across the model dimensions on a 5-Point Likert scale. Additionally, importance of the items within maturity dimensions is rated on a 4-point scale. The average importance rating was applied as the weighting factor for maturity items.

Defense industry is crucial for the autonomy and power of a country. A self-sufficient defense industry can strengthen a country's stance against potential military conflicts or challenges [20]. Particularly emerging countries need a robust defense industry to minimize their dependence to the foreign suppliers, and to the countries exporting defense systems, which in the long run give the supplier countries ability to influence the recipient country's policies and actions in the international arena [20]. Thus, establishing, sustaining, and constantly improving the defense industry domestically has been a high priority for emerging countries, and Turkey is no exception. In 2017, three Turkish defense and aerospace companies placed among the top 100 defense companies globally in US-based Defense News 2017 Top 100 list. To our knowledge, solely Bibby and Dehe [21] developed a I4MM for the defense industry which encompassed three dimensions; factory of the future (i.e., technologies necessary for Industry 4.0, such as additive manufacturing, cyber physical systems), people and culture, and strategy. They built and applied their framework in a large-scale defense manufacturing company. The model content and items were reviewed in collaboration with company employees during a workshop, and the model items were rated by the experts in the company. The maturity level was identified on a four-level scale (i.e., minimal, development, defined and excellence). They pinpointed the strengths (e.g., Manufacturing Execution Systems) and improvement areas (e.g., use of sensors) for the company. Then, an expert from the company used the same model to assess the maturity levels of twelve supply network partner firms. They further cautioned that their framework could favor technology driven companies against those performing better in people and culture, and strategy dimensions.

Few studies applied FAHP techniques to the maturity models in different fields. It should be noted that applying HFAHP to maturity model evaluation is a new approach. Related literature is reviewed as follows: Mendes et al. [22] applied AHP to rate the priorities of the components in the maturity model for demand-driven supply chains. Tan et al. [23] developed a maturity model for implementation of Cyber Physical Systems (CPS) in manufacturing environment. They applied group AHP to assign importance weights to the elements of the model and a fuzzy grading technique in evaluating the linguistic variables used to measure the responses given to the maturity assessment questionnaire. They aimed to handle the subjectivity in responses by fuzzy grading and applied different weights to the model elements. Amalia and Sarno [24] applied a maturity index for the risk management in IT-Governance context. They used fuzzy AHP to establish the weights of the criteria defined in the model, and these weights were used to rank the values taken from the risk factors.

According to the presented literature, there is lack of studies that handled HFAHP method in Industry 4.0 maturity models. To best of our knowledge, the current study is the first research that applies HFAHP to Industry 4.0 maturity models. Based on this motivation, in this study HFAHP method is applied to an Industry 4.0 maturity model in a company in Turkey.

3 Proposed Model

In the related literature there is no study that handled Industry 4.0 maturity evaluation in a defense industry using HFAHP method. Based on this motivation, in the proposed model HFAHP method is applied for prioritizing the identified and interpreted Industry 4.0 maturity parameters.

In the proposed model, all main and sub-criteria are defined by considering the maturity models previously discussed in the literature and the opinions of experts working in the defense industry. Also, the hierarchy of the Industry 4.0 maturity model for defense industry is illustrated as in Fig. 1.

3.1 Sub-criteria of Strategy

Quick response: Due to the nature of the defense industry, customer requests should be responded as quickly as possible. In case of any technical problem or when additional features are required, the solution should be implemented in a minimum time.

Product diversity: The change in military field needs is quite dynamic. The capabilities of the product should be adjusted accordingly, designed to support diversity, and this increases the services that offer solutions.

Using Industry 4.0 tools: The product designed and serving for a specific purpose should be examined in detail during the transformation phase. How and which points will be intervened by the experts should be planned within the framework of the protocols.

Setting goals and roadmaps: The projects that the company will take part in in the future should be determined by evaluating the features that the company has gained so far. Although its technical capacity is sufficient, the pros and cons of addressing every sector and every customer should be evaluated in detail by experts.

3.2 Sub-criteria of Management and Organization

Establishment of Industry 4.0 working groups: Labor force and time calculations should be made for the completion of existing projects. In addition, by whom, when and by which tools the necessary integrations will be made with a focus on transformation should be evaluated as a separate project by the subject experts.

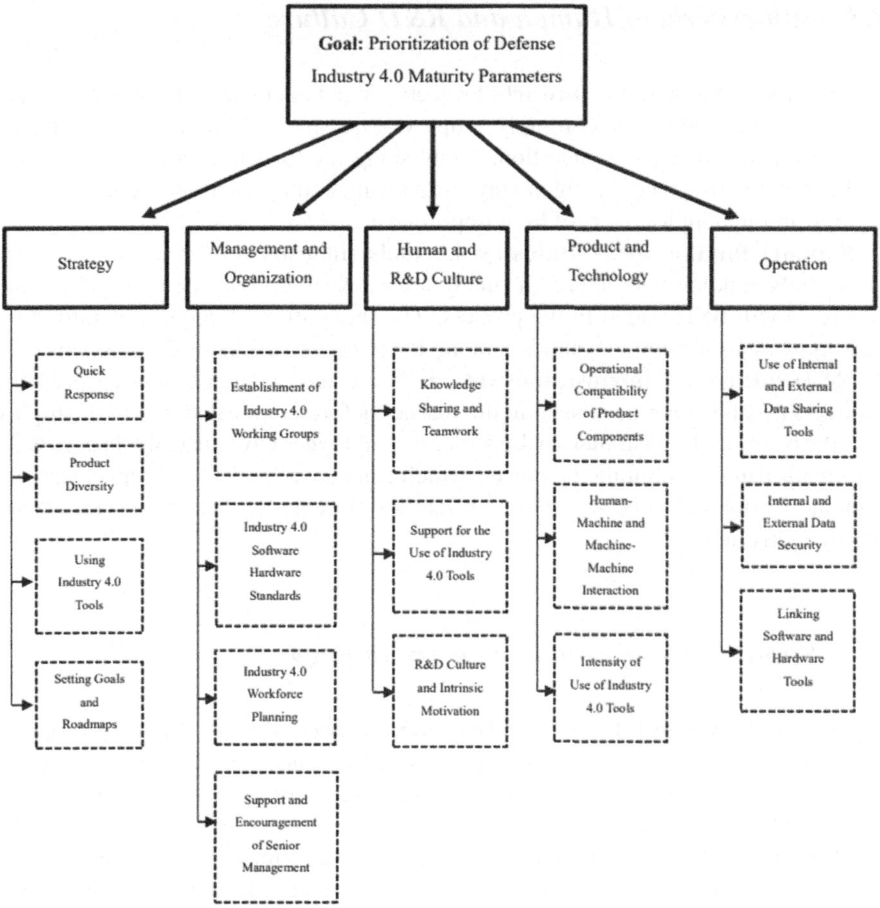

Fig. 1 The hierarchy of the Industry 4.0 maturity model for defense industry

Industry 4.0 software hardware standards: In cases such as adding additional features to the product and similar cases, every process must be designed in accordance with the standards, both on the hardware side and the software side.

Industry 4.0 workforce planning: Depending on the changes brought about by digital transformation, the relations between the employee and the work he/she does should be re-evaluated and arrangements should be made accordingly.

Support and encouragement of senior management: Learning a new knowledge can be a difficult process for the employee from time to time. In order to ensure maximum efficiency in this process, it would be appropriate for the management to support its employees with additional programs and trainings.

3.3 Sub-criteria of Human and R&D Culture

Knowledge sharing and teamwork: Projects carried out in the defense industry can be very broad in scope. Accordingly, employees are included in the process related to the division of labor. In addition, establishing a connection between processes will contribute to seeing the big picture. As a result, communication and information sharing among employees is of high importance.

Support for the use of Industry 4.0 tools: Industry 4.0 brings with it new, completely unknown structures in many areas. At this point, new programs and protocols will be included in the process. The integration of employees into these structures in a short time should be among the main objectives of the company.

R&D culture and intrinsic motivation: As a result of defense industry projects, products that have not been seen in the market before are created due to the highly specific needs of the customer. In this process, employees have difficulty in accessing information and combining the pieces, which can take a long time. Employees with high motivation can cope with these difficulties more easily than those who are not highly motivated.

3.4 Sub-criteria of Product and Technology

Operational compatibility of product components: Defense industry products must be able to operate in harsh environmental conditions. It must pass these tests successfully before it can be put on the field. A product without operational compatibility cannot be integrated into the field in the defense industry.

Human–machine and machine-machine interaction: It is important in terms of system management in studies on the system in the laboratory environment. Information transfers and user-to-user transfer will increase speed and efficiency.

Intensity of use of industry 4.0 tools: The digitalization process reduces the workforce due to its scope. Accordingly, employee productivity will increase, and this will be reflected in the product. It can be said that the more points Industry 4.0 tools are in the product, the more mature a product has been created.

3.5 Sub-criteria of Operation

Use of internal and external data sharing tools: It is indispensable for the company to prepare the documents or make the modeling of each stage of the projects carried out. In a company with a rich document repository, employees can easily access the required document thanks to special tools. This also applies to external sharing.

Internal and external data security: Every company wants to protect and preserve its own data. But in the defense industry, this requires more serious effort.

Data security must be fully ensured in scope. Company employees are also required to fulfill their responsibilities accordingly.

Linking software and hardware tools: Technically, it can be said that hardware and software work together in a product that contains electronic tools. At this point, within the scope of Industry 4.0, the characteristics of both parties and what they can do should be evaluated and a design should be made by considering the results.

4 Research Methodology

In today's business environment, there are many problems that have different types of uncertainties because of vague and imprecise information. As in the past, people strive to minimize these ambiguities and confusion today. As a matter of fact, it is aimed to obtain solution methods by developing various theories for the solution of problems in many different areas encountered in life. Fuzzy set theory is one of these methods [25].

Fuzzy sets are defined as a class of objects with membership degrees in a continuous array by Zadeh [26]. In this type of sets, each object has a membership function that assigns a degree of membership. Here, number 0 means that the related object is not a member of that membership function. On the other hand, number 1 corresponds that the related object is a fully member of that membership function and a number between 0 and 1 corresponds to the membership degree or partial membership of this object. Also, fuzzy set have been used in many MCDM approaches to cope with vagueness in problems. In this study, hesitant fuzzy sets (HFS), which are frequently used from fuzzy set theories, will be used in an integrated manner with AHP, which is one of the traditional MCDM methods and this method is called HFAHP.

Although the classical fuzzy set theory developed by Zadeh [26] considers uncertainties of human decisions with triangular or trapezoidal membership values, it is insufficient in hesitant situations where the decision maker has difficulty in making single choice. Therefore, Torra [27] proposed HFSs and after that study, Rodriguez et al. [28] examined these sets and proposed the HFS of linguistic terms (HFLTS) in order to enrich the content of linguistic expressions. In this way, the experts were able to use more flexible and expressively richer expressions when comparing two alternatives. In the classical fuzzy linguistic terms, alternatives can be compared with linguistic terms such as "Very High Important", "Absolutely Low Important", "Exactly Equal". However, in HFAHP fuzzy linguistic terms can be expressed in more flexible terms such as "Between Exactly Equal and Low Important", "Between Very High Important and High Important". Öztayşi et al. [29] recommended to use ordered weighted averaging (OWA) operator for the assessments of experts and OWA operator is applied in this problem.

HFSs determine the degree of membership of an element, and this comes from possible values that hesitate about which one is correct. It is a linguistic range obtained by the $env(H_s)$ envelope, upper bound and lower bound of the hesitant fuzzy linguistic term set (HFLTS) [28, 30].

$$env(H_s) = [H_{s^-}, H_{s^+}], H_{s^-} \leq H_{s^+} \tag{1}$$

Upper bounds and lower bounds are as follows:

$$H_{s^+} = \{S_i\} = S_j, S_i \leq S_j \text{ and } S_i, S_j \in H_s, \forall i \tag{2}$$

$$H_{s^-} = \{S_i\} = S_j, S_i \geq S_j \text{ and } S_i, S_j \in H_s, \forall i \tag{3}$$

n dimensional OWA operator: $R^n \rightarrow R$ and from here;

$$OWA(a_1, a_2, \ldots, a_n) = \sum_{(j=1)}^{n} w_j b_j \text{ happens} \tag{4}$$

b_j corresponds the j th largest aggregated argument. a_1, a_2, ..., a_n and $W = (w_1, w_2, \ldots . w_n)^T$ are the height vectors associated with $\sum_{i=1}^{n} w_i = 1$.

Parameters a and d are trapezoidal fuzzy membership functions indicated by $\tilde{A}$= (a, b, c, d) [29].

$$a = min\left\{a_L^i, a_M^i, a_M^{i+1}, \ldots, a_M^j, a_R^j\right\} = a_L^i \tag{5}$$

$$d = max\left\{a_L^i, a_M^i, a_M^{i+1}, \ldots, a_M^j, a_R^j\right\} = a_R^j \tag{6}$$

$$b = \left\{a_M^i, if\ i+1 = j \quad OWA_{w^2\left(a_m^j, \ldots, a_m^{(i+j)/2}\right), i+j\ çiftse} \quad OWA_{w^2\left(a_m^j, \ldots, a_m^{(i+j+1)/2}\right), i+j\ tekse}\right\} \tag{7}$$

$$c = \left\{a_M^{i+1}, if\ i+1 = j \quad OWA_{w^2\left(a_m^j, a_m^{j-1}, \ldots, a_m^{(i+j)/2}\right), i+j\ çiftse} \quad OWA_{w^2\left(a_m^j, a_m^{j-1}, \ldots, a_m^{(i+j+1)/2}\right), i+j\ tekse}\right\} \tag{8}$$

$\alpha_2 = \frac{(j-i)-1}{g-1}\alpha_1 = \frac{g-(j-i)}{g-1}$ Suppose that the value of α is in the unit interval ([0, 1]). The types of the first and second weights are expressed in Eq. (9) using α.

In equations $\alpha_1 = \frac{g-(j-i)}{g-1}\alpha_1 = \frac{g-(j-i)}{g-1}$ and $\alpha_2 = \frac{(j-i)-1}{g-1}$ (g is the top-ranking number of rating scores, j is the highest rating, and i is the lowest rating), the first weight $\left(W^1 = \left(w_1^1, w_2^1, \ldots, w_n^1\right)\right)$ is calculated using Eq. (9).

$$w_1^1 = \alpha_2, w_2^1 = \alpha_2(1-\alpha_2), \ldots, w_n^1 = \alpha_2(1-\alpha_2)^{n-2} \tag{9}$$

Second weight $\left(W^2 = \left(w_1^2, w_2^2, \ldots, w_n^2\right)\right)$ is calculated using Eq. (10).

$$w_1^2 = \alpha_1^{n-1}, w_2^2 = \alpha_2(1-\alpha_1)\alpha_1^{n-2}, \ldots, w_n^2 = 1-\alpha_1 \tag{10}$$

With $\tilde{C}$, the pairwise comparison matrix is expressed.

$$\tilde{c}_{ij} = (c_{ij_l}, c_{ij_{m1}}, c_{ij_{m2}}, c_{ij_u})$$

$$\tilde{C} = \left[1 \cdots \tilde{c}_{ln} \vdots \ddots \vdots \tilde{c}_{nl} \cdots 1 \right] \tag{11}$$

$$\tilde{c}_{ji} = \left(\frac{1}{c_{ij_u}}, \frac{1}{c_{ij_{m2}}}, \frac{1}{c_{ij_{m1}}}, \frac{1}{c_{ij_l}} \right) \tag{12}$$

Matrix values are ambiguous to check for consistency. Considering the number $\tilde{d} = (l, m_1, m_2, u)$ the net number $\left(\mu_{\tilde{d}}\right)$ is calculated as in Eq. (13).

$$\mu_{\tilde{d}} = \frac{l + 2m_1 + 2m_2 + u}{6} \tag{13}$$

Lastly,

$$CI = \frac{\lambda_{max} - n}{n - 1} \tag{14}$$

$$CR = \frac{CI}{RI} \quad CR >= 0.1 \tag{15}$$

CI is the consistency index, λ_{max} is the eigenvector of the matrix, n is the number of criteria, RI is the random index.

Each $\tilde{r}_i$ is calculated by the following equation.

$$\tilde{r}_i = (\tilde{c}_{i1} \otimes \tilde{c}_{i2} \cdots \otimes \tilde{c}_{in})^{1/n} \tag{16}$$

From this it can be seen that the $\tilde{r}_1\tilde{r}_2 \ldots \tilde{r}_n$ value is calculated to reduce the deviation in the weights.

$$\tilde{w}_i = \tilde{r}_i \otimes (\tilde{r}_1\tilde{r}_2 \ldots \tilde{r}_n)^{-1} \tag{17}$$

$$D = \frac{c_l + 2c_{m1} + 2c_{m2} + c_u}{6} \tag{18}$$

Normalization is applied to the fuzzy weights to obtain the local weights of the criteria. To calculate the global weights of the sub-criteria, their local weights are multiplied by the weights of the main criteria to which they belong. Finally, these steps are applied to calculate the score for each criterion as in Eq. (19).

$$S_i = \sum_{j=1}^{n} w_j s_j \quad \forall i \tag{19}$$

Here, w_j is the global weight of value j.

5 Real-life Case Study

Considering the given Industry 4.0 maturity criteria, surveys were designed according to the needs of HFAHP application, and these were replied by experts who have worked in various projects in the defense industry for many years. Linguistic scales used in these surveys is given in Table 1. Linguistic terms address how a criterion is prior to the other. In this case study, three experts were contacted, and their opinions were taken for determining the priorities of criteria over the others. The characteristics of some of these experts are given below:

Expert-1: Marine Communication Systems Software Design Manager (19 years of experience).

Expert-2: Low Level Software Design Manager (18 years of experience).

Expert-3: Marine Communication Systems Software Team Leader (11 years of experience).

The ranges obtained as a result of the evaluation carried out by the experts are given in Table 1. In addition, experts' integrated fuzzy envelope values are appointed as in Table 2.

By applying the OWA operator to the fuzzy envelope values, trapezoidal linguistic values represented as the set (a, b, c, d) were obtained. Trapezoidal fuzzy sets of each main criterion compared with other criteria are shown in Table 3.

The matrix obtained by defuzzification of the integrated fuzzy envelope values is given in Table 3. Considering this matrix, the consistency index was calculated and found to be 0.02. Since this value is less than 0.1, it can be said that the evaluation is consistent. For each main criteria, the geometric mean for each of the matrix elements

Table 1 Linguistic scales for HFAHP ([31]: 132)

Linguistic term	Abbreviation	Triangular fuzzy number
Absolutely high importance	AHI	(7, 9, 9)
Very high importance	VHI	(5, 7, 9)
Essentially high importance	ESHI	(3, 5, 7)
Weakly high importance	WHI	(1, 3, 5)
Equally high importance	EHI	(1, 1, 3)
Exactly equal	EE	(1, 1, 1)
Equally low importance	ELI	(0.33, 1, 1)
Weakly low importance	WLI	(0.2, 0.33, 1)
Essentially low importance	ESLI	(0.14, 0.2, 0.33)
Very low importance	VLI	(0.11, 0.14, 0.2)
Absolutely low importance	ALI	(0.11, 0.11, 0.14)

Table 2 Main criteria range values and integrated fuzzy envelope values

Criteria	C1	C2	C3	C4	C5
Strategy/C1	55/ EE	78/ WHI-ESHI	68/ EHI-ESHI	66/EHI	67/ EHI-WHI
Management and organization/ C2		55/EE	46/ELI-EHI	34/ WLI-ELI	35/WLI-EE
Human and R&D culture/C3			55/EE	35/ WLI-EE	46/ ELI-EHI
Product and technology/C4				55/EE	56/EE-EHI
Operation/C5					55/EE

Table 3 Trapezoidal fuzzy values of main criteria

Criteria	C1	C2	C3	C4	C5
C1	(1,1,1,1)	(1,3,5,7)	(1,2.77,3.22,7)	(1,1,1,3)	(1,1,3,5)
C2	(0.14,0.2,0.33,1)	(1,1,1,1)	(0.33,1,1,3)	(0.2,0.33,1,1)	(0.2,0.92,1.07,1)
C3	(0.14,0.31,0.36,1)	(0.33,1,1,3)	(1,1,1,1)	(0.2,0.92,1.07,1)	(0.33,1,1,3)
C4	(0.33,1,1,1)	(1,1,3,5)	(1,0.93,1.08,5)	(1,1,1,1)	(1,1,1,3)
C5	(0.2,0.33,1,1)	(1,0.93,1.08,5)	(0.33,1,1,3)	(0.33,1,1,3)	(1,1,1,1)

is calculated and it can be seen in Table 3. Normalization is done by dividing the defuzzified numbers by 9, which is the highest linguistic value and the results are as in Table 3. The weights are found by taking the average of each row of the normalized matrix values. The calculated weights of main criteria are shown in Tables 4 and 5.

In the evaluation made by the experts, it was seen that the strategy criteria is more important than the others, according to the results of the main criteria prioritization. The main criteria determinations were carefully considered and determined in line with the opinions of the experts. Each of them is essentially important and they play a major role in the realization of the stages leading to the conclusion.

Due to the needs of the defense industry, the products serve detailed areas and contain intensive technology. In this process, from the beginning of the project, every detail should be evaluated and interpreted using the necessary Industry 4.0 tools. To

Table 4 Defuzzified values, geometric mean of main criteria

Criteria	Defuzzified values					Geometric mean			
	C1	C2	C3	C4	C5				
C1	1	4	3.33	1.33	2.33	1	1.53	2.17	3.74
C2	0.25	1	1.22	0.64	0.86	0.29	0.57	0.81	1.25
C3	0.3	0.81	1	0.86	1.22	0.32	0.78	0.83	1.55
C4	0.75	1.55	1.15	1	1.33	0.80	0.99	1.27	2.37
C5	0.42	1.15	0.81	0.75	1	0.47	0.79	1.02	2.14

Table 5 Normalized matrix values and weights of main criteria

Criteria	Normalized matrix values				Weights
C1	0.11	0.17	0.24	0.42	0.343
C2	0.03	0.06	0.09	0.14	0.119
C3	0.04	0.09	0.09	0.17	0.145
C4	0.09	0.11	0.14	0.26	0.216
C5	0.05	0.09	0.11	0.24	0.175

do this, the distribution of technical processes, design activities, implementation processes and testing processes should have a unique strategy and should reflect these studies when the targets determined as a result are taken into consideration. At this point, what is wanted to be mentioned is how successfully the determined targets were achieved. This reflects the success of the project. In a structure that needs to be completely result-oriented as required by the sector, strategy determination activities are a subject that needs to be evaluated in detail on its own. The results also show this.

Product and technology criteria, by nature, reflect the product implemented at the end of the project. This stage directly affects the product. Accordingly, it appears as the second important criterion. It is very important to reflect both software and hardware elements on the product by using industry 4.0 tools in parallel. Considering the conditions of use, products that need to work for many years and that minimum failure is expected in this process should be designed in accordance with quality factors in terms of content, and test processes should be kept comprehensive.

After the weight values and consistency calculations of the main criteria are made, the same method steps are applied for the sub-criteria and the results are shown in Table 6.

Considering the results of the main criteria, the importance ranking is strategy, product and technology, operation, human and R&D culture and management and organization. According to the results, the strategy criterion became the most important criterion, surpassing the other criteria. In this case, it may be appropriate for a defense industry company to act according to the projects that have been done before and are suitable to be done while determining the strategy.

Products with advanced technology, produced for a certain purpose, should have the ability to change and update in a short time, considering the features and components they contain. Operational compatibility, human–machine interaction, and the intensity of use of Industry 4.0 tools directly affect product quality and capability. Additionally, in operation processes, software and hardware elements need to be associated effectively and efficiently. These two elements, which literally make up the whole of the product, contain many standards and protocols within themselves. Standards and protocols should be accessible at the maximum level with data sharing tools. At this point, internal and external data security is one of the most important features that the company policy should pay attention to.

Table 6 Local and global weights of sub-criteria

Main criteria	Sub-criteria	Local weights	Global weights
Management and organization	Establishment of industry 4.0 working groups	0.192	0.023
	Industry 4.0 software hardware standards	0.186	0.022
	Industry 4.0 workforce planning	0.192	0.023
	Support and encouragement of senior management	0.428	0.051
Human and R&D culture	Knowledge sharing and teamwork	0.083	0.012
	Support for the use of industry 4.0 tools	0.486	0.070
	R&D culture and intrinsic motivation	0.430	0.062
Product and technology	Operational compatibility of product components	0.245	0.053
	Human–machine and machine–machine interaction	0.328	0.071
	Intensity of use of industry 4.0 tools	0.425	0.092
Operation	Use of internal and external data sharing tools	0.300	0.052
	Internal and external data security	0.333	0.058
	Linking software and hardware tools	0.365	0.064

Defense industry companies aim to unite their employees under the same culture in order to increase efficiency, depending on the scope and diversity of the projects carried out in-house. Keeping information sharing dynamic, support efforts to increase the use of Industry 4.0 tools and motivation are the criteria that form the basis of this culture. Culture shows its importance in a structure where continuous learning is compulsory and necessary in a long and strenuous work tempo. In addition, while determining the management and organizational steps, it should be aimed to follow and implement innovations by carrying out studies for digital transformation. Industry 4.0 tools should take their place in the processes and be used to implement the transformation.

The ranking of the sub-criteria according to the global weights is given in Table 7.

The strategy criterion has the highest weight among the main criteria. Accordingly, the sub-criteria of this criterion appear in the upper ranks. The evaluation made by the experts was made by considering the digital transformation requirements in line with the situations within the company, the studies, and the works to be done.

Table 7 Ranking of sub-criteria with respect to global weights

Sub-criterion	Main criterion	Global weight
Product diversity	Strategy	0.125
Intensity of use of industry 4.0 tools	Product and technology	0.092
Using industry 4.0 tools	Strategy	0.083
Quick response	Strategy	0.073
Human–machine and machine-machine interaction	Product and technology	0.071
Support for the use of industry 4.0 tools	Human and R&D culture	0.070
Linking software and hardware tools	Operation	0.064
R&D culture and intrinsic motivation	Human and R&D culture	0.062
Setting goals and roadmaps	Strategy	0.061
Internal and external data security	Operation	0.058
Operational compatibility of product components	Product and technology	0.053
Use of internal and external data sharing tools	Operation	0.052
Support and encouragement of senior management	Management and organization	0.051
Establishment of industry 4.0 working groups	Management and organization	0.023
Industry 4.0 workforce planning	Management and organization	0.023
Industry 4.0 software hardware standards	Management and organization	0.022
Knowledge sharing and teamwork	Human and R&D culture	0.012

Considering the demands of the defense industry, the product diversity, quick response, and the intensity of use of Industry 4.0 tools have taken an important place compared to other sub-criteria. The scope of the product produced and the quickness of the intervention to this product when necessary have gained importance. Considering that the increase in diversity varies depending on the yield, the use and integration of the necessary tools for digitalization is at the forefront. Companies also need to update their working structures accordingly and speed up by integrating industry 4.0 tools in applying solutions to the product.

5.1 Industry 4.0 Maturity Level Measurement of a Real Company

In this section, a real-life case study is applied to check the validity of this methodology. A group of experts from a defense company in Turkey is selected to apply these steps. This company is Turkey's leading and most experienced national defense

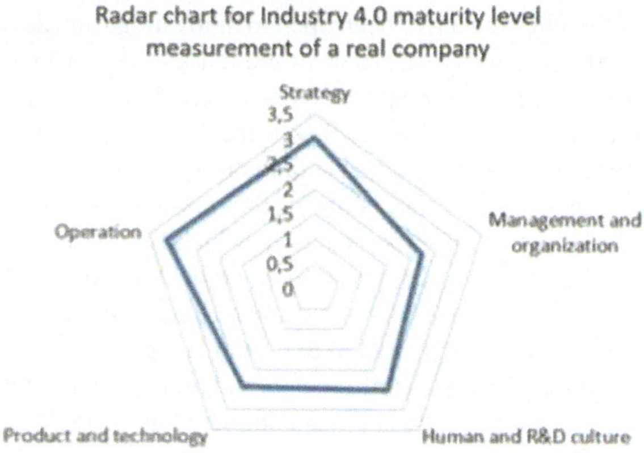

Fig. 2 Radar chart for the evaluation of main criteria in real company

industry company. In addition, it is Turkey's land systems manufacturer with vehicles and turret systems of different types and versions. 35 countries are actively using their products today. A survey was conducted with 10 experts from this company and expert gave scores for all our sub-criteria between 0 and 5. With the help of weighted mean method, scores for HFAHP is evaluated. The calculation of the weighted mean method is formulated as in Eq. (20). After calculating the scores, they are multiplied by weights that calculated with HFAHP of these sub-criteria. Finally, an overall score is calculated for checking how well the company can implement Industry 4.0 and for the company's state, and the company's situation is interpreted accordingly.

$$\underline{x} = \frac{\sum_{i=1}^{n} (x_i * w_i)}{\sum_{i=1}^{n} w_i} \tag{20}$$

According to the results of the experts, overall score is calculated as 2.734 out of 5. It can be concluded that this company has score higher than 50% of the maximal score. Therefore, we can say that this company can be in Level 3, which means experienced and in leaders group based on IMPULS ranking system. In addition, each main criteria are also compared among themselves to decide which aspects of the company should improve. In Fig. 2, the radar chart for each main criterion can be seen and it can be concluded that the company should improve the product and company criterion the most, followed by product and technology and people and R&D as the criteria to be developed.

6 Conclusions and Future Research Directions

Industry 4.0 is a very necessary transformation for companies to develop themselves and adapt to innovations. The speed of transformation and the accuracy of the transformation directly affect the competition of companies among themselves. Because, companies that have been open to innovations and developed themselves have always carried themselves further in the mark et According to this information, this study has drawn a roadmap that shows the experts how and in what order of importance they will apply Industry 4.0 in the defense industry. In this study, an Industry 4.0 maturity model for defense industry is proposed and then it is applied to a company in Turkey. Also, the nature of this problem is uncertain because of the differences of opinion of the experts and their inability to give a precise value. Therefore, fuzzy set theory was integrated into this model and the HFAHP method was used. Finally, various comments were made according to the order of importance of the main and sub-criteria and presented to users/managers.

The prioritization of Industry 4.0 maturity parameters determined and interpreted for the defense industry was completed using the HFAHP method. The results show that the strategy parameter is the most important criterion, surpassing the others. It may be considered appropriate for a defense industry company to act according to the projects that have been done before and that are appropriate to be done while determining the strategy. The strategy criterion should be evaluated significantly to maintain the ability to create specialized products that are not deformed by the attitude over time. When the requirements of the defense industry are taken into consideration, product diversity, quick response and the intensity of the use of Industry 4.0 tools are more important than other sub-criteria. The scope of the product produced and the quickness of intervention to this product when necessary have gained importance. Considering that the increase in diversity varies depending on efficiency, the use and integration of the tools required for digitalization are at the forefront. Companies also need to update their working structures accordingly and accelerate by integrating industry 4.0 tools in the application of solutions to the product. By evaluating all the results, companies need to organize their industry 4.0 routes.

In the future, we plan to apply new methodologies such as different MCDM approaches for measuring the Industry 4.0 maturity models. Also, we plan to test the applied method on different industrial sectors. Then, the comparisons can be done between MCDM approaches and various industrial sectors.

Acknowledgements We thank Assoc. Prof. Dr. Oğuzhan Erdinç for his assistance in constructing the literature review, and Dr. Aykut Duyguvar for his useful support on application.

References

1. Chen JY, Tai KC, Chen GC (2017) Application of programmable logic controller to build-up an intelligent industry 4.0 platform. Proc CIRP 63:150–155
2. Alcácer V, Cruz-Machado V (2019) Scanning the industry 4.0: a literature review on technologies for manufacturing systems. Eng Sci Technol Int J 22(3):899–919
3. Temur GT, Bolat HB, Gözlü S (2018) Evaluation of industry 4.0 readiness level: cases from Turkey. In: The international symposium for production research. Springer, Cham, pp 412–425
4. Basl J, Doucek P (2019) A metamodel for evaluating enterprise readiness in the context of industry 4.0. Information (Switzerland) 10(3):1–13

5. Mogos MF, Eleftheriadis RJ, Myklebust O (2019) Enablers and inhibitors of industry 4.0: results from a survey of industrial companies in Norway. Proc CIRP 81(2013):624–629
6. Castelo-Branco I, Cruz-Jesus F, Oliveira T (2019) Assessing Industry 4.0 readiness in manufacturing: evidence for the European Union. Comp Ind 107:22–32
7. Frank AG, Dalenogare LS, Ayala NF (2019) Industry 4.0 technologies: implementation patterns in manufacturing companies. Int J Prod Econ 210(September 2018):15–26
8. Kohnová L, Papula J, Salajová N (2019) Internal factors supporting business and technological transformation in the context of industry 4.0. Bus Theory Pract 20:137–145
9. Büyük AM, Ateş G, Burghli S, Yılmaz D, Temur GT, Sivri Ç (2020) Digital maturity assessment model for smart agriculture. In: The international symposium for production research. Springer, Cham, pp 289–301
10. Horvat D, Stahlecker T, Zenker A, Lerch C, Mladineo M (2018) A conceptual approach to analysing manufacturing companies' profiles concerning Industry 4.0 in emerging economies. Proc Manuf 17:419–426
11. PwC Report, Industry 4.0: Building the digital enterprise, PwC Report, 2016, 2015; 15(11):1136–1138
12. Rockwell Automation. The connected enterprise maturity model, International commission on illumination, p 12 (2014)
13. Yagiz Akdil K, Ustundag A, Cevikcan E, Akdil K, Ustundag ÁE, Cevikcan ÁA, Ustundag A, Cevikcan E (2018) Maturity and readiness model for industry 4.0 strategy implementation of industry 4.0 strategies require wide applications in companies. Springer International Publishing
14. Lichtblau K, Stich V, Bertenrath R, Blum M, Bleider M, Millack A, Schmitt K, Schmitz E, Schröter M (2015) Industrie 4.0 readiness IMPULS MODEL
15. Temur GT, Bolat HB, Gözlü S (2019) Evaluation of industry 4.0 readiness level: cases from Turkey. In: Proceedings of the international symposium for production research. Springer, Cham, pp 412–425
16. Schumacher A, Nemeth T, Sihn W (2019) Roadmapping towards industrial digitalization based on an Industry 4.0 maturity model for manufacturing enterprises. Proc CIRP 79:409–414
17. Schumacher A, Erol S, Sihn W (2016) A maturity model for assessing industry 4.0 readiness and maturity of manufacturing enterprises. Proc CIRP 52:161–166
18. Colli M, Berger U, Bockholt M, Madsen O, Møller C, Wæhrens BV (2019) A maturity assessment approach for conceiving context-specific roadmaps in the Industry 4.0 era. Ann Rev Control 48:165–177
19. Schuh G, Anderl R, Gausemeier J, Ten Hompel M (2017) Industrie 4.0 maturity index managing. In: Acathec, p 60
20. Kurç Ç, Neuman SG (2017) Defence industries in the 21st century: a comparative analysis. Def Stud 17(3):219–227
21. Bibby L, Dehe B (2018) Defining and assessing industry 4.0 maturity levels–case of the defence sector. Prod Plan Control 29(12):1030–1043
22. Mendes P, Leal JE, Thomé AMT (2016) A maturity model for demand-driven supply chains in the consumer product goods industry. Int J Prod Econ 179:153–165
23. Tan Q, Tong Y, Wu S, Li D (2017) Evaluating the maturity of CPS in discrete manufacturing shop-floor: a group AHP method with fuzzy grade approach. Mechanika 24(1):100–107
24. Amalia PR, Sarno R (2018) Measuring maturity index of risk management for IT-governance using fuzzy AHP and fuzzy topsis. In: Proceedings—2018 international seminar on application for technology of information and communication: creative technology for human life, ISemantic, pp 17–22
25. Ross TJ (1995) Fuzzy logic with engineering applications. McGraw-Hill, USA
26. Zadeh LA (1965) Fuzzy sets. Inf Control 8:338–353
27. Torra V (2010) Hesitant fuzzy sets. Int J Intell Syst 25:529–539
28. Rodriguez RM, Martinez L, Herrera F (2011) Hesitant fuzzy linguistic term sets for decision making. IEEE Trans Fuzzy Syst 20(1):109–119

29. Öztaysi B, Onar SÇ, Boltürk E, Kahraman C (2015) Hesitant fuzzy analytic hierarchy process. In: 2015 IEEE international conference on fuzzy systems (FUZZ-IEEE). IEEE, pp 1–7
30. Rodriguez RM, Martínez L, Herrera F (2012) Group decision making with comparative linguistic terms. In: International conference on information processing and management of uncertainty in knowledge-based systems. Springer, Berlin, Heidelberg, pp 81–190
31. Başar A (2017) Hesitant fuzzy pairwise comparison for software cost estimation: a case study in Turkey. Turk J Electr Eng Comput Sci 25(4):2897–2909
32. US-based Defense News 2017 Top 100 List, Defense News

A Reinforcement Learning-inspired Estimation of Antibiotic Resistance in Poultry Environments

Ahmet Eren Tomurcuk, Elif Bozcali, Furkan Dogancan Baytemur, Caglar Sivri, and Ece Gelal Soyak

Abstract Antibiotic resistance occurs when bacteria evolve to withstand the effects of antibiotics, reducing the effectiveness of treatment medication. Estimation of antibiotics resistance holds importance, as antibiotic-resistant infections can spread rapidly and knowledge in antibiotic resistance allows healthcare providers to make informed decisions. However, this is a challenging task because there are very limited open data sets to work on, containing systematic measurements. In this work, we propose machine learning-inspired estimation of resistance of five specific types of antibiotics on different poultry environments and created confidence percentages in which tell us how accurate the program is for a specific type of antibiotic. The analysis in this work highlights that a variety of factors impact antibiotic resistance in poultry; however, an informative prediction using a model such as the one proposed in this paper would assist in proper planning and usage of antibiotics in terms of effectiveness and necessity.

Keywords Machine learning · Antibiotics · Resistance · Poultry

1 Introduction

Public health has been influenced by bacterial infections to a great extent [1]. Antibiotics are the most common treatment method for dealing with bacterial infections. Antibiotics have saved many lives since they were discovered. However, with the overuse of antibiotics, bacteria gained the ability to alter themselves and become resistant to antibiotics. This situation is called antibiotic resistance [2]. Antibiotic resistance renders antibiotics less effective, dampening their ability to treat infectious and bacterial diseases [3]. It is one of the most serious threats to public health in the

A. E. Tomurcuk · F. D. Baytemur · E. G. Soyak
Computer Engineering, Bahcesehir University, Istanbul, Turkey

E. Bozcali · C. Sivri (✉)
Management Engineering, Bahcesehir University, Istanbul, Turkey
e-mail: caglar.sivri@bau.edu.tr

F. Çebi et al. (eds.), *Engineering and Technology Management in Challenging Times*, Lecture Notes in Networks and Systems 1137,
https://doi.org/10.1007/978-3-031-70935-7_7

twenty-first century [4]; causing over 50,000 deaths in 2015, in Europe and the USA, with estimated death toll up to 10 million per year worldwide by 2050 [5].

Multiple factors contribute to antibiotic resistance, namely globalization, the absence of adequate antimicrobial stewardship, and the excessive use of antibiotics in aquaculture and in animal breeding. Specifically, for the last item, the amount of antibiotic usage continues to increase due to changing farming practices worldwide. Large amounts of antibiotics are used while breeding animals, to avoid and treat diseases as well as support their growth [6]. It is revealed that more than 80% of antimicrobials that are crucial to human medicine were sold or delivered to be utilized in food-producing animals in 2014. However, failure in the treatment process that might result in economic loss can arise due to antimicrobial-resistant animal pathogens. Besides, antibiotic resistance in poultry may cause the emergence of resistant bacteria/genes that are connected to human health problems [7]. Direct or indirect communication, food consumption and feeding or environment can cause the resistant bacteria or resistance genes to transfer between humans and animals [8]. When we think about how big this percentage is and the relation of these food-producing animals with humans and the environment, it can be said that there are growing concerns about antimicrobial usage. Although this is an important global concern, there is limited information about the usage of antimicrobials for food-producing animals and resistance surveillance in the food chain [6]. This situation points out that further research about the subject is required. As a result, identification of the resistance issue has great importance in terms of economic loss and evaluation of their usage for treatment from both human and animal health perspectives.

Escherichia coli (E. coli) has been accepted as a possible resistance exposure source for humans within the animal commensal flora [8]. E. coli is a Gram-negative, facultative bacterium which belongs to the Enterobacteriaceae family. E. coli has been commonly used for observing AMR in food animals which also includes poultry because it is omnipresent in the gastrointestinal tract of warm-blooded animals. Additionally, AMR genes that might be transferred to humans can be caused by E. coli strains hosted by poultry [7].

Reinforcement learning is a useful tool for predicting future trends because it can model the interactions with many variables and the outcomes. In this work, we use an analysis that is inspired by reinforcement learning, to estimate the antibiotics resistance of soil samples. We analyze the resistance of E. coli against five antibiotics used in animal production, specifically ciprofloxacin, tetracycline, cephalexin, streptomycin and amoxicillin; and evaluate the consequences from both human and animal health perspectives. The results of this work can help relevant institutions to predict their costs and assess the effectiveness of antibiotic usage considering resistance.

2 Related Work

In the literature, different studies for resistance to different types of antibiotics exist. Machine learning has the potential to be a decision-support tool for the advancement of clinical practice. Today, diagnosis and treatments are primarily carried out based on clinical guidelines and the experience of doctors. However, in certain scenarios a more detailed analysis of correlations among factors may be needed [9]. Machine learning can analyse patient characteristics quickly from patient information and can reach a conclusion by synthesizing these data thanks to the algorithm it is trained [10]. The widespread use of Health Information Systems (HIS) in the world allows access to medical information, enabling meaningful results.

It is estimated that 1.27 million human deaths in 2019 were caused by antimicrobial resistance [11]. It was stated that the use of AI for antibiotic resistance would improve the healthcare system manyfold [12]. An AI-based, offline smartphone application for antibiogram analysis was developed in [13], utilizing machine learning and the image of the bacteria-free area around the cellulose, to categorize the susceptibility of bacteria to the tested antibiotics. The use of artificial intelligence (AI) for antimicrobial resistance prediction was reviewed in [14, 15], which argued that the use of AI for practical diagnosis and treatment would be beneficial.

To the best of our knowledge, this is the first work that formulates a machine learning driven estimation of antibiotic resistance, based on the different factors that may potentially impact the resistance.

3 Background

3.1 Antibiotics Resistance in Poultry

Poultry is one of the most common types of consumed meat. Large quantities of antibiotics are used in their cultivation to support their growth as well as to prevent and treat diseases. However, failure in the treatment process that might result in economic loss can arise due to antibiotic-resistant poultry pathogens. Besides, antibiotic resistance in poultry may cause the emergence of resistant bacteria/genes that are connected to human health problems [7]. In addition to problems related to animals, regarding the products like meat, egg, and milk are important parts of the human diet, the antimicrobial residue existence in those products can cause concerns about human health. In general, evolution and prevalence assumptions/knowledge about AMR in an animal production system is restricted to bacterial species such as Escherichia coli and Staphylococcus aureus that are non-commensal in poultry [7]. To get ahead of AMR, the conscious use of antibiotics is crucial for both human and animal health. It is predicted the effect of antimicrobials on veterinary medicine will also loosen as seen in human medicine [8]. Direct or indirect communication like food/feed consumption or environment can cause the resistant bacteria or resistance genes to transfer between humans and animals [8].

3.2 *Escherichia Coli (E. Coli)*

E. coli has been accepted as a possible resistance exposure source for humans within the animal commensal flora [8]. Escherichia coli is a Gram-negative, facultative bacterium which belongs to the Enterobacteriaceae family. E. coli has been commonly used for observing AMR in food animals which also includes poultry because it is omnipresent in the gastrointestinal tract of warm-blooded animals. Additionally, AMR genes that might be transferred to humans can be caused by E. coli strains hosted by poultry [7].

4 Methodology

Our methodology builds on predicting antibiotic resistance in poultry data sets using machine learning. Our approach is to use reinforcement learning (RL) with an algorithm that computes a numerical score for each sample in our data. These scores are used for predicting if the sample is resistant to the specified antibiotics or not. We treat the dataset as the *environment* in RL formulation, the studied soil sample's resistance as the state, and the score that is calculated based on how many samples of a characteristic is found with antibiotic resistance, as the reward. Figure 1 represents the ML-derived formulation.

Figure 2 represents the steps taken in our proposed algorithm. We first step is to acquire a data fit for our aim of identifying key features to determine the bacteria

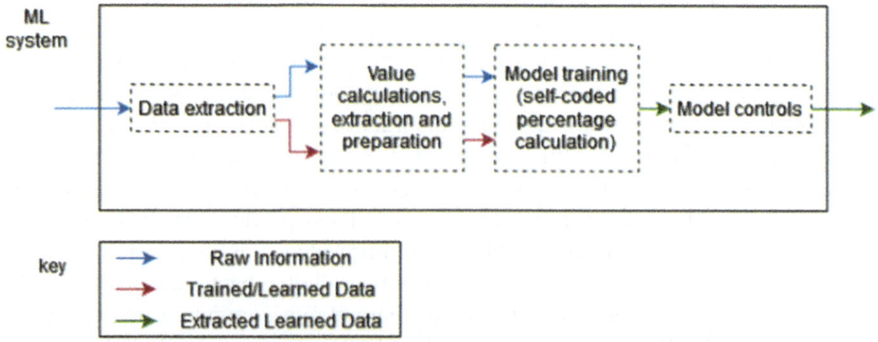

Fig. 1 The flow of data in the system

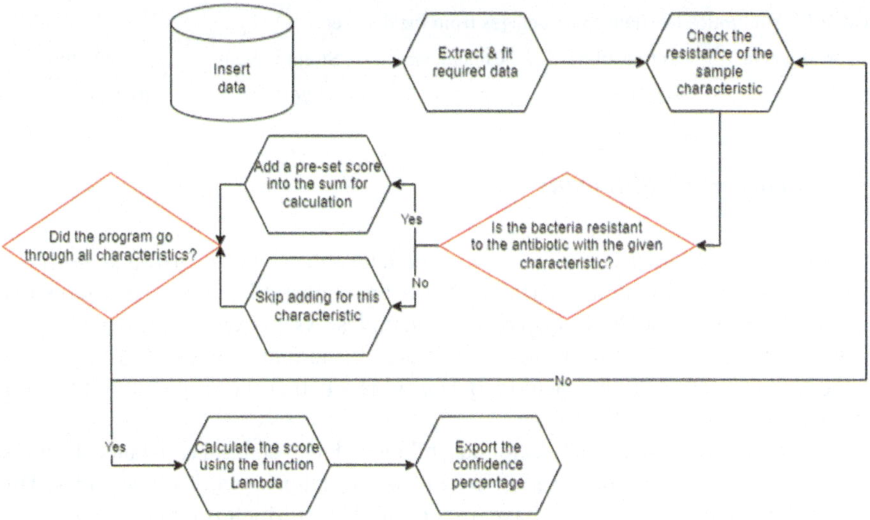

Fig. 2 Process chart for the system

resistance. This part is critical, as the data can be focused on different elements, e.g., effectiveness of the antibiotic or the number of bacteria in the sample.

We use the *Farm management and longitudinal data on antibiotic use and antibiotic resistant E. coli* data set [16], comprising over 4000 contaminated environmental samples collected over 2 years (2017–2019) across 53 dairy farms in England. We chose this data set because it has various features representing farm soil samples. The records of samples include date, location of farm, types of animals in location, answers to questionnaires administered to farmers while data was collected, and microbiological analyses to identify E. coli resistant to amoxicillin, cephalexin, tetracycline, ciprofloxacin and/or streptomycin. The data set contains 4000 samples. Listing 1 shows a sample of the features in the data set. For the whole set of features, we refer the reader to [16].

The data set contained binary, ordinal, categorical and continuous types of data. On this data set, little preprocessing was needed, since the data set did not contain rows or columns with null values. In addition, binary features such as forage type were derived as binary variables created from categorical features in the data set.

Fig. 3 Selected features of the data set

```
database = {
    "Amoxicillin Resistance": amoxResList,
    "Herd size": herdSizeList,
    "Organic": organicList,
    "Poultry": poultryList,
    "Waste Milk": wasteMilkList,
    "Wean": weanList
  }
```

Table 1 Antibiotics resistance percentages from the data set

Ciprofloxacin	Tetracycline	Cephalexin	Streptomycin	Amoxicillin
5.77%	62.91%	27.65%	32.26%	60.38%

4.1 *Feature Engineering*

With the given characteristic features in the data set, we performed a *score-based* training paradigm. In our approach, all values are converted to a score based on their occurrences (shown in Table 1) in resistant samples. As an example, for any bacteria A that has been affected from factors *a*, *b*, and *c*, and that was resistant to *x*, *y*, and *z*, we map the data with the bacteria to determine whether they are resistant towards that factor.

The scoring captures which factor has influenced the bacterial resistance the most. For example, characteristic q is seen in 55 resistant samples out of 110 samples. This means that characteristic q will have a score of 55/110 which is equal to 0.5.

The formulation of the *confidence ratio*, λ, is shown in Eq. 1,

$$\lambda = \frac{\sum_{i=1}^{n} x_i}{\sum_{j=1}^{m} y_j} \tag{1}$$

where x, y are the elements of given array (i.e., the features of the measurement from the poultry data), n is the number of resistant samples of the given characteristic, and m is the number of all resistant samples.

These scores are used to determine a threshold that helps to identify whether the sample is resistant. For example, assume the algorithm outputs a threshold value of 40% resistance to Ciprofloxacin. Our algorithm determines the score for each characteristic and these points will be summed up. If the summed score exceeds the threshold, then AI will estimate that it will be resistant. The threshold will be picked from resistant samples which have the lowest score but are still resistant to antibiotics. To prevent borderline miscalculations, threshold will be picked as lower quartile point from points of resistant samples.

4.2 *Score Based Resistance Confidence Calculation (SBRCC) Algorithm*

Our SBRCC algorithm comprises the following steps.

- The score for each characteristic is calculated based on Eq. 1.
- A general score for each sample is computed by summing the score for all characteristics.

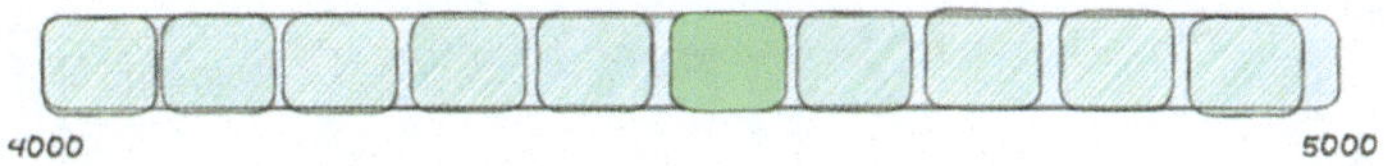

Fig. 3 Sample range filtering example

- Based on the total score of each sample, we determine whether the sample is resistant to specific antibiotics.
- Finally, the 10% average of the scores is computed to set a threshold that is used for defining resistance.

As explained in the steps above, we derive a threshold using the data set, that is used to specify whether the sample is resistant to the specified antibiotic. Accordingly, if the score exceeds the threshold, our algorithm outputs that this sample is resistant to the specified antibiotics. Using this approach, we can identify, for new samples, whether they are resistant. However, this is not applicable for samples with low scores such as Ciprofloxacin or Streptomycin, since their range of scores is small and threshold scoring is not able to filter results very effectively. For such samples, instead, we develop a new method based on **range filtering**. We divide our array of scores into 30 sub-arrays, and determine which sub array contains the most weight of scores. This part constitutes our award system. If a sample's score is inside this sub array's range, we can classify it as *resistant*. Range filtering is depicted in Fig. 3.

After examining the score results of each antibiotic, we proceed with a simple calculation to use this range filtering for scores below 20,000 and threshold filtering for above.

5 Experiments

We used Python v.3.11.5 with four librares, namely csv (for reading the data set), collections (for sorting), matplotlib (for plotting graphs) and tabulate (for creating tables in the terminal emulator).

We evaluate each antibiotic by its scores and use a specific filtering method depending on the scores as explained above. Figure 4 demonstrates the resistance per score for each of the five antibiotics that were tested. We observe that the score per resistance is smaller for Ciprofloxacin compared to the other four antibiotics.

Table 2 shows the calculated confidence ratios for each antibiotic.

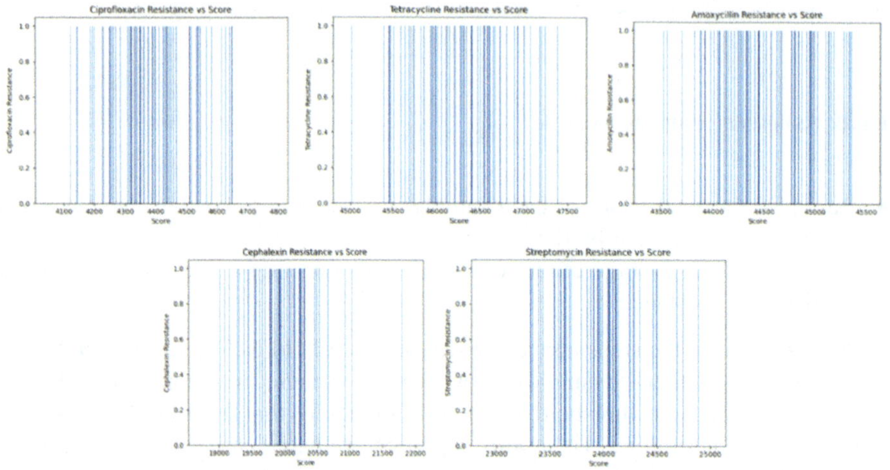

Fig. 4 Resistance versus score tables for each antibiotic

Table 2 Calculated confidence percentages of each antibiotic type

Antibiotic name	Confidence percentage (%)	Calculation method
Ciprofloxacin	94.62	Range filtering
Tetracycline	68.82	Threshold scoring
Cephalexin	68.10	Threshold scoring
Streptomycin	67.03	Range filtering
Amoxicillin	65.23	Threshold scoring

6 Conclusions and Future Work

Bacterial infections have a significant impact on public health and antibiotics have played an important role in saving lives since their discovery. However, it is clear that antibiotic resistance seriously hinders treatment efficacy and leads to significant economic losses. Our study aimed to analyze bacterial antibiotic resistance in order to guide preventive measures to reduce economic losses and improve treatment efficacy in the future.

Our data analysis was performed by assigning percentages within a specialized parameter system, resembling a form of reinforcement learning. Instead of going through epochs, we utilized the data set itself to enhance the model's performance.

Though it would take several years to practically measure the accuracy of our predictions in relation to actual resistance that E. coli presents, sharing our results with relevant institutes could allow them to assess the need for action concerning specific antibiotics or bacteria based on emerging trends.

References

1. Doron S, Gorbach SL (2008) Bacterial infections: overview. In: International encyclopedia of public health, p 273
2. Antibiotic resistance (2020). [Online]. https://www.who.int/news-room/fact-sheets/detail/antibiotic-resistance
3. Habboush Y, Guzman N (2023) Antibiotic resistance. Treasure Island (FL)
4. Crowther-Gibson P, Govender N, Lewis D, Bamford C, Brink A, Von Gottberg A, Klugman K, Du Plessis M, Fali A, Harris B et al (2011) Part IV. Human infections and antibiotic resistance. South Afr Med J 101(1):567–578
5. Langdon A, Crook N, Dantas G (2016) The effects of antibiotics on the microbiome throughout development and alternative approaches for therapeutic modulation. Genome Med 8(1):1–16
6. Weatherbie L (2017) Antibiotic use in Canadian livestock. [Online]. https://saltyisland.com/antibiotic-use-in-canadian-livestock/
7. Nhung NT, Chansiripornchai N, Carrique-Mas JJ (2017) Antimicrobial resistance in bacterial poultry pathogens: a review. Front Vet Sci 4:126
8. Joosten P, Ceccarelli D, Odent E, Sarrazin S, Graveland H, Van Gompel L, Battisti A, Caprioli A, Franco A, Wagenaar JA et al (2020) Antimicrobial usage and resistance in companion animals: a cross-sectional study in three european countries. Antibiotics 9(2):87
9. Barnett K, Mercer SW, Norbury M, Watt G, Wyke S, Guthrie B (2012) Epidemiology of multimorbidity and implications for health care, research, and medical education: a cross-sectional study. Lancet 380(9836):37–43
10. Rajpurkar P, Chen E, Banerjee O, Topol EJ (2002) AI in health and medicine. Nat Med 28(1):31–38
11. Salehi B, Abu-Darwish M, Tarawneh A, Cabral C, Gadetskaya A, Salgueiro L, Hosseinabadi T, Rajabi S, Chanda W, Sharifi-Rad M (2022) Antimicrobial resistance collaborators global burden of bacterial antimicrobial resistance in 2019: a systematic analysis. Lancet 399:629–655
12. Rabaan AA, Alhumaid S, Mutair AA, Garout M, Abulhamayel Y, Halwani MA, Alestad JH, Bshabshe AA, Sulaiman T, AlFonaisan MK et al (2022) Application of artificial intelligence in combating high antimicrobial resistance rates. Antibiotics 11(6):784
13. Pascucci M, Royer G, Adamek J, Asmar MA, Aristizabal D, Blanche L, Bezzarga A, Boniface-Chang G, Brunner A, Curel C et al (2021) AI-based mobile application to fight antibiotic resistance. Nat Commun 12(1):1173
14. Lv J, Deng S, Zhang L (2021) A review of artificial intelligence applications for antimicrobial resistance. Biosaf Health 3(01):22–31
15. Ali T, Ahmed S, Aslam M (2023) Artificial Intelligence for antimicrobial resistance prediction: challenges and opportunities towards practical implementation. Antibiotics 12(3)
16. Reyher KK, Avison MB (2021) Farm management and longitudinal data on antibiotic use and antibiotic resistant e. coli for 53 dairy farms, south west England, 2017–2019. [Online]. https://doi.org/10.5285/c9bc537a-d1c5-43a0-b146-42c25d4e8160

References

A Review of Machine Learning for Additive Manufacturing

Beyza Gavcar

Abstract Additively manufactured parts can present lower mechanical properties because of the anisotropy and defects. Also reliability, repeatability, durability as well as the forecast of characteristics can remain difficult. Artificial intelligence techniques can prevent these challenges. They can analyze complex design requirements and generate optimized designs. These algorithms can be applied for real-time monitoring and quality control. By analyzing high amounts of data and simulating properties with regard to material, artificial intelligence can allow describe new materials with specified features. Integrating artificial intelligence with additive manufacturing processes can lead to increased efficiency, quality, and innovation, ultimately driving advancements in manufacturing. On the other hand, artificial intelligence driven additive manufacturing may face hurdles in ensuring consistent quality, scalability, and generalization across materials and processes. These algorithms require extensive computational resources and time to find optimal solutions, limiting their applicability. In the research regarding with additive manufacturing, machine learning has become a significant field of artificial intelligence, which deals with the development of algorithms and statistical models that allow computers to operate without specific programming. This study provides a review of the recent research on machine learning methods in the additive manufacturing in terms of analysis of strengths, weaknesses, opportunities, and threats (SWOT) analysis, which is crucial in determining if a specific machine learning model suits a specific additive manufacturing process. Conducting a SWOT analysis on the integration of artificial intelligence with additive manufacturing can provide valuable insights into its potential benefits, challenges, technologic and strategic considerations.

Keywords Additive manufacturing · Artificial intelligence · Machine learning

B. Gavcar (✉)
Yildiz Technical University, Istanbul, Turkey
e-mail: beyza.gavcar@gmail.com

F. Çebi et al. (eds.), *Engineering and Technology Management in Challenging Times*,
Lecture Notes in Networks and Systems 1137,
https://doi.org/10.1007/978-3-031-70935-7_8

1 Introduction

Artificial Intelligence (AI) is a field of computer science dedicated to exploring the essence of intelligence and developing innovative machines capable of exhibiting intelligent behavior [1, 2]. It can be considered as the simulation of human intelligence described by machines, which depends on mathematical algorithms and statistical analysis of data. A wide variety of scientific fields and industries has been considerably affected by this technology since the introduction of this topic. There is a great attention to AI techniques from both industry and academics thanks to their advantages [3]. AI technology provides decision-making, prediction, clustering, classification, and optimization based on available data instead of a problem's analytical model relying on assumptions. The characteristics of this digital technology go beyond the natural abilities of learning, reasoning, adaptative qualities, and its ability to mimic the human brain. It also combines high efficiency with autonomy, adaptability, cooperation, and creativity [4]. These qualities help the usage of artificial intelligence more prevalent, such as finance, health care, education, logistics, transportation and other fields [5–9]. AI techniques involving automation and robotics, big data, data mining, expert system, machine learning (ML), and machine vision have been confirmed in the manufacturing industry [10].

Additive manufacturing (AM), is also called as three dimensional printing (3DP), allows to produce parts by adding material in the form of layers from an STL file [11]. AM has a number of outstanding properties, including a short preparation cycle between the design and manufacturing, the capacity to effectively fabricate intricate components with high degree of freedom, customer-oriented production, the ability to fabricate complex shapes in one part, restricting the waste of raw material and the capacity to simulate light bioinspired structures [12]. It is regarded as an encouraging solution for achieving environmentally friendly and sustainable manufacturing due to its ability to reduce raw material requirements, resulting in less waste generation [13, 14]. Therefore, AM has emerged as a rapidly evolving technology in both academic and industrial research over the last two decades. It has been widely embraced across various industrial sectors, including automotive, aerospace, and medical applications [15–17].

Researchers apply AI data-based algorithms to evaluate the relation between output and input variables of AM. AI facilitates the collection, processing, and analysis of vast quantities of data generated during the AM process. It intelligently oversees the manufacturing process, ensuring personalized production, high-quality output, and the safety of the products. The combination of AI and AM makes available the sustainable growth and innovation. AI-based AM has possible uses in design of part, plan of manufacturing path, and quality control to allow customization, short production time, and improved quality. This combination can result in the digital transformation of the industry and enable more various part options to customers [18].

Moreover, 3DP contributes the development of AI technologies. By fabricating AM models with high complexity and accuracy, more and improved data samples can be obtained, so assisting AI algorithms to learn and predict better; AM technology is employed to produce different wearable devices, sensors, etc., increasing the applications of AI technology [19].

As an interdisciplinary approach Machine Learning, which is the branch of AI, is the most commonly used low-cost algorithm in AM processes [20]. The main focus of ML technology is the utilization of available information for improving the system's performance by computation. The predictions performed for making decisions accurately are based on the models consistent with the empirical data and learning algorithms [21]. ML is a large-scale algorithm, where data is extracted. ML models are trained by huge amounts of data to estimate the output based on input [22]. However, further AI technologies with potential applications in AM exist, like the utilization of data mining for detecting the available process instabilities automatically or the use of machine vision for comparing the geometry of the designed model with the obtained model [23, 24]. ML models provide opportunities for automation, cost reduction, enhanced productivity, and predictive maintenance of machinery tools to AM. ML predicts optimal process parameters, mechanical properties, quality of manufactured parts, and topology optimization of design in an efficient manner [25–29]. Compared to the experimental methods, which are only aimed and designed for a particular situation, ML algorithms have more flexibility and can be easily adjusted upon application. These models like Adaptive Boosting (AdaBoost), Artificial Neural Network (ANN), Decision Tree (DT), Support Vector Regression (SVR), Random Forest (RF), and Linear Regression (LR) are employed for the prediction of the output response, optimization of the process parameters, and decreasing the process time [30, 31].

In various scientific and industrial areas, there is a growing attention to SWOT analysis for different applications [32]. This method assesses the existence of intrinsic outstanding features of a part or process (strengths), which support performing the aim, as well as intrinsic adverse aspects that prevent to carry out the aim (weaknesses). Extrinsic components, which support enhancing the properties (opportunities) as well as extrinsic components, which prevent the use of a part or process (threats), are present [33]. ML presents the optimal operating conditions for the different cases having sufficient training data from the databases about the materials, parameters, processes as well as precision in optimization [34]. This method has a high potential to learn models from data and to support decision-making according to the available dataset. Analyzing the strengths of ML describes the properties necessary for accurate simulation modelling. Weaknesses identifies the areas that need improvement. Finally, opportunities and threats clarify the particular attributes that may promote or prevent the application of simulation tools. This study reviews a SWOT analysis of ML, for AM applications from the standpoint of internal (strengths and weaknesses) and external (opportunities and threats) points of view.

2 Artificial Intelligence and Additive Manufacturing

Additive manufacturing brings innovation to the manufacturing sector due to its original manufacturing concept. It provides various opportunities like decentralized, flexible, and on-demand manufacturing and mass customization. There is a growing demand for AM products that are highly complex and quality considering the design principles, quality, and standardization. Despite the advantages of AM, the use of AM processes also brings several challenges, which are related to the process and system's complexity. AM parts' quality is influenced by different constituents, such as the process duration and parameters. If these factors are not suitable to meet the requirements, process errors and defects take place. These defects and errors, which stem from the lack of knowledge about the process or use of unsuitable process parameters as input, in AM products hinder the use of parts, decrease durability of the parts and lead to the increase of cost due to the waste of time and material. Thus, to detect defects in the product and process as well as tackle other challenges, a model is required. In this respect, data-driven models have verified their success as powerful and reliable approaches to predict the properties of AM products, to optimize the process parameters, and to provide the mechanisms of defects that would have on the other hand been very challenging or too costly to find out by in situ experiments [35]. AI methods have importance in removing the shortcomings of AM since they can present effective approaches for the management of sources, modeling of highly complex systems, optimization of process, and control of quality. With the rapid advance and adoption of AI technology, some AI models used for AM can accurately remove its drawbacks via data analysis, collection, management, processing, and storage [36].

One of these advanced numerical methods is the machine learning tool introduced by [37]. ML is described as developing an algorithm to ensure performance criteria optimization utilizing programming. ML is a widely applied branch of AI for AM to enhance the process' and product's quality, accelerate the manufacturing, and decrease the costs and defects by achieving optimal process parameters [38]. The major ML models used in AM can be summarized as in Table 1.

ML is utilized to comprehend the AM process by predictive analysis [59, 60]. This requires relating 3D model and optical tomography images of in-process product as input or training data with X-ray computer tomography (CT) imaged of finished product as output to control and improve the quality. This integration provides the determination of defects formed during AM process and optimization of process parameters [61]. Successful AM processes require optimization before, during, and after manufacturing, and there are various factors to regard (Fig. 1).

Before the process of AM, AI is used to develop predictive models and to predict suitability between the process and materials to decrease errors by choosing the accurate material, process, and parameters. Generally, to achieve 3D models, AI techniques are applied to assist with image segmentation and following optimization of these models. AM is a dynamic manufacturing process, which also entails different stages, hence some problems can occur at anywhen, and errors can still take place

Table 1 Summary of studies of the last 5 years where ML is used in AM

Machine learning approach	Description	Research aim	References
Linear regression	Based on a collection of input data, it generates a best-fit line to predict the output	• To enhance the strength • To achieve optimal process parameters • To foresee hardness for further process parameters • To determine the correlation among energy consumption, part quality, process parameters • To achieve a forecast model for the tensile properties of additively manufactured parts	Veeman et al. [30], Malviya and Desai [39], Tian et al. [40], Kusano et al. [41]
Decision tree	Offers solutions for cases requiring both classification and regression	• To predict compressive strength • To achieve optimal process parameters • To foresee hardness for further process parameters • To predict the dimensional variation • To predict the tensile ductility • To predict the tensile and flexural strengths	Veeman et al. [30], Agarwal et al. [42], Sharma et al. [43], Yao et al. [44], Sharma et al. [45]
Random forest	Includes various decision trees trained on arbitrary sub-samples	• To estimate the optimal deposition angle for different categories of geometrical shapes • To predict the compressive strength • To determine the surface roughness • To achieve optimal process parameters • To foresee hardness for further process parameters	Veeman et al. [30], Agarwal et al. [42], Parsazadeh et al. [46], Hooda et al. [47], Barrios and Romero [48]
AdaBoost	Provides a combination of its weak learners to achieve one strong learner	• To achieve optimal process parameters • To foresee hardness for further process parameters • To evaluate the wear rate • To predict the tensile and flexural strengths • To perform classification at part density level according to the remelting process parameters	Veeman et al. [30], Sharma et al. [45], Barrionuevo et al. [49], Hu et al. [50]

(continued)

Table 1 (continued)

Machine learning approach	Description	Research aim	References
Support vector regression	A powerful approximation method for regression cases to obtain the best-fit line by providing a broad interval band with numerous sample points if possible	• To estimate form of melt pool • To maximize the estimation performance by utilizing different models and parameters • To predict compressive strength • To predict deposited height • To find relation between part properties and process parameters • To predict tensile properties	Zhu et al. [21], Agarwal et al. [42], Yao et al. [44], Parsazadeh et al. [46], Sharma et al. [51], Ding et al. [52]
Artificial neural networks	This nonlinear optimization technique can distinguish and estimate the complicated relation between output and input datasets	• To predict specimens' distortion • To determine correlation between parts and their properties • To predict mechanical properties • To predict fatigue life • To predict residual stress • To reveal the correlation between the mechanical properties and microstructures	Gairola et al. [53], Karna et al. [54], Suo et al. [55], Zhan and Li [56], Hajializadeh and Ince [57], Peng et al. [58]

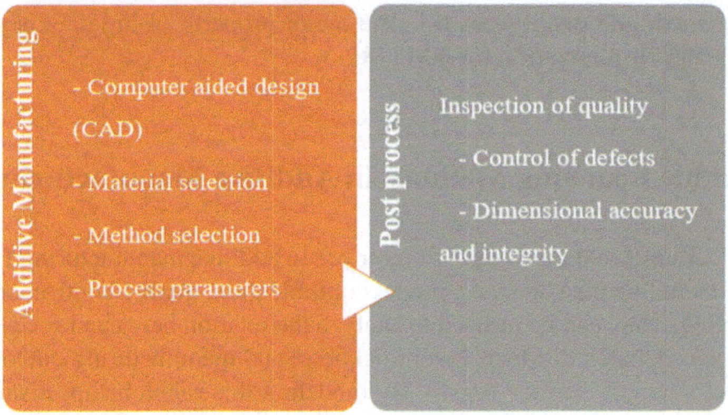

Fig. 1 The stages of AM and the factors to regard at each step

even after the forecast of the suitability and the optimization of the Computer-Aided Design (CAD) models. For instance, materials' properties are affected by the changes in the environment, and AM part's quality is affected by the utilized AM technology and the performance of AM system [62].

During the manufacturing process, challenges, for example, the uniformity of AM part over the build plate to cause failures; material's uncertainty (e.g., distribution, linear expansion coefficient, and viscosity) to induce plugging of the nozzles; unsuitable support structures to result in part's distortion. These situations cause a waste of time and material if there is not any inspection [63]. In large-scale manufacturing, detection of single-layer defects is challenging visually, thereby while the process is in progress, AI tools are needed to effectively distinguish defects, remove these defects, and improve process performance by selecting optimal process parameters based on databases.

For the post-process step of AM, AI applications are limited since the part is currently present at this stage. When the manufacturing process is finished, quality control of the part is required to avoid training with defective models to provide dimensional accuracy and integrity. After manufacturing, the conservation of parts is also an important problem as working conditions (such as humidity, light, PH, and temperature, PH) influence the product properties and quality, especially when AM is used for biomedical applications. To tackle these issues in AM, AI is the most appropriate technology.

The usage of AI technology in AM is also investigated in terms of the aspect of the big data. The big data analysis can provide the optimal part quality, accurately manufacturing paths, higher manufacturing efficiency, optimize the AM parameters and the cost of process as well as control the AM process. Thanks to the gathered and analyzed the process data, AI is applied to generate related algorithms to find the suitable process conditions and new materials to guarantee that the product seems to has good variations according to the needs. Big data can allow real-time data in the AM process. The early interference of big data offers the opportunity of pre-arrangement and quality control. The major property of big data is to achieve innovation and customization for AM [18].

3 Machine Learning Methods in Additive Manufacturing

Machine learning techniques can be employed in additive manufacturing for various purposes, including data analysis, process optimization, quality control, and predictive modeling. They can be utilized to analyze the relation between process parameters (e.g., speed, hatch distance, layer thickness) and manufacturing outcomes (e.g., mechanical strength, surface finish). By modeling this relationship, manufacturers can identify optimal parameter settings to improve process efficiency and product quality. They can be employed to predict material properties of 3D-printed parts based on input parameters such as material composition, printing conditions, and geometry. This allows manufacturers to optimize material usage and ensure the desired performance of printed components. These techniques can be utilized to estimate the performance of 3D-printed designs based on geometric features, material properties, and environmental conditions. This enables engineers to optimize designs for specific functional requirements, such as hardness, heat resistance, or aerodynamic performance. ML learns from training data, detects connections and models among properties, and create nonlinear correlations within the data. The amount and quality of the data have importance in ML, they directly affect the performance and accuracy of predictive algorithm. To reduce the manufacturing time and the cost, combining available data verifies to be the most efficient strategy when building datasets [64].

3.1 Linear Regression (LR)

Linear regression presents the relation between a dependent variable and at least independent variable statistically [30]. Linear regression models can be trained on historical data to identify patterns associated with defects or anomalies in manufactured parts. By analyzing sensor data or image features, these models can predict the likelihood of defects during AM process, enabling real-time quality control and intervention. They can analyze sensor data from AM system to predict equipment failures or maintenance needs. By identifying trends and patterns indicative of potential issues, predictive maintenance strategies can be implemented to minimize downtime and optimize equipment utilization. LR can be applied to analyze experimental data obtained from designed experiments (e.g., factorial designs) in AM. By modeling the effects of different factors and interactions, manufacturers can find out the underlying processes and identify significant variables affecting the outcomes. LR serves as a significant approach in AM for data analysis, modeling, and decision-making, allowing manufacturers to optimize processes, enhance quality, and drive innovation.

Danish et al. used linear regression model for successfully prediction of tool wear of additively manufactured parts [65].

3.2 *Decision Tree (DT)*

Decision trees find solutions for use cases including both classification and regression [30]. They can be trained on historical data to classify manufactured parts as defective or non-defective based on inputs such as sensor data or image characteristics. By recursively partitioning the feature space, decision trees can effectively identify patterns associated with defects and enable real-time quality control during the AM process. By partitioning the feature space into decision regions, DTs can capture complex relationships between input variables and predict the desired output properties of manufactured components. By recursively splitting the data based on sensor readings and operational parameters, DTs can identify conditions indicative of potential equipment failures and trigger maintenance actions to prevent downtime. DTs can be integrated into closed-loop control systems to adaptively arrange process parameters based on real-time sensor feedback. By continuously updating the decision rules based on observed data, DTs can optimize printing processes to maintain desired quality standards and minimize variability. Zhu et al. applied Gradient-Boosting Decision Trees (GBDT) to analyze the influences of AM process parameters on the surface quality [66].

3.3 *Random Forest (RF)*

Random forest is a supervised ensemble method that operates as a clustered model. For the RF algorithm, particular decision tree models (n) are allocated as base learners. Later, the training dataset is sampled haphazardly into various subsets, and each subset is considered as input for each base learner. This approach is called as bootstrap aggregation or Bagging, a commonly utilized ensemble strategy in the recent machine learning area [30]. By aggregating predictions from multiple DTs, RF can provide robust and accurate defect detection capabilities, enabling real-time quality control during the AM process. Roy et al. trained random forest to estimate the detect formation in AM process [67].

3.4 *Adaptive Boosting (AdaBoost)*

AdaBoost relates multiple weak learners (typically decision trees) to obtain a strong learner, with a focus on improving predictive accuracy [30]. By iteratively arranging the weights of misclassified constituents, AdaBoost focuses on difficult-to-classify samples, resulting in improved defect detection capabilities. AdaBoost iteratively adjusts the model to prioritize misclassified instances, enabling the detection of subtle deviations that may indicate abnormal conditions in the manufacturing process. This allows for timely intervention to maintain quality and reliability in additively manufactured parts. Li et al. utilized AdaBoost to forecast the surface roughness in AM process [68].

3.5 Support Vector Regression (SVR)

Support vector regression, in which an insensitive loss function is presented, is an effective approximation approach for regression problems [69]. This powerful machine learning algorithm works by finding the hyperplane that best fits the data in a high-dimensional space while maximizing the margin around it. By analyzing the correlations between the input parameters of AM and the desired material features, SVR can accurately model complex nonlinear patterns in the data. It can identify optimal parameter settings that maximize the desired outcome, such as minimizing defects, cost or manufacturing time.

3.6 Artificial Neural Network (ANN)

Artificial neural networks are inspired by the function and structure of the neural networks in the human brain. ANNs include interconnected nodes (neurons) arranged in layers, with each neuron performing simple computations and transmitting signals to neurons in the next layer. A neural network can learn from data, thus it can be trained to identify models, categorize data, and predict future cases [70]. In the AM, ANNs can be utilized for various applications owing to their ability to learn complex patterns from data. ANNs can detect anomalies or abnormalities in the manufacturing process, such as variations in temperature or pressure. By alerting operators to potential issues, ANNs help ensure the quality and consistency of AM products.

4 SWOT Analysis

To process vast amounts of data and to reveal patterns in data, the applications of AI algorithms in AM field have a crucial role because of their strengths and offered opportunities, which will be evaluated in future studies. The performances of models are considerably associated with the training data amount and type. There are also weaknesses and threats of AI tools that restrict the applications of them. To perform

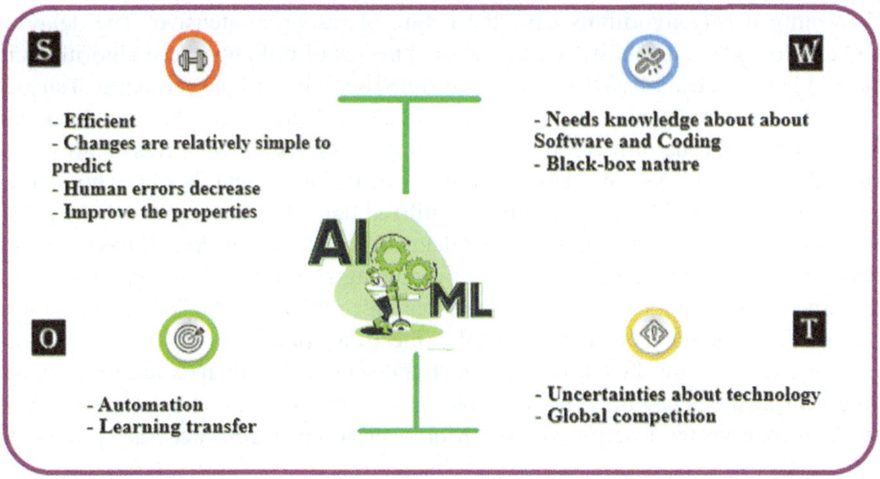

Fig. 2 SWOT analysis for AI and ML

a main comprehension of AI and ML in AM, a summary of the SWOT analysis is presented here, as illustrated in Fig. 2.

4.1 *Strengths*

AI, and especially ML applications are utilized in various areas of AM thanks to their superior features. ML models possess the property of learning from large datasets without human interference [71]. To enhance their efficiency, they can be integrated with analytical, empirical, or finite element models. They are not particularly programmed to solve a specific problem. The accessibility of them is not difficult thanks to their packages and open-source extension. Approaches for improvement of efficiency and product quality, management of the sources as well as process optimization are major strengths of ML in AM. Its flexibility, operating speed, and user-friendly applications are other outstanding properties [72]. Well-trained algorithms can forecast accurately in a fraction of a second. Accurate predictions propose easier solutions to be employed. These models simulate the complicated relation between the dependent and independent variables.

4.2 *Weaknesses*

AI models in the AM heavily rely on high-quality data, and the availability of suitable data can be a challenge, especially for emerging applications or industries.

Conventional ML algorithms have the nature of resource-intensive. The delay of find and low generalizability are major weaknesses of utilizing these algorithms in the AM [73–75]. Data-based methods developed applying simple geometrical shapes may not be generalizable to complex, functional AM products. This drawback can be removed with the utilization of physics-informed ML (PIML) model [46]. The other drawbacks of ML are data gathering, extrapolation, and its black-box structure. One of the critical shortcomings is utilized data. The data amount is high, so accurate utilization and interpretation of data are needed. When dataset has too many systematic errors, the network will fail by causing systematic errors itself, of what is defined as the principle of "garbage in–garbage out" (GIGO) [76]. One potential solution for systematic errors is removal of these data points manually. Other option is application of algorithms for detection. To present physically meaningful explanations for specific behaviors is not possible. Therefore, ML approaches are unsuitable modeling options for descriptive cases. Moreover, some models' accuracy is not high sufficiently.

4.3 Opportunities

ML algorithm's strengths provide different opportunities that are offered to increase the researches about ML for AM. ML approaches are prospective for automatic detection, categorization, and quantification of defects as well as learning transfer. They are well-suited for AM processes where the speed and accuracy are significant. Also, data training is performed by cloud-edge synergy. ML for AM offers sustainability from the point of energy and cost management. For energy modeling, it is complex to consider several factors based on conventional approaches like mathematical methods. Therefore, ML models are preferred to analyze and model the energy consumption of AM systems. Decrease of costs, which is related to material, post-processing, and process duration, has an important role in AM sustainability. Through ML technologies, the time and cost of AM applications can be reduced. The successful estimation of parameters and material properties utilizing models decreases the consumption in discovering materials and manufacturing of the products to meet process needs by determining optimal AM parameters to combine parameters and features. These opportunities lead to decreasing capital investment, enhancing of automation levels, and strategy development [77].

4.4 Threats

Adversarial attacks are one of the most significant threats for ML, in particular concerning AM for highly crucial components. They force ML models to output inaccurate predictions. AI-driven AM systems are vulnerable to cybersecurity threats, such as data breaches or malicious attacks on AM processes. This threat has become prior for cybersecurity experts as well as manufacturers. To hinder this threat, ML algorithms can be hardened and robust neural network optimization approach can be developed to reduce their susceptibility. Inaccurate evaluation of input data brings improper forecasts that are not simply revealed and defined because of ML's black-box property. Psychological threats also exist. There are concerns about the reliability of ML models because the uncertainties and their potential effects are present. Rapid advancements in AI and AM technologies may lead to the risk of investing in outdated or obsolete systems if not carefully managed. Increasingly global competition needs reliable ML models and drives continuous advancement. Increased adoption of AI in AM may intensify market competition, requiring organizations to continuously innovate and differentiate their offerings. Although it is considered as a strength, access to ML models is another important threat when the user has only basic programming skills. In this case, ML can be misused and the outputs can be misinterpreted in the context of the complicated AM processes because of the lack of knowledge. Users should have accurate knowledge, thus specialized training is required [78]. Even for users who are experienced and well-informed, ML can be considered as a trial-and-error model.

5 Conclusion

Additive manufacturing and artificial intelligence have become technologies that greatly affect humanity and assist each other. Thanks to the improvements in these two technologies, the interaction and integration between them will become even closer, generating more beneficial applications and products. With the rapid advancement of AI recently, researchers also focus on innovative approaches to strengthen AI as well as to integrate with AM. There is an extensive interest in machine learning in AM applications, which is crucial in the development of automated technologies thanks to its outstanding properties. This branch of AI is generally used as a tool for the prediction of properties, in situ correction, and parameter optimization. ML-based AM applications provide high efficiency and quality parts through improved image processing, perfectly foreseen the convenience between the materials and process and inspected AM process. Accurate predictions provide to expansion of the dataset and increase the accuracy and reliability of the model. The usage of ML processes in AM should follow a well-organized strategy, which typically includes phases such as data gathering, categorizing, dividing, parameter selecting, implementation, simulation, and assessment. This structured methodology ensures effective usage of ML models to optimize AM processes. A promising future research direction would be to completely comprehend ML methods and integrate them with the specific knowledge domains in AM to develop dedicated models for AM. Furthermore, the possible applications of ML for sustainability of AM have not been in detail studied in which more research about utilizing ML to save energy, cost, and material as well as to improve the efficiency of the process is essential.

References

1. Cai D, Aziz G, Sarwar S, Alsaggaf MI, Sinha A (2023) Applicability of denoising-based artificial intelligence to forecast the environmental externalities. Geosci Front 101740
2. Schramm S, Wehner C, Schmid U (2023) Comprehensible artificial intelligence on knowledge graphs: a survey. J Web Semant 79:100806
3. Wieczorowski M, Kucharski D, Sniatala P, Pawlus P, Krolczyk G, Gapinski B (2023) A novel approach to using artificial intelligence in coordinate metrology including nano scale. Measurement 217:113051
4. Li J, Herdem MS, Nathwani J, Wen JZ (2023) Methods and applications for artificial intelligence, big data, internet of things, and blockchain in smart energy management. Energy AI 11:100208
5. Chen XQ, Ma CQ, Ren YS, Lei YT, Huynh NQA, Narayan S (2023) Explainable artificial intelligence in finance: a bibliometric review. Financ Res Lett 104145
6. Chakrabarty N, Mahajan A (2023) Imaging analytics using artificial intelligence in oncology: a comprehensive review. Clin Oncol
7. Chiu TK, Xia Q, Zhou X, Chai CS, Cheng M (2023) Systematic literature review on opportunities, challenges, and future research recommendations of artificial intelligence in education. Comput Educ Artif Intell 4:100118
8. Bouanba N, Barakat O, Bendou A (2022) Artificial intelligence & agile innovation: case of moroccan logistics companies. Proc Comput Sci 203:444–449
9. Yang B (2023) Research on vehicle detection and recognition technology based on artificial intelligence. Microprocess Microsyst 104937
10. Sarker IH (2022) AI-based modeling: techniques, applications and research issues towards automation, intelligent and smart systems. SN Comput Sci 3(2):158
11. ISO A (2015) ISO/ASTM 52900: 2015 additive manufacturing–general principles–terminology: ASTM F2792-10e1, vol 1, pp 1–19
12. Bacciaglia A, Falcetelli F, Troiani E, Di Sante R, Liverani A, Ceruti A (2023) Geometry reconstruction for additive manufacturing: from G-CODE to 3D CAD model. Mater Today Proc 75:16–22
13. Chen D, Heyer S, Ibbotson S, Salonitis K, Steingrímsson JG, Thiede S (2015) Direct digital manufacturing: definition, evolution, and sustainability implications. J Clean Prod 107:615–625
14. Mani M, Lyons KW, Gupta SK (2014) Sustainability characterization for additive manufacturing. J Res Nat Inst Stand Technol 119:419
15. Guo N, Leu MC (2013) Additive manufacturing: technology, applications and research needs. Front Mech Eng 8:215–243
16. Liu R, Wang Z, Sparks T, Liou F, Newkirk J (2017) Aerospace applications of laser additive manufacturing. In: Laser additive manufacturing. Woodhead Publishing, pp 351–371
17. Javaid M, Haleem A (2018) Additive manufacturing applications in medical cases: a literature based review. Alexandria J Med 54(4):411–422
18. Yu Q, Zhang M, Mujumdar AS, Li J (2024) AI-based additive manufacturing for future food: potential applications, challenges and possible solutions. Innov Food Sci Emerg Technol 103599
19. Dabbagh SR, Ozcan O, Tasoglu S (2022) Machine learning-enabled optimization of extrusion-based 3D printing. Methods 206:27–40
20. Jordan MI, Mitchell TM (2015) Machine learning: trends, perspectives, and prospects. Science 349(6245):255–260
21. Zhu X, Jiang F, Guo C, Wang Z, Dong T, Li H (2023) Prediction of melt pool shape in additive manufacturing based on machine learning methods. Opt Laser Technol 159:108964
22. Dias RD, Gupta A, Yule SJ (2019) Using machine learning to assess physician competence: a systematic review. Acad Med 94(3):427–439
23. Oleff A, Küster B, Stonis M, Overmeyer L (2021) Process monitoring for material extrusion additive manufacturing: a state-of-the-art review. Prog Addit Manuf 6(4):705–730

24. Johnson MV, Garanger K, Hardin JO, Berrigan JD, Feron E, Kalidindi SR (2021) A generalizable artificial intelligence tool for identification and correction of self-supporting structures in additive manufacturing processes. Addit Manuf 46:102191
25. Wang C, Tan XP, Tor SB, Lim CS (2020) Machine learning in additive manufacturing: state-of-the-art and perspectives. Addit Manuf 36:101538
26. Li X, Zhang M, Zhou M, Wang J, Zhu W, Wu C, Zhang X (2023) Qualify assessment for extrusion-based additive manufacturing with 3D scan and machine learning. J Manuf Process 90:274–285
27. Xu K, Lyu J, Manoochehri S (2022) In situ process monitoring using acoustic emission and laser scanning techniques based on machine learning models. J Manuf Process 84:357–374
28. Jiang J, Yu C, Xu X, Ma Y, Liu J (2020) Achieving better connections between deposited lines in additive manufacturing via machine learning. Math Biosci Eng 17(4):3382–3394
29. Zhou Y, Lu H, Wang G, Wang J, Li W (2020) Voxelization modelling based finite element simulation and process parameter optimization for Fused Filament Fabrication. Mater Des 187:108409
30. Veeman D, Sudharsan S, Surendhar GJ, Shanmugam R, Guo L (2023) Machine learning model for predicting the hardness of additively manufactured acrylonitrile butadiene styrene. Mater Today Commun 35:106147
31. Song K, Yan F, Ding T, Gao L, Lu S (2020) A steel property optimization model based on the XGBoost algorithm and improved PSO. Comput Mater Sci 174:109472
32. Gürel E, Tat M (2017) SWOT analysis: a theoretical review. J Int Soc Res
33. Noguerol TM, Paulano-Godino F, Martín-Valdivia MT, Menias CO, Luna A (2019) Strengths, weaknesses, opportunities, and threats analysis of artificial intelligence and machine learning applications in radiology. J Am Coll Radiol 16(9):1239–1247
34. Zahidin MR, Yusof F, Rashid SHA, Mansor S, Raja S, Jamaludin MF, Hussein NIS, et al (2023) Research challenges, quality control and monitoring strategy for Wire Arc Additive Manufacturing. J Mater Res Technol
35. Bayat M, Zinovieva O, Ferrari F, Ayas C, Langelaar M, Spangenberg J, Hattel J (2023) Holistic computational design within additive manufacturing through topology optimization combined with multiphysics multi-scale materials and process modelling. Progr Mater Sci 101129
36. Ma L, Yu S, Xu X, Amadi SM, Zhang J, Wang Z (2023) Application of artificial intelligence in 3D printing physical organ models. Mater Today Bio 100792
37. Ko H, Witherell P, Lu Y, Kim S, Rosen DW (2021) Machine learning and knowledge graph based design rule construction for additive manufacturing. Addit Manuf 37:101620
38. Selvam PP, Prabhakaran S, Vinod B, Jishnu T (2024) A review of energy efficiency and machine learning analysis for additive manufacturing of direct laser metal deposition. Mater Today Proc
39. Malviya M, Desai K (2019) Build orientation optimization for strength enhancement of FDM parts using machine learning based algorithm. Comput Des Appl 17:783–796
40. Tian W, Ma J, Alizadeh M (2019) Energy consumption optimization with geometric accuracy consideration for fused filament fabrication processes. Int J Adv Manuf Technol 103:3223–3233
41. Kusano M, Miyazaki S, Watanabe M, Kishimoto S, Bulgarevich DS, Ono Y, Yumoto A (2020) Tensile properties prediction by multiple linear regression analysis for selective laser melted and post heat-treated Ti-6Al-4V with microstructural quantification. Mater Sci Eng A 787:139549
42. Agarwal R, Singh J, Gupta V (2022) Predicting the compressive strength of additively manufactured PLA-based orthopedic bone screws: a machine learning framework. Polym Compos 43(8):5663–5674
43. Sharma P, Vaid H, Vajpeyi R, Shubham P, Agarwal KM, Bhatia D (2022) Predicting the dimensional variation of geometries produced through FDM 3D printing employing supervised machine learning. Sens Int 3:100194
44. Yao Z, Jia X, Yu J, Yang M, Huang C, Yang Z, Liu X et al (2023) Rapid accomplishment of strength/ductility synergy for additively manufactured Ti–6Al–4V facilitated by machine learning. Mater Des 225:111559
45. Sharma S, Gupta V, Mudgal D, Srivastava V (2023) Predicting biomechanical properties of additively manufactured polydopamine coated poly lactic acid bone plates using deep learning. Eng Appl Artif Intell 124:106587

46. Parsazadeh M, Sharma S, Dahotre N (2023)Towards the next generation of machine learning models in additive manufacturing: a review of process dependent material evolution. Progr Mater Sci 101102
47. Hooda N, Chohan JS, Gupta R, Kumar R (2021) Deposition angle prediction of Fused Deposition Modeling process using ensemble machine learning. ISA Trans 116:121–128
48. Barrios JM, Romero PE (2019) Decision tree methods for predicting surface roughness in fused deposition modeling parts. Materials 12(16):2574
49. Barrionuevo GO, Walczak M, Ramos-Grez J, Sánchez-Sánchez X (2023) Microhardness and wear resistance in materials manufactured by laser powder bed fusion: machine learning approach for property prediction. CIRP J Manuf Sci Technol 43:106–114
50. Hu F, Liu Y, Li Y, Ma S, Qin J, Song J, Tang Q, et al (2023) Task-driven data fusion for additive manufacturing: framework, approaches, and case studies. J Ind Inf Integr 34:100484
51. Sharma R, Paul AR, Mukherjee M, Vadali SRK, Singh RK, Sharma AK (2023) Forecasting of process parameters using machine learning techniques for wire arc additive manufacturing process. Mater Today Proc 80:248–253
52. Ding D, He F, Yuan L, Pan Z, Wang L, Ros M (2021) The first step towards intelligent wire arc additive manufacturing: an automatic bead modelling system using machine learning through industrial information integration. J Ind Inf Integr 23:100218
53. Gairola S, Singh G, Jayaganthan R, Ajay J (2023) High temperature performance of additively manufactured Al 2024 alloy: constitutive modelling, dynamic recrystallization evolution and kinetics. J Market Res 25:3425–3443
54. Karna M, Kakalis C, Chatzidai N, Kousiatza C, Tambouratzis T, Karalekas D (2023) A combined experimental and artificial neural networks study of distortion in 3D printed beam specimens. Mater Today Proc
55. Suo S, Yang Y, Wang Z, Rao WF (2023) The property palette: a rapid printing of performance-tunable blended polymers guided by artificial neural network. Appl Mater Today 32:101837
56. Zhan Z, Li H (2021) Machine learning based fatigue life prediction with effects of additive manufacturing process parameters for printed SS 316L. Int J Fatigue 142:105941
57. Hajializadeh F, Ince A (2021) Integration of artificial neural network with finite element analysis for residual stress prediction of direct metal deposition process. Mater Today Commun 27:102197
58. Peng J, Liu B, Li W, Liaw PK, Li J, Fang Q (2023) Data-driven investigation of microstructure and surface integrity in additively manufactured multi-principal-element alloys. J Alloy Compd 937:168431
59. Rachmawati SM, Putra MAP, Lee JM, Kim DS (2023) Digital twin-enabled 3D printer fault detection for smart additive manufacturing. Eng Appl Artif Intell 124:106430
60. Li C, Cabrera D, Sancho F, Cerrada M, Sánchez RV, Estupinan E (2021) From fault detection to one-class severity discrimination of 3D printers with one-class support vector machine. ISA Trans 110:357–367
61. Feng S, Chen Z, Bircher B, Ji Z, Nyborg L, Bigot S (2022) Predicting laser powder bed fusion defects through in-process monitoring data and machine learning. Mater Des 222:111115
62. Go J, Hart AJ (2017) Fast desktop-scale extrusion additive manufacturing. Addit Manuf 18:276–284
63. Paraskevoudis K, Karayannis P, Koumoulos EP (2020) Real-time 3D printing remote defect detection (stringing) with computer vision and artificial intelligence. Processes 8(11):1464
64. Wang H, Li B, Zhang W, Xuan F (2024) Microstructural feature-driven machine learning for predicting mechanical tensile strength of laser powder bed fusion (L-PBF) additively manufactured Ti6Al4V alloy. Eng Fract Mech 295:109788
65. Danish M, Gupta MK, Irfan SA, Ghazali SM, Rathore MF, Krolczyk GM, Alsaady A (2024) Machine learning models for prediction and classification of tool wear in sustainable milling of additively manufactured 316 stainless steel. Results Eng 102015
66. Zhu J, Su Z, Wang Q, Lan Z, Chan FSF, Han Z, Ngan ACF, et al (2024) Surface quality prediction and quantitative evaluation of process parameter effects for 3D printing with transfer learning-enhanced gradient-boosting decision trees. Expert Syst Appl 237:121478

67. Roy A, Swope A, Devanathan R, Van Rooyen IJ (2024) Chemical composition based machine learning model to predict defect formation in additive manufacturing. Materialia 102041
68. Li Z, Zhang Z, Shi J, Wu D (2019) Prediction of surface roughness in extrusion-based additive manufacturing with machine learning. Robot Comput Integr Manuf 57:488–495
69. Roy A, Chakraborty S (2022) Reliability analysis of structures by a three-stage sequential sampling based adaptive support vector regression model. Reliab Eng Syst Saf 219:108260
70. Kang M, Elbel S (2023) Novel regenerator design for caloric cycles using artificial neural network—genetic algorithm method and additive manufacturing. Energy Rep 9:4257–4274
71. Goh GD, Hamzah NMB, Yeong WY (2023) Anomaly detection in fused filament fabrication using machine learning. 3D Printing Addit Manuf 10(3):428–437
72. Dobbelaere MR, Plehiers PP, Van de Vijver R, Stevens CV, Van Geem KM (2021) Machine learning in chemical engineering: strengths, weaknesses, opportunities, and threats. Engineering 7(9):1201–1211
73. Liu S, Stebner AP, Kappes BB, Zhang X (2021) Machine learning for knowledge transfer across multiple metals additive manufacturing printers. Addit Manuf 39:101877
74. Gunasegaram DR, Murphy AB, Barnard A, DebRoy T, Matthews MJ, Ladani L, Gu D (2021) Towards developing multiscale-multiphysics models and their surrogates for digital twins of metal additive manufacturing. Addit Manuf 46:102089
75. Mani M, Lane BM, Donmez MA, Feng SC, Moylan SP (2017) A review on measurement science needs for real-time control of additive manufacturing metal powder bed fusion processes. Int J Prod Res 55(5):1400–1418
76. Bininda-Emonds OR, Jones KE, Price SA, Cardillo M, Grenyer R, Purvis A (2004) Garbage in, garbage out: data issues in supertree construction. In: Phylogenetic supertrees: combining information to reveal the tree of life, pp 267–280
77. He F, Yuan L, Mu H, Ros M, Ding D, Pan Z, Li H (2023) Research and application of artificial intelligence techniques for wire arc additive manufacturing: a state-of-the-art review. Robot Comput Integr Manuf 82:102525
78. Guo S, Agarwal M, Cooper C, Tian Q, Gao RX, Grace WG, Guo YB (2022) Machine learning for metal additive manufacturing: Towards a physics-informed data-driven paradigm. J Manuf Syst 62:145–163

Assessment of Emerging Digital Technology in Renewable Energy Sector

Melisa Ozbiltekin Pala and Yesim Deniz Ozkan Ozen

Abstract Economic prosperity, energy security, and environmental sustainability are all greatly impacted by the use of renewable energy sources. Renewable energy sources are vital in the battle against climate change and do not affect the environment like energy production based on fossil fuels does. Furthermore, more efficiency, sustainability, and safety are guaranteed in the energy industry through the integration of digital technology and the renewable energy sector. By ensuring the energy supply and promoting environmental sustainability goals, this improves energy security. As a result, the significance of this integration will be crucial to the future of the energy industry. Finding the emerging digital technology that the market requires also helps with managerial and real-world application decision-making processes. Determining which technology is more significant in the field of renewable energy is the goal of this study. This study presents eight distinct new digital technologies, all of which are evaluated using the fuzzy Delphi technique. According to the results, Internet of Things, Artificial Intelligence (AI) and Machine Learning (E2), and Big Data Analytics are found as the most important technologies for the renewable energy sector.

Keywords Emerging digital technologies · Renewable energy · Fuzzy Delphi · Digital transformation

---

M. O. Pala (✉) · Y. D. O. Ozen
Logistics Management, Yasar University, Bornova, Turkey
e-mail: melisa.ozbiltekin@yasar.edu.tr

Y. D. O. Ozen
e-mail: yesim.ozen@yasar.edu.tr

F. Çebi et al. (eds.), *Engineering and Technology Management in Challenging Times*,
Lecture Notes in Networks and Systems 1137,
https://doi.org/10.1007/978-3-031-70935-7_9

1 Introduction

Renewable energy is energy obtained from energy sources that are produced indefinitely by natural processes and continue without human influence [1]. These sources are considered environmentally friendly because they minimize the release of greenhouse gases and other environmental pollutants during energy production. Renewable energy sources represent a sustainable approach to energy production and provide cleaner and environmentally friendly energy compared to fossil fuels [2]. Therefore, it is increasingly used in response to important global challenges such as combating climate change, energy security and environmental protection.

In the globalizing world, digital technologies are increasing their importance day by day and emerging digital technologies have begun to be used in every sector [1]. It is extremely important to adopt emerging digital technologies, especially in the renewable energy sector, where decision mechanisms must operate flawlessly and sustainability must be ensured in the processes [3]. The renewable energy sector has become heavily dependent on emerging digital technologies to make energy production and distribution more efficient, sustainable and secure [4].

Recently, many different emerging digital technologies are emerging and each of them offers sector-specific benefits. The technology that every sector should take into consideration and invest in is shaped according to the needs of the sector [3]. Especially in the renewable energy sector, emerging digital technologies enable renewable energy production facilities and networks to operate more efficiently and reliably, and enable the performance of the facilities to be monitored in real time [4]. Additionally, emerging digital technologies are making energy storage and distribution smarter. Energy storage systems regulate energy supply by storing energy from variable sources such as solar or wind energy [5]. Moreover, emerging digital technologies also increase data analysis and prediction capabilities. It is aimed to predict future energy demand by analysing energy production and consumption data [4]. This helps energy companies plan their resources more effectively and better adapt to energy demand fluctuations. In addition, emerging digital technologies facilitate energy trading and sustainability monitoring and reporting [3]. Furthermore, it is important to know how to produce renewable energy resources, which technologies to use and how to distribute them, as well as the use of renewable energy resources [6]. Therefore, research question of this study is planned as follows;

- **Research Question:** What is the importance order of emerging digital technologies in renewable energy sector?

To give an answer of the research question, firstly, emerging technologies in renewable energy sector are defined by considering the literature. Then, fuzzy Delphi method is applied to determine the importance order of emerging digital technologies in renewable energy sector. The main motivation and contribution of the study is to reveal which technologies should be used more in the renewable energy sector and which technologies should be given importance. Realizing which technology is most important in this field in terms of managerial and practical industry applications will be guiding for decision analysis. Therefore, this study aims to determine the importance orders of emerging digital technologies in renewable energy sector. The sections of this study are organized as follows; firstly, theoretical background about renewable energy and emerging technologies is explained. Then, methodology is given. Section 4 covers implementation and result and lastly, discussions and conclusions are given.

2 Theoretical Background

Rapidly increasing population, globalization and industrial growth put environmental, social and economic pressure on the world. In order to ensure sustainability, renewable energy has become more prominent today compared to previous years [7]. They play a very important role in solving some of the most pressing global problems, and above all, renewable energy sources such as solar, wind and hydroelectricity are environmentally friendly alternatives to fossil fuels. Renewable energy resources are sustainable and inexhaustible and provide a continuous and reliable energy supply for the future [8]. Additionally, the renewable energy sector supports energy security by diversifying energy sources, reducing dependence on fossil fuel imports, and increasing resilience to supply disruptions [1]. It also fosters innovation and technological advancement, spurring economic growth and job creation. Overall, the renewable energy sector is vital to a cleaner, more sustainable and resilient energy future [5].

The issue of "digitalization", which is encountered in almost every sector and all business models are organized within this framework, is also extremely critical for the renewable energy sector [3]. In other words, the effects of emerging digital technologies on the renewable energy sector are highly transformative. First of all, the efficiency of power plants is increasing thanks to the ability of these technologies to monitor and control renewable energy production. Sensors, smart measuring devices and internet of things (IoT) devices enable renewable energy sources such as solar panels, wind turbines and hydroelectric power plants to be continuously monitored and data collected [7]. This provides important information for optimizing energy production, predicting maintenance requirements and predicting failure conditions.

Moreover, emerging digital technologies increase efficiency in the renewable energy sector and also contribute to making energy production more sustainable and reliable [7]. With greater adoption of these technologies, there is wider use of renewable energy sources and reduced dependence on fossil fuels, increasing environmental sustainability and energy security. Emerging digital technologies make the energy sector more innovative and enable the development of new business models [4]. Remote monitoring and management of energy facilities provides energy companies with greater control and data. In addition, emerging digital technologies increase efficiency in the renewable energy sector, optimizing energy production and making energy facilities more sustainable [5]. Widespread adoption of these technologies contributes to the more widespread and effective use of renewable energy and the future sustainability of the energy sector [6].

It is extremely critical that emerging digital technologies can be adopted in the renewable energy sector and that the right technology can be applied in the right process [3]. Choosing the right emerging digital technologies in the renewable energy sector significantly affects the efficiency, reliability and sustainability of energy production and distribution [6]. When selecting and implementing these emerging digital technologies in the renewable energy sector, it is important to consider factors such as cost effectiveness, scalability, interoperability, and the specific needs of the renewable energy project or facility [9]. In addition, choosing the right technology significantly affects the efficiency, reliability and sustainability of energy production and distribution. For this reason, emerging technologies that can be used in the renewable energy sector are listed as follows (Table 1).

Particularly in the renewable energy sector, it has become highly dependent on emerging digital technologies to make energy production and distribution more efficient, sustainable and safe [8]. When academic researchers are examined, it is seen that these studies on this subject are focused on a single technology, not on multiple technologies, and focus on practical applications such as [15–19]. Determining which technologies are at the forefront in the renewable energy sector is important to develop further understanding and broader strategic approaches. As a research gap, there is no study in which technologies are evaluated with a focus on the renewable energy sector with expert opinions and ranked in order of importance to guide decision-making mechanisms.

3 Methodology

To evaluate the emerging technologies for renewable energy sector, fuzzy DELPHI method is employed in this study. Fuzzy DELPHI method is derived from traditional DELPHI method, which integrates traditional DELPHI method with fuzzy theory [20]. Classical DELPHI method is a very well-known system that is based on consensus of decision makers [21]. On the other hand, in fuzzy DELPHI fuzzy linguistic variables are used for evaluation of the concepts. Furthermore, in fuzzy DELPHI method number participants required for the application is reduced, requirement for multiple round of interviews eliminated, and consequently investigation time is shortened [22]. From this point of view, it can be said that Fuzzy DELPHI method can be utilized to translate experts' viewpoints into specific numerical values, thus increase the efficiency of decision-making process [23]. In this study triangular fuzzy numbers are used and showed in Table 2.

The procedure of the of fuzzy DELPHI method is summarized from [24] as follows:

Step 1: Collect expert opinions by using fuzzy linguistic scale presented in Table 2. Fuzzy triangular numbers are shown for attribute k as follows:

Table 1 Emerging technologies in renewable energy sector

Emerging digital technologies	Explanation
Internet of things (E1)	IoT devices, such as sensors and smart meters, are used to monitor renewable energy systems and collect real-time data on energy production and consumption [9]. This data is important to help optimize energy production, improve maintenance plans and improve grid management [7]
Artificial intelligence (AI) and machine learning (E2)	AI and machine learning algorithms are used for predictive modelling and optimization in renewable energy systems [10]. It helps predict energy production, recognize performance issues and optimize energy production in real time [11]
Digital twins (E3)	Digital twin technology creates virtual copies of physical renewable energy assets, enabling real-time monitoring, simulations and predictive maintenance [12]. This optimizes the performance of renewable energy plants
Big data analytics (E4)	Big data analytics tools process and analyse large amounts of data generated by renewable energy sources [13]. This data is used to optimize predictive maintenance, demand forecasts and energy production, increasing efficiency and cost savings [14]
Blockchain (E5)	Blockchain technology increases transparency and traceability in the renewable energy sector and it is used to track the source of renewable energy generation and facilitate secure peer-to-peer energy trading, making renewable energy markets more efficient and reliable [15]
Cybersecurity solutions (e6)	As the renewable energy sector becomes more digital, ensuring the security of digital infrastructure is extremely important [16]. Strong cybersecurity solutions are vital to protect renewable energy systems against cyber threats [17]
Edge computing (E7)	Edge computing brings data processing closer to the data source, reducing latency and increasing real-time decision making [18]. This is particularly valuable for monitoring and controlling distributed renewable energy systems
Augmented reality (AR) and virtual reality (VR) (E8)	AR and VR technologies are used for training, maintenance and remote support in the renewable energy sector [19]. It allows technicians to visually inspect and interact with the equipment, increasing efficiency and safety

Table 2 Fuzzy linguistic scale

Linguistic variable	Triangular fuzzy number
Extremely unimportant (EU)	(0.1, 0.1, 0.3)
Unimportant (U)	(0.1, 0.3, 0.5)
Medium (M)	(0.3, 0.5, 0.7)
Important (I)	(0.5, 0.7, 0.9)
Extremely important (EI)	(0.7, 0.9, 0.9)

$$\tilde{W}_k = (x_k, y_k, z_k) \tag{1}$$

where, x_k is the minimum value, y_k is the medium value and z_k is the maximum value.

Step 2: Defuzzification of the fuzzy numbers is conducted by using centre of gravity method as presented in Eq. 2, where D_k reprents the defuzzified value.

$$D_k = \frac{x_k + y_k + z_k}{3} \tag{2}$$

Step 3: Aggregation of expert opinions by calculating the average for each attribute.

Step 4: Selection of threshold value, λ based on the research needs and evaluate the attributes by using following principles.

a. If $D_k \geq \lambda$ then accept the attribute k.
b. If $D_k < \lambda$ then eliminate the attribute k.

The following section presents the implementation and results of the study.

4 Implementation and Results

This study aims to reveal the most critical emerging digital technologies and their importance order for renewable energy sector. With this aim, implementation of this study is conducted with the participation of experts, who works in different energy related firms. On the contrary of the surveys that require certain sample size, number of decision makers needed in group decision making approaches may vary according the features of the study. Characteristics of experts, such as experience, skills, consistency and consensus are the determinants of selection of the experts in the study [25]. From this point of view, in total 6 senior employees participated to the study, who has at least 5-year experience in the sector. They work in different positions including research and development engineer, information technologist, data analyst and 2 project managers. These experts were asked to evaluate 8 emerging digital technologies according to their importance level by using the linguistic scale. The evaluation of experts is shown in Table 3.

After the data collection part, firstly by using Eq. 2 defuzzified values are found, then by calculating the average for each emerging digital technology a final score is found. In this study, threshold value, λ, is determined as 0.5, which is the average of medium values minimum and maximum number. The results are presented in Table 4 with the ascending order.

According to the results, Internet of Things (E1), Artificial Intelligence (AI) and Machine Learning (E2), Big Data Analytics (E4), Cybersecurity Solutions (E6) and Blockchain (E5) are found above λ, therefore accepted for the evaluation. The reason behind this result is that companies are more familiar with these technologies, as well

Table 3 Evaluation of experts

	Expert 1	Expert 2	Expert 3	Expert 4	Expert 5	Expert 6
E1	EI	I	EI	EI	I	M
E2	I	I	M	EI	I	M
E3	U	EU	U	U	EU	U
E4	I	EI	M	M	I	I
E5	M	M	I	I	M	I
E6	EI	EI	M	I	M	M
E7	EU	EU	U	U	U	EU
E8	U	U	M	EU	EU	EU

Table 4 Results

Emerging digital technologies	Score (D_k)
Internet of things (E1)	0.733
Artificial intelligence (AI) and machine learning (E2)	0.656
Big data analytics (E4)	0.656
Cybersecurity solutions (E6)	0.644
Blockchain (E5)	0.600
Augmented reality (AR) and virtual reality (VR) (E8)	0.267
Digital twins (E3)	0.256
Edge computing (E7)	0.233

as they can be adopted easier to the current system. Furthermore, return on investment for these technologies, especially Iot, AI and Machine Learning, big data analytics, might be relatively lower since they mostly aim increasing the efficiency, optimizing systems, demand forecasting etc. Moreover, the importance of cybersecurity and blockchain shows the critical role of transparency, traceability and data security in renewable energy sector. On the other hand, emerging digital technologies including Augmented Reality (AR) and Virtual Reality (VR) (E8), Digital Twins (E3) and Edge Computing (E7) are found below the threshold value. Hence, they are omitted for this study. Not being familiar to these technologies as well as not having enough background knowledge related to their potential applications in renewable energy sector could be the main reasons for these results. In addition, higher infrastructure needs as well as relatively higher return on investment are some other reasons to eliminate these technologies. However, for future applications the potential benefits of these technologies should be considered by organizations.

5 Conclusion

Applying emerging digital technologies in renewable energy sector is very important to achieve sustainability in energy production systems. Emerging digital technologies are useful in terms of increasing efficiency, reducing errors, better forecasting, providing data security and improving traceability. Therefore, it is significant to evaluate different emerging digital technologies according to their importance order for renewable energy sector to provide better understanding for future investments as well as guiding the strategical decisions. From this point of view, this study aimed to present the critical emerging digital technologies for renewable energy sector with their applications, and to select the most important technologies by using expert opinion.

In total eight emerging digital technologies are presented in this study, and fuzzy DELPHI method is applied to evaluate these technologies. Results showed that, Internet of Things, Artificial Intelligence (AI) and Machine Learning, Big Data Analytics, Cybersecurity Solutions and Blockchain are the technologies, which should be considered for renewable energy sector. On the other hand, Augmented Reality (AR) and Virtual Reality (VR), Digital Twins and Edge Computing are omitted for this study.

High investment costs, duration of return on investment, employee capabilities, impacts on operational efficiency, familiarity to the technologies, technological background related to these technologies and knowing the potential benefits are some of the factors that are affected the results. This study can be seen as an initial attempt to evaluate different emerging technologies for renewable energy sector. In future studies, these technologies can be extended, and comparative studies can be conducted to reveal sector-specific features.

References

1. Razzaq A, Sharif A, Ozturk I, Skare M (2023) Asymmetric influence of digital finance, and renewable energy technology innovation on green growth in China. Renew Energy 202:310–319
2. Sharif A, Kartal MT, Bekun FV, Pata UK, Foon CL, Depren SK (2023) Role of green technology, environmental taxes, and green energy towards sustainable environment: insights from sovereign Nordic countries by CS-ARDL approach. Gondwana Res 117:194–206
3. Lin B, Huang C (2023) Promoting variable renewable energy integration: the moderating effect of digitalization. Appl Energy 337:120891
4. Mendoza JMF, Ibarra D (2023) Technology-enabled circular business models for the hybridisation of wind farms: integrated wind and solar energy, power-to-gas and power-to-liquid systems. Sustain Prod Consum
5. Bähr K, Fliaster A (2023) The twofold transition: framing digital innovations and incumbents' value propositions for sustainability. Bus Strateg Environ 32(2):920–935
6. Adebayo TS, Kartal MT, Ağa M, Al-Faryan MAS (2023) Role of country risks and renewable energy consumption on environmental quality: evidence from MINT countries. J Environ Manage 327:116884

7. Ren Y, Li B, Liang D (2023) Impact of digital transformation on renewable energy companies' performance: evidence from China. Front Environ Sci 10:2702
8. Wei C, Li C, Löschel A, Managi S, Lundgren T (2023) Digital technology and energy sustainability: recent advances, challenges, and opportunities. Resour Conserv Recycl 190:106803
9. Zhao N, Zhang H, Yang X, Yan J, You F (2023) Emerging information and communication technologies for smart energy systems and renewable transition. Adv Appl Energy 100125
10. Entezari A, Aslani A, Zahedi R, Noorollahi Y (2023) Artificial intelligence and machine learning in energy systems: a bibliographic perspective. Energ Strat Rev 45:101017
11. Ahmad T, Zhang D, Huang C, Zhang H, Dai N, Song Y, Chen H (2021) Artificial intelligence in sustainable energy industry: status quo, challenges and opportunities. J Clean Prod 289:125834
12. Agostinelli S, Cumo F, Nezhad MM, Orsini G, Piras G (1817) Renewable energy system controlled by open-source tools and digital twin model: zero energy port area in Italy. Energies 15(5):2022
13. Mostafa N, Ramadan HSM, Elfarouk O (2022) Renewable energy management in smart grids by using big data analytics and machine learning. Mach Learn Appl 9:100363
14. Giest S, Mukherjee I (2018) Behavioral instruments in renewable energy and the role of big data: a policy perspective. Energy Policy 123:360–366
15. Juszczyk O, Shahzad K (2022) Blockchain technology for renewable energy: principles, applications and prospects. Energies 15(13):4603
16. Cali U, Kuzlu M, Pipattanasomporn M, Elma O, Reddi R (2021) Cybersecurity of renewable energy data and applications using distributed ledger technology. arXiv:2110.11354
17. Pandey V, Sircar A, Bist N, Solanki K, Yadav K (2023) Accelerating the renewable energy sector through Industry 4.0: Optimization opportunities in the digital revolution. Int J Innov Stud 7(2):171–188
18. Li W, Yang T, Delicato FC, Pires PF, Tari Z, Khan SU, Zomaya AY (2018) On enabling sustainable edge computing with renewable energy resources. IEEE Commun Mag 56(5):94–101
19. Erten B, Oral B, Yakut MZ (2022) The role of virtual and augmented reality in occupational health and safety training of employees in PV power systems and evaluation with a sustainability perspective. J Clean Prod 379:134499
20. Saffie NAM, Rasmani KA (2016) Fuzzy Delphi method: Issues and challenges. In: 2016 international conference on logistics, informatics and service sciences (LISS). IEEE, pp 1–7
21. Holsapple CW, Joshi KD (2002) Knowledge manipulation activities: results of a Delphi study. Inf Manage 39(6):477–490
22. Bui TD, Tsai FM, Tseng ML, Ali MH (2020) Identifying sustainable solid waste management barriers in practice using the fuzzy Delphi method. Resour Conserv Recycl 154:104625
23. Lee CH, Wu KJ, Tseng ML (2018) Resource management practice through eco-innovation toward sustainable development using qualitative information and quantitative data. J Clean Prod 202:120–129
24. Shen YC, Chang SH, Lin GT, Yu HC (2010) A hybrid selection model for emerging technology. Technol Forecast Soc Chang 77(1):151–166
25. Blagojević B, Nordstroem EM, Lindroos O (2023) A framework for defining weights of decision makers in group decision-making, using consistency between different multicriteria weighting methods. Int J For Eng 34(2):130–142

Battery Electric Vehicles in Freight Transportation

Levent Özlü and Dilay Çelebi

Abstract Historically, electric freight vehicles faced skepticism due to perceived impracticality and high costs for long-distance journeys. However, recent research suggests a shifting paradigm. Owing to advancements in electric vehicle technology and mass production, the future may reveal a different narrative. Electric vehicles, once considered less suitable for extended transport, are now emerging as a promising solution to mitigate transportation-induced air pollution. Numerous studies, conducted globally, yield varying insights into the impact of electric freight vehicles on logistics, spanning technological, managerial, and operational dimensions, as well as key parameters influencing their adoption. This article provides an extensive literature review to reveal the factors fostering the widespread adoption of electric freight vehicles. Our objective is to raise awareness of their unemployed potential within the long-haul transportation sector, highlighting their crucial role in emissions reduction and the advancement of sustainable transportation solutions.

Keywords Electric freight vehicles · Greenhouse gas emissions · Life cycle cost · Transportation

L. Özlü (✉)
Department of Business Engineering, Istanbul Technical University, Istanbul 34469, Turkey
e-mail: ozlule17@itu.edu.tr

D. Çelebi
South-East European Research Centre, 54622 Thessaloniki, Greece

City College, University of York Europe Campus, 54626 Thessaloniki, Greece

D. Çelebi
e-mail: dgonidis@seerc.org

F. Çebi et al. (eds.), *Engineering and Technology Management in Challenging Times*, Lecture Notes in Networks and Systems 1137,
https://doi.org/10.1007/978-3-031-70935-7_10

1 Introduction

In a period characterized by rising environmental concerns and an urgent need to address climate change, the transportation industry emerges as a pivotal domain for significant and far-reaching changes. The rise of electric freight vehicles (EFVs) has attracted considerable attention as one of the key developments in the field of sustainable transportation. EFVs provide the potential to mitigate greenhouse gas emissions, reduce air pollution, and transform the logistics sector. The understanding of the function of electric vehicles in the logistics sector receives increased attention due to growing concerns of governments, industries, and consumers regarding sustainability.

This study investigates the integration of electric vehicles, particularly electric trucks, in the logistics industry, with the goal of offering an extensive understanding of their implications and challenges. The use of electric trucks implies a fundamental change in the way goods are transported, presenting the potential for decreased emissions, less operational expenses, and improved environmental sustainability. Nevertheless, this shift involves several complexities, including technological challenges, economic factors, and the requirement for supportive infrastructure.

We provide a thorough analysis of the available literature to explore the current state of electric vehicles, various variables which impact their adoption, and their potential implications for the logistics industry. Furthermore, this research delves into the theoretical and managerial implications of this substantial change, offering perspectives for policymakers, industry stakeholders, and academics. Our research attempts to contribute to the current academic discussion regarding sustainable mobility and explore the possibility of this paradigm shift in promoting a more sustainable future.

This study significantly contributes to the field of freight transportation by offering a comprehensive review of the latest research, including emerging trends, technological advancements, and regulatory developments. Our detailed examination of recent literature not only highlights the current state of EFVs but also identifies critical gaps in the existing body of knowledge, thereby setting a clear agenda for future research. By integrating insights from diverse geographic and regulatory contexts, we provide an understanding of the factors influencing EFV adoption, ranging from technological challenges to economic incentives.

The chapter is structured as follows: Sect. 2 presents the research approach, employing a wide-ranging literature review to underpin the study's findings. Section 3 presents a critical analysis of current emission levels and the performance benchmarks of EFVs. Following this, Sect. 4 evaluates the challenges and opportunities facing the adoption of EFVs, informed by global insights. Section 5 summarizes the findings. The paper concludes in Sect. 6, which synthesize the research outcomes and outline both the managerial implications and future research directions, emphasizing the pivotal role EFVs could play in achieving sustainable freight transportation.

2 Methodology

In conducting the literature review, we employed the Google Scholar search engine to explore a broad spectrum of studies related to the keywords: "electric vehicles," "electric trucks," "battery electric trucks," "battery electric vehicles," "battery electric logistics vehicles," "life cycle emissions," "lifetime costs," "life cycle costs," "transportation," and "long haul transportation." This search strategy allowed us to comprehensively survey the available scholarly works on these topics. We systematically screened and selected relevant articles, scrutinized their content, and synthesized the findings to gain insights into the intersection of electric vehicles, particularly trucks and logistics vehicles, with considerations such as life cycle emissions, lifetime costs, and their implications for long-haul transportation.

The selection of appropriate studies was determined by the implementation of strict inclusion and exclusion criteria. These criteria took into account several factors, such as the relevance of the research topic and the peer-reviewed status of the sources. This careful process played a crucial part in concluding the selection of studies that met the desired requirements. Through the use of this methodical methodology, we have assured that our literature evaluation is firmly grounded in a comprehensive exploration of reputable scholarly sources, so establishing a strong basis for our analysis and synthesis.

3 Transportation Sector Greenhouse Gas Emissions and Electric Vehicle Performance

Articles showing both the CO_2 emissions caused by the transportation sector and the share of medium and heavy-duty vehicles in the emissions were examined. It has been stated that the transportation sector is responsible for 19% of CO_2 emissions in Latin America, and Battery Electric Trucks (BET), which have the potential to reduce freight transport emissions, represent a modern alternative to diesel trucks [1]. Plötz et al. [2] states that the transportation sector is responsible for 25% of greenhouse gas emissions in Europe, and heavy-duty vehicles are the cause of approximately 25% of transportation sector emissions. International Energy Agency [3] stated that approximately 65% of freight activity is carried out by heavy-duty trucks with a gross weight of more than 15 tons, and heavy-duty trucks consume half of the oil in the road cargo transportation sector. Kast et al. [4] cites that medium-duty and heavy-duty vehicles are responsible for approximately 26% of carbon dioxide emissions in the US transportation sector, and heavy-duty vehicles contribute 36% to national nitrogen oxide emissions.

Articles on greenhouse gas emissions of electric freight vehicles were reviewed within the scope of the study. Çabukoğlu et al. [5] stated that electric trucks transported with a well-developed battery swapping infrastructure would halve emissions, while innovations in battery technology or infrastructure could lead to further improvements. Sen et al. [6] stated that, in comparison of life cycle greenhouse gas emissions, battery electric heavy-duty trucks perform slightly better than their closest competitors, hybrid and biodiesel trucks. Sen et al. [6] also stated that battery electric trucks perform much better than their competitors when comparing the total conventional air pollutants released throughout their life cycle, and if more electrical energy is obtained from renewable energy sources, the performance of battery electric trucks can increase even more. Sabet and Farooq [7] perform an extensive microsimulation of traffic and emissions and finds that battery electric vehicles have the lowest greenhouse gas emissions compared to hybrid electric vehicles and renewable fuels. Syre et al. [8] conducted a life cycle assessment and stated that greenhouse gas emissions for battery electric vehicles are consistently lower than fuel cell electric vehicles. Rogstadius et al. [9] expressed that biofuels, hydrogen and e-fuels do not have the ability to significantly contribute to greenhouse gas emissions before 2035, given technological limitations and scalability. It has been stated that battery electric vehicles are the only feasible solution to achieve zero tailpipe emissions on a large scale.

Wang and Yu [10] reveal the detailed life cycle process of Lithium-ion batteries (LiB) of electric vehicles in China. The environmental impact of each type of LiB is then estimated using the life cycle assessment (LCA) method. It states that with the evolution of LiBs, their environmental impact has decreased. Additionally, if waste LiBs can be properly recycled, their life cycle environmental impacts will be significantly reduced.

4 Potential of Electric Freight Vehicles

Wang and Thoben [11] analyze the effective factors in urban freight transportation for electric commercial vehicles. It has been stated that long charging times, short driving range and high initial investments are the dominant obstacles to the diffusion of electric vehicles.

Cheng and Lin's study [12] identifies major hurdles to the adoption of long-haul electric trucks by fleet carriers in the United States, highlighting high upfront costs, extended charging times, a scarcity of highway charging stations, and insufficient policy support. Lin et al. [13] emphasize the necessity for government intervention through the implementation of supportive policies, funding for charging infrastructure, and financial incentives to facilitate the adoption of long-haul electric trucking among fleet carriers. Bae et al. [14] explore the viewpoints of heavy-duty vehicle fleet operators on various alternative fuel vehicles, including electric, hydrogen, compressed natural gas, and hybrid options. The research points out concerns such as prohibitive initial costs, uncertain investment returns, operational inadequacies, immature infrastructure, vehicle unavailability, and lackluster state government support. Chowdhury et al. [15] examine how behavioral factors influence the adoption of low-carbon transportation options, revealing that financial hurdles significantly deter smaller fleet organizations, whereas technical challenges are more daunting for larger entities. The study suggests that policymakers could foster a more rapid adoption of low-carbon transportation by removing these obstacles and creating a supportive ecosystem. Tanco et al. [16] identify the main barriers to the adoption of Battery Electric Trucks (BETs) in Uruguay, including concerns about driving range, purchase cost, charging infrastructure availability, and insufficient incentives. Lastly, Konstantinou and Gkritza [17] discuss how factors such as charging duration, product availability, business models, grid resilience, operational reliability, and government backing critically influence the adoption of electric trucks in the United States.

Samet et al. [18] states that battery electric trucks have a potential of large-scale adoption and it will reduce the greenhouse gas emission in freight transport sector if the minimum requirements are met for operational range restrictions and access an electricity power network for the battery charging.

Wang and Thoben [19] state that alternative fuel vehicles (electric vehicles, hybrid electric vehicles and natural gas vehicles) constitute 1% of the total light commercial vehicles as of 2013, and investigates the reason why the spread is low although the intention to adopt electric commercial vehicles is high. It has been stated that low operating and environmental costs, public participation, low emissions and noise and energy consumption, fast acceleration and high energy efficiency are the factors that encourage electric commercial vehicles. It has been determined that the problems that negatively affect the spread of electric commercial vehicles are investment costs, battery replacement costs, drivers' costs, accessibility, security, training, use of second batteries, technological limitations of batteries and vehicles and charging infrastructures.

Taefi et al. [20] conducted interviews with policy makers and electric freight vehicle users and asked them to rate 23 policy measures suggested in the literature. They found that there were large differences between the ratings of policy makers and electric freight vehicle users. While financial supports are seen as the main solution in the literature to close the gap between the total cost of ownership between the internal combustion engine and the electric freight vehicle, the study results show that supporting the installation of company charging infrastructure and special legal measures are not given enough importance by policy makers.

Taefi et al. [21] aimed to discuss case study examples of electric freight vehicle initiatives in Denmark, Germany, the Netherlands, Sweden and the United Kingdom and identify supports and barriers for common trends. Common findings for all countries are listed as follows; the expectation of increasing the range to ensure a higher usage rate, load loss due to the weight of the batteries, lack of quality in after-sales services (lack of spare parts, limited information for repairs, etc.), electric freight vehicle users do not see it as profitable but can only gain experience and create future policies and strengthen their image in the eyes of customers regarding sustainability. They note that the lack of after-sales services shows that truck manufacturers are not strongly involved in the development of markets for electric trucks.

Quak et al. [22] discusses the obstacles and opportunities for the use of electric freight vehicles in daily city logistics operations, based on experiences from a number of implemented applications (FREVUE and the Cargohopper case in Amsterdam).

Although using electric freight vehicles compared to conventional vehicles suffers from high vehicle purchase price and uncertainty in second-hand value, it has been noted that it has good technical performance in most of the cases where new generation electric freight vehicles are used in daily operations. It has been stated that companies using electric freight vehicles are generally satisfied with these vehicles. The following elements were stated in the SWOT analysis of electric cargo vehicles compared to internal combustion engine vehicles;

- Strengths: low fuel costs, good environmental performance, silence of the vehicle, efficiency of operation with government support, positive acceptance by the public.
- Weaknesses: high supply costs, limited loading capacity, limited and unreliable and expensive after-sales support, network issues with large fleet, limited vehicle availability.
- Opportunities: newer vehicles have higher range, good fit for specific niche jobs, availability of public charging points, innovative vehicle/battery rental programs, reduction in battery price.
- Risks: unclear regulations regarding certification, low oil prices and rising energy prices, better environmental performance of vehicles running on alternative fuels.

Quak et al. [23] examined the feasibility of using electric vehicles in urban freight transportation from a carrier's perspective. The study examined all relevant factors affecting the business situation, including carriers' attitudes towards electric freight vehicles; technological features, existing restrictive and incentive policies, financial and non-financial incentives, types of activities, urban environments and logistics organization were taken into account. The results confirm the positive social attitude towards electric freight vehicles. Undoubtedly, the environmental performance of electric freight vehicles is one of their main strengths. It is emphasized that the purchase price of electric vehicles is significantly higher than that of conventional vehicles, and the operational advantages of electric freight vehicles are not sufficient to turn the commercial situation into a positive one. Therefore, operators continue to seek new forms of ownership and an active role is still expected from local authorities.

Rizet et al. [24] predicts that carbon dioxide rates will decrease if electric freight vehicles replace internal combustion engine vehicles in urban transportation, but they stated that range and loading areas pose an obstacle to the use of electric freight vehicles.

There are studies that reveal different results for different countries regarding the effects of electric freight vehicles on the future of long-distance transportation, the critical parameters of this change, the threshold values of these parameters and their effects on the usage rates of electric vehicles in long-distance transportation [1].

Higueras-Castillo et al. [25] finds that cultural considerations are important in determining electric vehicle adoption.

Smith et al. [26] stated that electric vehicles need to achieve lower total cost of ownership than conventional trucks in order to increase their share in the medium-duty and heavy-duty vehicles market. It is thought that R&D studies addressing maintenance, reliability, performance at extreme temperatures, efficiency and component life will reduce uncertainties in cost analysis. It has been stated that chargers and infrastructure that can provide fast charging for larger batteries used by medium-duty and heavy-duty goods vehicles need to be developed.

Zhou et al. [27], in their Canadian comparison of class 6 (medium-duty class) battery electric trucks and diesel trucks, showed that battery electric trucks are an alternative to diesel trucks in both life cycle greenhouse gas emissions and life cycle total cost of ownership. It has been stated that the comparison results are affected by factors such as air temperature, electricity prices, diesel prices, frequency of stops during travel, and that battery-electric trucks or diesel trucks can take the lead in both greenhouse gas emission and total cost of ownership comparisons in different scenarios.

Tanco et al. [1] highlights initial investments in battery electric trucks as the main barrier to adoption, which is the main reason for the difference in total cost of ownership between electric vehicles and diesel vehicles. The difference in electricity and fuel costs of battery-electric trucks has been identified as a factor that can cover the difference in initial investment. Fuel, electricity, maintenance, insurance and import tariffs are factors that make electric vehicles and diesel vehicles break-even in terms of total cost of ownership. It has been stated that different factor prices in different Latin American countries create significant country-by-country differences in the break-even point, and this changes the time horizon in which electric vehicles are expected to become widespread.

De La Peña et al. [28] states that the availability of charging infrastructure and route optimization are important, thus the relationship between infrastructure system development and vehicle operation planning is ignored.

Çabukoğlu et al. [5] concludes that electric trucks cannot be replaced by diesel trucks, although it says that electric trucks transported with a well-developed battery replacement infrastructure would halve emissions, and that innovations in battery technology or infrastructure could lead to further improvements. Although he reaches this conclusion by citing the many obstacles he identified with the data from Switzerland, he also notes that there are still many opportunities for further research.

Liimatainen et al. [29] states that electric trucks could cover 71% of tonne-kilometers in Switzerland and 38% in Finland. This study shows that electric trucks are already a viable solution for a large share of road freight transport with medium-duty trucks. They state that improvements in battery capacity and charging infrastructure could make electric trucks a viable option for heavy-duty rigid trucks and semi-trailers.

Sen et al. [6] states that battery electric and mild hybrid trucks have the best overall performance when compared on the basis of total costs incurred over the life cycle.

Burke and Sinha [30] stated that, in the long-term economic evaluation, electric freight vehicles appear to be more suitable than fuel cell vehicles if range needs are met. It was emphasized that the difference depends on the relative costs of electricity and hydrogen. Similarly, it has been stated that the competitive costs of electric freight vehicles with diesel vehicles depend on the relative prices of electricity and diesel fuels.

One of the most obvious obstacles to the use of electric vehicles in long-distance transportation is the distances that electric vehicles can travel on a single charge. How much of the transportation can be done on a single charge seems to be one of the basic thresholds for the widespread use of electric vehicles in long-distance transportation.

One of the findings that Çabukoğlu et al. [5] reached with the data from the Swiss example is that the majority of vehicles travel under 800 km per day. A significant amount of vehicle mileage is 400 km.

The finding that electric trucks can cover 71% of ton-kilometers in Switzerland and 38% in Finland [29] is an important example of how different the expectations in terms of deployment can be between the two countries.

Earl et al. [31] makes it possible to make a technical analysis, total cost of ownership comparison between diesel trucks and battery electric trucks, taking into account wages, maintenance, insurance, fuel and electricity prices and on-road charging. The results show that the greatest sensitivity to cost competitiveness is the price of electricity and, to a lesser extent, the road charging discount that can be applied to zero-emission trucks on highways. This finding runs counter to some doubts about both the technical and economic feasibility of battery-electric heavy-duty trucks.

Mareev et al. [32] shows that, on a life cycle cost basis, battery electric trucks can perform at the same cost level as diesel trucks in certain scenarios.

Tesla, which plans to produce Battery Electric Trucks, claims that battery electric trucks will provide competitive figures with diesel alternatives in terms of initial investment costs and operating costs. Tesla has announced the following information for the Semi model (Class 8—Heavy Duty Truck) [33]:

Price: $180,000 (Estimated).

Range: 500 mile (800 km) (Fully loaded, Battery 900–1100 kWh).

Energy Consumption < 2 kWh/mile.

Cost: 1.26 $/mile (General Operating Cost, estimation: 7cent/kWh).

Tesla has disclosed the following information about the costs of Class 8 competing diesel trucks:

Price: $120,000.

Cost: 1.51 USD/mile (General Operating Cost, estimation: 2.50 $/galon).

Tesla states that 80% of the routes of battery electric trucks are shorter than 250 miles (400 km) and predicts that they will be able to go to these destinations and return without charging.

The cost and route data regarding the battery electric truck Semi and diesel vehicles shared by Tesla needs confirmation. The details of the data and calculation method used when calculating these costs were not disclosed. Since the Semi is not a mass-produced vehicle, assumptions are made when calculating life cycle costs, but the assumptions made for the calculation are unknown.

There are also electric vehicle alternatives that use different technologies than battery-electric trucks.

Plötz et al. [2] examined catenary hybrid electric trucks fed by overhead lines. It states that the additional burden it will impose on the European electricity system will have a limited impact on the total energy system.

Kast et al. [4] states that fuel cell electric vehicles (FCEV) provide a scalable route to clean hydrogen fuel to be used for transport power generation and are applicable for many truck classes.

Quaranta et al. [34] makes a comparison between a diesel truck (a 450 HP 6-cylinder diesel motor and a gross vehicle weight of 44 tonnes), an electric truck and a hydrogen-powered truck. It states that hydrogen trucks have lower Total Cost of Ownership and Well to Wheel emissions.

De Wolf and Smeers [35] state the main advantages of hydrogen vehicles as autonomy and fast recharging and the main advantages of battery electric vehicles as lower price and the wide availability of the electricity grid. It estimates that battery electric vehicles will be used for short-distance travel and fuel cell electric vehicles will meet the demand for long-distance and heavy-duty transport.

Göckeler et al. [36] states that battery electric heavy-duty vehicles will dominate the transport market. The potential of fuel cell trucks and overhead catenary trucks are estimated to be low.

Burke and Sinha [30] aim to project the entry of battery electric vehicles and fuel cell hydrogen technologies into the medium-duty (MD) and heavy-duty (HD) vehicle markets and to determine which markets will be most suitable for each technology and the factors that will be most critical for their successful entry. The study considers trucks and buses of various types (delivery vans, transit buses, intercity buses, long-haul and short-haul tractor trailers, and heavy-duty pickup trucks). In summary, they note that if the range requirement (miles) for the truck can be met using batteries, the long-term economics of battery-electric buses and trucks appear to be more favorable than the fuel cell/hydrogen option. This is primarily due to significantly lower energy operating costs ($/mi) using electricity from hydrogen. The difference, of course, depends on the relative costs of electricity and hydrogen.

Canese [37], based on technical and economic analysis, finds that fuel cell electric vehicle weight and cost are much less sensitive to increased range than batteries. The main reason for the finding is that in battery electric trucks, the battery is responsible for both power and energy, while in fuel cell electric trucks, the power is provided by the fuel cell system and the energy is related to the amount of hydrogen fuel stored on the vehicle. For these reasons, the study identifies fuel cell electric trucks as a promising technology for future truck decarbonization scenarios.

Hovi and Ruben [38] aim to provide an overview of the technology state, user experiences of early adopters of battery electric vehicles and the potential for operating battery electric heavy-duty vehicles in Norway. In small-scale mass production, no other option appears to be competitively cost-effective against diesel rivals in heavy-duty trucks. It is predicted that if mass production is made with the same assumptions, battery electric vehicles can reduce costs that provide a serious competitive advantage over their diesel rivals in heavy-duty trucks. When mass produced, it is estimated that electric battery-powered vehicles will provide a serious cost advantage over both diesel vehicles and hydrogen fuel cell vehicles, when compared with the same assumptions about tractors for semi-trailers.

Mareev and Sauer [39] state that the overhead catenary truck is an interesting technology for long-distance transportation with heavy-duty trucks because it can combine the advantage of providing energy via the catenary while driving and the flexibility of a battery-powered truck on non-catenary routes using the traction battery. In their study, they investigate the energy consumption of catenary trucks on German roads and consider different configurations for the traction battery and catenary power system. Next, the life cycle costs of roof-mounted catenary trucks are calculated for a specific long-haul transportation scenario and the results are compared with the battery electric truck and diesel truck using the findings of the authors' previous study. The energy consumption of the roof catenary trucks considered is approximately equal to the energy consumption of a battery electric truck, but only about half the equivalent energy consumption of a conventional diesel truck. Based on the cost assumptions in this study, total life cycle costs for roof-mounted catenary trucks could be in the range of conventional diesel trucks, demonstrating the competitiveness of this alternative truck technology.

According to Speth and Funke [40], among other solutions such as overhead lines or hydrogen, battery swapping systems could also be interesting for use cases in heavy road transport with only short idle times that do not allow a battery charge. The paper makes a techno-economic comparison of battery replacement and electric road systems for a specific use case along the Berlin-Peine route in Germany. From a total cost of ownership perspective, both alternatives, battery swapping stations and catenary vehicles, appear to be an option for electrifying heavy-duty road transportation.

Mourelo [41] shows that intermodal distribution networks, which can combine a train and several electric vehicles, are more efficient and environmentally friendly than single-modal networks for high-volume and long-distance transportation, regardless of customers' distribution. Such a combination results in an environmentally friendly distribution network that maximizes the capacity of the system, minimizes the total distance traveled, reduces road congestion and energy consumption, and contributes to the improvement of air quality and environmental conditions by reducing CO_2 emissions by more than 50%.

5 Findings

Greenhouse gases are an important factor in global warming. The transportation sector has a large share in greenhouse gas emissions [1, 2]. In the transportation sector, although the number of freight vehicles may seem small in the total vehicle park, they have a serious share in greenhouse gas emissions [2–4]. At this point, battery electric vehicles with low emission values [5, 6, 10], hydrogen fuel cell vehicles [4] and catenary vehicles fed from overhead lines [2] emerge as different alternatives.

It is clear that developments in future oil prices, electricity prices and hydrogen prices will be effective in lifetime costs among the alternatives. It is stated that battery electric vehicles can have lower lifetime costs than both internal combustion vehicles and hydrogen fuel cell vehicles, especially when mass produced [38].

The lifetime emission values of electric freight vehicles are lower than today's diesel freight vehicles, but they are not zero. The lifetime emission values of electric freight vehicles are higher than the emission values of railway systems. The combination of rail and electric freight vehicle appears to be the best option for an Eco-Friendly Distribution Network, which maximizes the capacity of the system, minimizes the total distance traveled, reduces road congestion and energy consumption, and contributes to the improvement of air quality and environmental conditions by reducing CO_2 emissions [41].

The prominent findings about electric freight vehicles in the literature review are stated below:

- The main obstacle for those who adopt electric freight vehicles is the high initial investment cost of electric freight vehicles [11–16, 19, 22, 23].
- The idea that electric freight vehicles are not suitable for long-distance transportation due to their high costs is changing. Some studies conducted with current data indicate that battery-electric trucks and internal combustion engine vehicles may be at par with life cycle costs in long-distance freight transportation [6, 27, 29–32, 38, 41].
- Although the fuel supply infrastructure of internal combustion engine vehicles does not require new investments, the magnitude of the infrastructure cost required for charging electric freight vehicles emerges as a significant disadvantage in lifetime cost comparisons [13, 16, 19, 20].
- The difference between the electricity cost of battery-electric trucks and the fuel cost of internal combustion engine vehicles can be a factor that can offset the disadvantage of the initial investment cost of battery-electric trucks. Electricity prices and diesel/gasoline prices play an important role in this regard [1, 19, 22, 39].
- Different factor prices in different countries create significant country-by-country differences in the break-even point for battery electric trucks and internal combustion engine vehicles, and this changes the time horizon over which electric vehicles are expected to become widespread [1].
- One of the most obvious obstacles to the use of electric vehicles in long-distance transportation is the distances that electric vehicles can travel on a single charge. The distances traveled by vehicles on a daily basis vary depending on the country [11, 16, 21, 24].
- The long charging time of electric freight vehicles is one of the important obstacles to their widespread use [11–13, 17].
- Load loss due to the weight of the batteries negatively affects the decision to purchase an electric freight vehicle [21, 22].
- The routing of electric freight vehicles has different constraints than the routing of internal combustion engine vehicles. The distance it can travel on a single charge, the availability of a charging station on the route, etc. constraints require a new planning approach for electric freight vehicles [28].
- Studies on batteries, engines and powertrains of electric freight vehicles are promising in increasing the distance they can travel on a single charge and reducing

the total cost of ownership. Once electric freight vehicles go into mass production, significant reductions in production costs are expected [38].

- The service quality and widespread maintenance network of electric freight vehicle manufacturers are an effective factor in carriers' decisions. It is important for electric freight vehicle manufacturers to take a strong initiative regarding maintenance [17, 21, 22, 26].
- Second-hand prices and scrap prices of electric freight vehicles, which have high initial investment costs, affect their life cycle costs. Uncertainty in second-hand prices plays an important role in carriers' decisions [14, 22].
- The lower greenhouse gas emissions of electric freight vehicles compared to internal combustion engine vehicles is one of the most important advantages in this period when environmental sensitivity is increasing in society. Carriers who want to positively influence their perception in society are showing interest in this issue as part of their public relations and social responsibility activities, but this factor alone is not sufficient for the widespread use of electric freight vehicles [14, 19, 21–23].
- It is important but insufficient for state institutions to set strategies and targets. Uncertainty is one of the major obstacles to the spread of electric freight vehicles. The fact that government institutions present a road map and a strong will is an important factor in carriers' decision to invest in electric freight vehicles [12–17, 20, 22].

6 Conclusions

This literature review highlights the potential of electric freight trucks to significantly revolutionize the freight transportation sector by offering a more sustainable and economically feasible substitute for traditional internal combustion engine vehicles. Nevertheless, it also shows the multiple obstacles that need to be overcome in order to accomplish this transformation. With the development in technology and the implementation of strict regulations by governments, it is possible that electric freight trucks might serve an essential part in developing a more environmentally sustainable and efficient transportation system. The understanding of the full potential of electric freight cars in decreasing greenhouse gas emissions and promoting the sustainability of freight transportation requires further research and real-world applications.

This study presents several managerial implications. Firstly, managers in the transportation industry should conduct comprehensive Total Cost of Ownership (TCO) analyses to evaluate the financial viability of transitioning to electric freight vehicles. Additionally, they should actively engage with government agencies to advocate for supportive policies and incentives. Secondly, managers should prioritize the development of a robust charging infrastructure and collaborate with stakeholders to ensure a smooth transition to electric vehicles. Thirdly, managers can capitalize on the environmental benefits of electric vehicles to align with sustainability objectives and enhance their public image. They should also explore route optimization strategies and risk mitigation plans to minimize operational disruptions. Finally, a gradual approach to fleet transition, as well as customer awareness programs, should be explored to attract environmentally concerned customers. It is essential to engage in regular monitoring of the electric vehicle market in order to remain up to date with technological progress and market dynamics. This practice enables managers to successfully adjust their strategies and operations in response to developments within the sector.

Several limitations should be acknowledged in this study. Firstly, the research primarily relies on existing literature, and while efforts were made to include a comprehensive range of sources, there may be relevant studies that were not considered. Secondly, the study's focus on electric freight vehicles does not cover all potential solutions or technologies for sustainable transportation, and factors such as hydrogen fuel cells or other alternative fuels may require further investigation. Finally, the study's assessment of managerial and theoretical implications is based on current knowledge and may evolve as new research emerges in this dynamic field. Consequently, one important aspect of future research direction includes the revision and expansion of the literature study to incorporate a comparative analysis conducted not only among electric freight vehicles but also by considering alternative sustainable transport technologies like hydrogen fuel cells, providing a comprehensive evaluation of sustainable transport solutions. Moreover, the examination of consumer behavior can provide light on the various elements that influence the acceptance of electric trucks. Additionally, conducting policy assessments can provide insights on the efficacy of government incentives in promoting desired outcomes.

A significant future research direction involves the development of a "system dynamics" simulation model. Such a model would help in comprehensively understanding the evolving role of electric trucks within the logistics market in the future. By simulating various scenarios, this model could project the long-term impacts of electric vehicle adoption, accounting for factors such as infrastructure development, adoption rates, environmental consequences, and economic considerations. This system dynamics approach would not only facilitate policymakers and industry stakeholders in making informed decisions but also offer valuable insights into the dynamic interplay of factors shaping the future of sustainable transportation in the logistics sector.

References

1. Tanco M, Cat L, Garat S (2019) A break-even analysis for battery electric trucks in Latin America. J Clean Prod 228:1354–1367
2. Plötz P, Gnanna T, Jochemb P, Yilmaz HÜ, Kaschub T (2019) Impact of electric trucks powered by overhead lines on the European electricity system and CO_2 emissions. Energy Policy 130:32–40

3. International Energy Agency (2017) The future of trucks. IEA, Paris, France. https://doi.org/10.1787/9789264279452-en. Accessed 22 Sept 2023
4. Kast J, Vijayagopal R, Gangloff JJ Jr, Marcinkoski J (2017) Clean commercial transportation: medium and heavy duty fuel cell electric trucks. Int J Hydrogen Energy 42:4508–4517
5. Çabukoğlu E, Georges G, Küng L, Pareschi G, Boulochos K (2018) Battery electric propulsion: an option for heavy-duty vehicles? Results from a Swiss case-study. Transp Res Part C 88:107–123
6. Sen B, Ercan T, Tatari O (2017) Does a battery-electric truck make a difference? Life cycle emissions, costs, and externality analysis of alternative fuel-powered Class 8 heavy-duty trucks in the United States. J Clean Prod 141:110–121
7. Sabet S, Farooq B (2023) Exploring sustainable pathways for urban traffic decarbonization: vehicle technologies, management strategies, and driving behaviour. arXiv:2308.14914
8. Syré AM, Shyposha P, Freisem L, Pollak A, Göhlich D (2024) Comparative life cycle assessment of battery and fuel cell electric cars, trucks and buses
9. Rogstadius J, Alaküla M, Plötz P, Márquez-Fernández FJ, Nordin L (2024) 2035 Joint impact assessment of greenhouse gas reducing pathways for EU road transport. Research Institutes of Sweden
10. Wang S, Yu J (2020) A comparative life cycle assessment on lithium-ion battery: case study on electric vehicle battery in China considering battery evolution. Waste Manage Res 0734242X20966637
11. Wang M, Thoben KD (2016) The influential factors for application of the electric commercial vehicle in the urban freight transport. In: Dynamics in logistics. Springer, Cham, pp 375–378
12. Cheng X, Lin J (2024) Is electric truck a viable alternative to diesel truck in long-haul operation? Transp Res Part D Transp Environ 129:104119
13. Lin J, Cheng YL, Cheng X, Shen H, Pawar A, Koziel K (2024) Development of commercial vehicle emission inventory and analysis. FHWA-ICT-24-002
14. Bae Y, Rindt CR, Mitra SK, Ritchie SG (2024) Fleet operator perspectives on alternative fuels for heavy-duty vehicles. Transp Policy
15. Chowdhury V, Tuli FM, Mitra SK (2024) Sustainable shift: analyzing drivers for low-carbon transportation adoption in California's heavy-duty and off-road sectors. Sustainability 16(5):1722
16. Tanco M, Cavallieri MS, Levy M, Rossi S, Jurburg D (2023) Barriers towards the adoption of battery electric trucks-an exploratory study in Uruguay. Int J Electr Hybrid Veh 15(3):292–317
17. Konstantinou T, Gkritza K (2023) Examining the barriers to electric truck adoption as a system: a grey-DEMATEL approach. Transp Res Interdiscip Persp 17:100746
18. Samet MJ, Liimatainen H, van Vliet OPR (2023) GHG emission reduction potential of road freight transport by using battery electric trucks in Finland and Switzerland. Appl Energy 347:121361
19. Wang M, Thoben KD (2017) Sustainable urban freight transport: analysis of factors affecting the employment of electric commercial vehicles. In: Dynamics in logistics. Springer, Cham, pp 255–265
20. Taefi TT, Kreutzfeldt J, Held T, Fink A (2016) Supporting the adoption of electric vehicles in urban road freight transport—A multi-criteria analysis of policy measures in Germany. Transp Res Part A Policy Pract 91:61–79
21. Taefi TT, Kreutzfeldt J, Held T, Konings R, Kotter R, Lilley S, Baster H, Green N, Laugesen MS, Jacobsson S, Borgqvist M, Nyquist C (2016) Comparative analysis of european examples of freight electric vehicles schemes—a systematic case study approach with examples from Denmark, Germany, the Netherlands, Sweden and the UK. In: Dynamics in logistics. Springer, Cham, pp 495–504
22. Quak H, Nesterova N, van Rooijen T (2016) Possibilities and barriers for using electric-powered vehicles in city logistics practice. Transp Res Proc 12:157–169
23. Quak H, Nesterova N, van Rooijen T, Dong Y (2016) Zero emission city logistics: current practices in freight electromobility and feasibility in the near future. Transp Res Proc 14:1506–1515

24. Rizet C, Cruz C, Vromant M (2016) The constraints of vehicle range and congestion for the use of electric vehicles for urban freight in France. Transp Res Proc 12:500–507
25. Higueras-Castillo E, Singh V, Singh V, Liébana-Cabanillas F (2023) Factors affecting adoption intention of electric vehicle: a cross-cultural study. Environ Dev Sustain 1–37
26. Smith D, Ozpineci B, Graves RL, Jones PT, Lustbader J, Kelly K, Mosbacher J et al (2020) Medium-and heavy-duty vehicle electrification: an assessment of technology and knowledge gaps (No. ORNL/SPR-2020/7). Oak Ridge National Lab (ORNL), Oak Ridge, TN (United States); National Renewable Energy Lab (NREL), Golden, CO (United States)
27. Zhou T, Roorda MJ, MacLean HL, Luk J (2017) Life cycle GHG emissions and lifetime costs of medium-duty diesel and battery electric trucks in Toronto, Canada. Transp Res Part D Transp Environ 55:91–98
28. De la Peña AG, Davendralingam N, Raz AK, DeLaurentis D, Shaver G, Sujan V, Jain N (2020) Projecting adoption of truck powertrain technologies and CO_2 emissions in line-haul networks. Transp Res Part D Transp Environ 84:102354
29. Liimatainen H, Vliet OV, Aplyn D (2019) The potential of electric trucks—an international commodity-level analysis. Appl Energy 236:804–814
30. Burke A, Sinha AK (2020) Technology, sustainability, and marketing of battery electric and hydrogen fuel cell medium-duty and heavy-duty trucks and buses in 2020–2040. University of California, Davis, Institute of Transportation Studies, USA
31. Earl T, Mathieu L, Cornelis S, Kenny S, Ambel CC, Nix J (2018) Analysis of long haul battery electric trucks in EU. In: Proceedings of the 8th commercial vehicle workshop, Graz, Austria, 17–18 May 2018
32. Mareev I, Becker J, Sauer DU (2018) Battery dimensioning and life cycle costs analysis for a heavy-duty truck considering the requirements of long-haul transportation. Energies 11(1):55
33. https://www.tesla.com/semi. Accessed 23 Sept 2023
34. Quaranta S, Aiello G, Inguanta R, Bonafede F, Abusohyon IAS, Venticinque M (2023) Mutli-objectve performance assessment of vehicles for logistic operations: methodology and case study. In: 2023 3rd international conference on electrical, computer, communications and mechatronics engineering (ICECCME). IEEE, pp 1–6
35. De Wolf D, Smeers Y (2023) Comparison of battery electric vehicles and fuel cell vehicles. World Electric Veh J 14(9):262
36. Göckeler K, Steinbach I, Görz WK, Hacker F, Blanck R, Mottschall M (2023) StratES—Scenarios for the electrification of road freight transport. Study based on market ramp-up modeling. In: Third report of the StratES research and dialogue project. OekoInstitut, Berlin
37. Canese L (2018) Decarbonization scenarios for medium and heavy-duty transportation. PhD Thesis, Politecnico di Torino, Turin, Italy
38. Hovi IB, Ruben D (2019) Erfaringer fra tidlige brukere av. Fuel 1:1–26
39. Mareev I, Sauer DU (2018) Energy consumption and life cycle costs of overhead catenary heavy-duty trucks for long-haul transportation. Energies 11(12):3446
40. Speth D, Funke SÁ (2019) A techno-economic comparison of battery swap and electric road systems for heavy road transport. A German case study. In: Proceedings of the 3rd electric road systems conference
41. Mourelo S (2016) The eco-friendly intermodal delivery network. Liberty University, USA

Capturing Small Space Debris: Inspired by the Frog Tongue's Strategy for Catching Its Prey

Ayşe Meriç Yazıcı and Erik Persson

Abstract The development of space technologies and the opportunities offered by the space economy are leading to a rapid increase in space traffic. Thousands of satellites are expected to be launched into space in the coming years. Metal, plastic, glass and paint scraps are small space debris objects. There are many studies in the literature on the capture of large space debris. This paper discusses a nature-inspired solution to clean up the increasing amount of micro-sized space debris in space. In particular, the prey capture strategy of the frog tongue is examined and how this strategy can be used to capture micro space debris. The sticky tongue mechanism used by frogs to capture their prey is translated into an effective model for capturing space debris. This mechanism utilizes a sticky and reversible adhesive layer that can rapidly capture micro-sized pieces of debris. Furthermore, the fast mobility of the linguistic mechanism provides maneuverability, which is critical for the capture of space debris. The study suggests that this nature-inspired solution is a promising method for effective collection and cleanup of micro-debris in space. This approach could offer an innovative and effective strategy for managing space debris in the future and improve the safety of activities in space.

Keywords Space debris · Frog tongue · Space technologies

A. M. Yazıcı (✉)
Istanbul Gelisim University, Istanbul 34310, Turkey
e-mail: ayazici@gelisim.edu.tr

E. Persson
Lund University, 192 221 00 Lund, Sweden
e-mail: erik.persson@fil.lu.se

F. Çebi et al. (eds.), *Engineering and Technology Management in Challenging Times*,
Lecture Notes in Networks and Systems 1137,
https://doi.org/10.1007/978-3-031-70935-7_11

1 Introduction

In the introduction to their book "The Ethics of Space Exploration", [1] asks: "Are we obliged to "clean-up" [space] debris if we can develop the requisite capabilities?" Considering our dependency on satellites for everything from mobile phone communication to environmental monitoring, however, it seems more reasonable that we are not just obliged to clean up if we manage to develop the requisite capabilities, but that we are obliged to actually make a real effort to develop these capabilities.

In recent years, space debris in Earth orbit has become a serious problem and is constantly increasing [2]. Collisions between satellites not only damage the satellites, but also increase the amount of space debris when parts from the damaged satellite join the existing debris. The same thing happens when debris collide with other debris and break into smaller parts, thus increasing the amount of debris even more [3]. It is very important to protect the space environment, reduce the risk of collisions between satellites, and remove active space debris from orbit [4, 5]. One of the key elements in the development of space debris removal missions is the ability to capture the debris, whose shape and motion are generally unknown. Therefore, the development of robotic devices capable of grasping, manipulating and capturing objects that are mostly uncooperative and not equipped with appropriate devices to improve their manipulation is the primary challenge for the near future [6]. Therefore, it is necessary to track the movement and trajectory information of space debris to enable effective prediction of future space debris activities and avoid satellite accidents [7].

In low Earth orbit, the orbit most intensively used, the debris travels at 7.8 km/s (initial cosmic velocity) and has an average speed of 10 km/s to enter the atmosphere [8]. Internationally, up to 30 satellite maneuvers per year are required to avoid debris impacts [9]. Debris smaller than 1 cm causes spatial impact and leads to loss of function and failure of components, subsystems and even the entire spacecraft [10]. Objects as small as 0.5 cm could pose a risk [4] to spacecraft. Such debris cannot be tracked and catalogued and can only be prevented passively by installing a protective structure. In addition to small debris, there are also different debris sizes. Debris larger than 10 cm can lead to explosion, fragmentation and complete failure of the spacecraft. No protection can be taken, but debris can be accurately identified, catalogued and managed. Spacecraft actively avoid such debris and debris removal is also possible [11].

The de-orbit of end-of-life rocket upper stages and satellites has only been carried out in a very small number of cases, and most remain in orbit. The explosion of residual propellants and collisions between satellite debris or rocket upper stages can produce large quantities of smaller debris, which greatly increases the probability of collision through a cascade effect. Due to such cascading collisions, the amount of space debris is predicted to increase at a large rate henceforth, eventually jeopardizing near-Earth space activities [12].

Space debris can be classified as detectable and non-detectable. Detectable space debris has detectable hardware, while undetectable space debris does not have detectable features. Different strategies can be used to capture space debris, one of which is to hold the object firmly [13]. Unlike nets, harpoons, and other proposed systems, a gripper robot can apply controlled forces and motions [14].

In this work, we present a new model for capturing small space debris in Low Earth Orbit. The work is inspired by frogs, which have developed various strategies to capture prey in nature. With their sticky tongues, frogs can quickly hunt prey in the blink of an eye. Frogs owe this ability to the physical properties of their tongue and the chemical properties of their saliva. Inspired by this strategy developed by frogs to capture their prey, we believe that this model, which we have developed to capture small objects, will be useful for the space industry and for space sustainability.

2 Concepts of Space Debris

Space or orbital debris, understood as any part of a non-functional man-made object, is often presented as one of the most significant challenges to the sustainable development of outer space. Increased space traffic, coupled with the irresponsible behavior of some of the actors due to involved in the context of the absence of legally binding rules, makes outer space vulnerable to many threats to the sustainable use of Earth's orbits. Recently, the gradual accumulation of space debris has increased the probability of collisions between debris and orbiting satellites. This situation threatens the long-term safety and stability of the space environment. Active debris removal (ADR) has become an important task for the entire spaceflight community [15]. Space debris is any man-made object that is no longer in active use and orbits the Earth. This includes any released object or instrument lost by astronauts during a space mission, space crafts, launch vehicles or parts thereof [16, 17]. Space debris comes in different sizes. Small space objects are objects, such as broken rocket parts, screwdrivers and flaking paint, that pose a lethal threat to spacecraft [18]. The harmful characteristic of space debris is due to its self-generating nature or cascading effect or Kessler syndrome, a phenomenon already predicted during the Cold War, describes a situation in which the number of space debris reaches a critical mass and will self-replicate through mutual interaction and fragmentation, even in the absence of further launches [19].

When orbital debris collides with other space objects, the result is more debris. Eventually, some chain reaction of collisions could potentially damage some of the most valuable orbits [20]. Judging by the growing threat of space debris, there are 130 million pieces of debris smaller than 1cm in orbit, millions of which are too small to be tracked [21]. 34,000 pieces of space debris larger than 10cm in diameter are currently being tracked. 8950+ satellites have been launched to date, 5000+ of which are currently in orbit, but only 1950 are operational [22].

There are growing concerns that we are creating an envelope of space debris in parts of Earth's orbit that could impede human space travel, space exploration and satellite utilization. This scenario, which continues as collisions between space objects create more debris, could have implications for global communications, navigation [4], electronic commerce, national security, telecommunications, GPS, news reporting and remote sensing [23]. There is an urgent need to address the problem of space debris in order to prevent a catastrophe later [24].

Therefore, the development of active debris removal technologies is important and urgent. So far, various strategies have been conceptualized to solve the problem of large space debris. However, our proposal is aimed at capturing small space debris, which is as important as the dangers posed by large space debris, and more difficult to solve. Our strategy for capturing small space debris is inspired by the nature-based strategy of frogs to capture prey with their sticky tongues.

3 Sticky Tongues of Frogs and Prey Capture Strategies

Frogs are well known for using their sticky tongues to catch elusive prey. Tongue movements of frogs are extremely fast. A single feeding event usually occurs within a tenth of a second [25]. It might be thought that the frog tongue is only successful in capturing light prey. However, a frog can pull up to 1.4 times its own body weight using its tongue [26] and can capture very large prey compared to its own body size [27]. In most frogs, the tongue is attached to the front of the mouth and moves forwards under its own inertia, often undergoing incredible acceleration as it moves out of the mouth to capture prey [28, 29].

Frogs are notable for having a tongue that is propelled through the mouth to strike, latch onto, and pull prey into the oral cavity. Although sometimes listed as a class-defining characteristic, among some salamanders and frogs the tongue appears to be turned outwards rather than moved by progressive control. The movement of the tongue thus resembles the bullet tongue of chameleons rather than the merely extensible tongue of anteaters, some birds and snakes [30].

Most vertebrates feed on a variety of prey species that differ in their physical and behavioral characteristics. Therefore, vertebrates have developed alternative strategies to capture diverse prey [31–33]. Frogs have an average adhesion force of 3:01 + 2:53 kPa and a maximum recorded adhesion force of 17.7 kPa. These values are not the highest in the animal kingdom. Geckos have an adhesion force of 16.5–100 kPa [26]. Chameleons have developed a highly specialized feeding system based on one to two body-length ballistic projections of the tongue with accelerations ranging from 300 to 1500 ms^2 depending on size [34]. Besides vertebrates, Philcoxia, a carnivorous plant, has rounded leaves covered with glands that produce a sticky substance. Philcoxia is the first species known to capture and digest earthworms underground using these sticky leaves. The secret of this unique food-capturing mechanism is hidden in the sand beneath the sticky leaves [35]. The reason for choosing the frog in this study is that it has the highest adhesion strength and is very fast in catching its prey.

Frogs have identified three non-exclusive mechanisms that they use to extend their tongues during feeding. These are mechanical traction, inertial elongation and hydrostatic elongation [36]. According to [26], the frog tongue behaves more like a car shock absorber than a pressure-sensitive adhesive. The viscoelastic structure of the frog tongue allows rapidly applied forces to dissipate in the tongue tissue. The stickiness of the tissue is due to the unique saliva that can flow onto textured surfaces and grip tightly during tongue retraction. Frog tongues have a number of characteristics that enable successful prey capture at high speed. Firstly, their tongues are one of the softest biological materials known, which allows the tongue to flatten on impact, facilitating a large contact area. Secondly, the tongue is highly damped, like the shock absorber of a car. When an insect is suddenly pulled at high speed, the inertia of the insect causes large separation forces with the tongue. These forces are reduced by internal damping in the tongue [26].

4 Overview of Existing Suggestions for Mechanisms for Capturing Large Scale Space Debris

Cleaning up small space debris in the atmosphere may sound very simple, but making it a reality is not so simple. Space debris mitigation is about minimizing the production of new debris and requires a man-made system or atmospheric drift [37]. Although space debris mitigation seems to be a simple and inexpensive solution, it requires complex and expensive technologies [25]. There are many proposals in the literature to clean up space debris. Some of these suggestions are as follows.

According to [18], the new sensor to be installed on the International Space Station (ISS) offers better control over the small debris population. The 1 m^2 detector comprises a network of thin wires supported by a doubly thin sensor. Impacting debris will break some of these wires, which correlates with the size of the particle. By detecting where the wires are damaged and how fast the particle enters the layers, the sensors will be able to track particle velocities and trajectories. A back plate will measure the intensity of the impact and help scientists estimate the density of the object. With this data, they can determine the trajectories of the particles, which provide clues to their origins.

However, the International Space Station (ISS) does have various sensors and systems in place to monitor space debris and ensure the safety of the astronauts and the station itself. These systems include radar and optical sensors that track larger debris objects, and the ISS can perform maneuvers to avoid potential collisions with known objects [38]. NASA and other space agencies are continuously working on improving debris tracking and mitigation techniques [25].

Srikrishnan et al. [39] describe a concept according to which a Semi-Autonomous Satellite Tracking Robot (SASTROBOT) is launched into space carrying different de-orbit rockets on its body. They suggest that this system is economically favorable for space debris removal.

The most favored space debris acquisition methods are harpooning and netting [40]. RemoveDEBRIS [41], proposes a net and a harpoon to capture space debris, as well as a Vision-Based Navigation (VBN) system to validate debris tracking techniques in orbit with cameras and LIDAR.

Piergentili et al. [42] describe a spray foam system developed to capture space debris.

Sun et al. [43] developed a new adaptive space debris capture approach based on the origami principle. In this approach, a gripper configuration is designed for different dimensions of the space debris target. Furthermore, considering that the disturbance in the capture process is nonlinear and difficult to obtain accurately, a new extended state observer-based sliding control is developed for the capture process control.

Trivailo and Kojima [44] developed a comprehensive nonlinear model of a web capture system in the form of a rapidly deployed system containing spider-like webs, which are launched towards the targeted debris at a distance.

Ru et al. [45] focused on capturing irregular and rotating debris through a flexible mesh that can be switched on and off repeatedly, which means it can capture many pieces of debris in a single mission. The flexible net can capture debris rotating at an angular velocity of 6.28 rad/s. The debris can be completely wrapped and dragged further along the expected trajectory.

Takeichi and Tachibana [46] propose a tethered plate satellite that can lift a large number of small space debris in succession only through collisions, without capturing them or performing large orbital maneuvers.

One method for active space debris removal described in the literature is harpooning [2]. 'Harpooning' here means that a projectile is ejected that penetrates the surface of a piece of space debris and roped back with the piece of space debris attached. A risk connected with this capture method is that if the velocity of the harpoon is too low it will not penetrate the target but just push it off course. If it is too high the target may break and the problem will be worsened [2]. Harpooning is also not effective for very small pieces of debris.

Another example is in the form of robotic arms with some kind of grip. A problem with this technology is that it is very difficult to secure a rigid grip of pieces of debris moving around in space [4, 12]. This problem is worsened for really small objects [4]. Nishida et al. [12] suggest that the robot arm needs to be constructed so as to buffer and brake the motion of the debris. This is difficult to do with rigid mechanics but easier with the frog tongue inspired solution suggested here.

One way of dealing with large debris would be to attach something to the debris that can be used to change its trajectory and steer it into the Earth atmosphere. Several such solutions have been suggested including balloons, sails and thrusting engines. In addition to only working for very large objects like satellites and large rocket parts, this solution is also very expensive [4].

Another suggestion for capturing space debris is to use a net. That solution is free of most of the problems mentioned above but it only works for large debris. For smaller debris that could fit through the holes of a net, some kind of membrane could be used but that would instead suffer from drag and may easily break [4].

The examples given above illustrate different ideas for cleaning up large space debris. As pointed out by St-Onge et al. [4], however, systems made for large debris removal are not useful for small debris removal.

Although most of the focus today is on large and medium sized debris, some work is also being done on active capture of small size debris [4]. St-Onge et al. [4] discuss using a large cupola to swipe the Low Earth Orbit. A risk with using a large structure like that is that it itself becomes vulnerable to collisions with debris. St-Onge et al. [4] suggest a solution for stabilizing the course of the cupola in cases of collisions with medium sized debris using flywheels but do not find it economically viable at present times. In addition to the risk of course change, however, we also need to acknowledge the risk of the cupola breaking and creating more debris if hit by a large piece of existing debris.

5 A New Model for Capturing Small Space Debris Inspired by Frogs

The model we propose is for cleaning up small space debris. The model is inspired by the frog tongue. With their sticky tongues, frogs can catch prey larger than their own size and weight very quickly and manoeuvrably. Figure 1 shows the tongue thrust of the frog.

Figure 2 shows a small space debris removal model inspired by the frog's tongue thrust.

The small space debris removal model shown in Fig. 2 is inspired by the biological model of the frog pushing its tongue while catching it prey. A detailed description of the model is given below.

Number 1 in the model represents the frog's n-shaped tongue. Frogs use the sticky property of their tongues to quickly capture their prey. For the purpose of the model, number 1 is the main mechanism for capturing small space debris. In this section, small debris can stick to this mechanism, which acts as a glue, and small space debris can be captured by this mechanism. Number 2 in the model is the mechanism that acts as a kind of stabilizer of the main mechanism in number 1. More precisely, it is the mechanism that helps the main mechanism in No 1 to catch space debris without being thrown to the left or right. These long curved arms can also prevent small space debris from flying around. Number 3 in the model is the mechanism that connects mechanism number 1, with mechanism number 2. Number 4 in the model is the lever to be placed inside mechanism number 1. The main purpose of mechanism No 4 is to connect and fix together the cable No 6, the collection box No 5 for capturing small space debris and the main mechanism No 1 and the long curved arms No 2. Number 5 in the model is the storage (collection) box for capturing small space debris. After the small space debris is attached to the main mechanism No 1, it will be transferred to the space debris storage (collection) box No 5 via cable No 4 and No 6. Capturing small space debris is indeed a significant challenge, an done proposed method is the use of a cable-based system. The length of the cable required for effective capture depends on various factors, including the specific mission objectives and the characteristics of the debris being targeted. Number 6 in the model is the cable connecting No 4

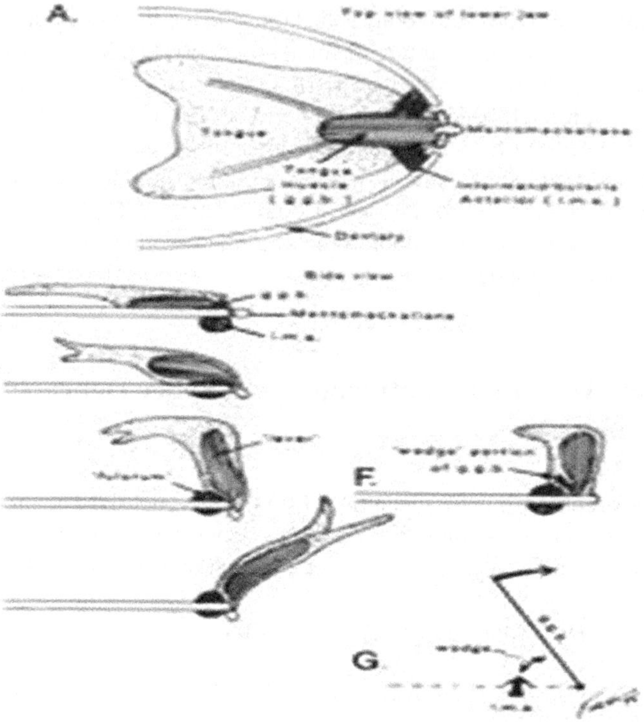

Fig. 1 Tongue thrust pattern of the frog. *Source* Regal and Gans [36, 719]. Figure courtesy of NN

Fig. 2 Small space objects debris removal model

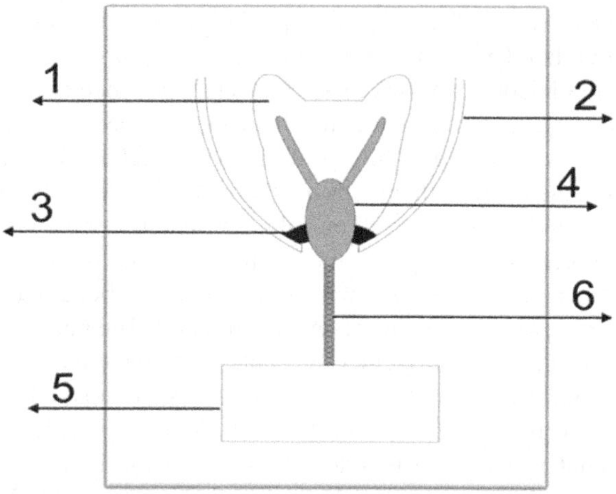

and Number 5. This cable can be made of carbon tubes, which are flexible, hard to break and 200 times stronger than steel. The cable should be between 150 and 200 m long. While there is no universally agreed-upon length for a debris-capturing cable, some studies and proposals have suggested lengths in the range of 150–200 m. This length allows for sufficient coverage and maneuverability to intercept and capture debris. However, it is important to note that different designs and concepts may have varying cable lengths based on their specific requirements and capabilities.

6 Advantages and Disadvantages with Our Approach

A general problem for solutions that target each piece of debris individually, including robotic arms and lasers, and also our proposed solution, is that the amount of debris is substantial and it will take a long time to deal with each piece of debris individually [4]. This problem also affects our approach.

The most important advantage of the frog tongue model is that the strong adhesively created by a combination of chemical and physical properties, can catch and hold on to debris of any shape and material moving in high speed with no risk of breaking it apart in smaller pieces and thus create more debris.

Modeling the frog tongue for capturing small space junk has both advantages and disadvantages.

6.1 Advantages

Adhesive properties: Frog tongues have a unique adhesive mechanism that allows them to capture their prey quickly and efficiently. By modelling this mechanism, it could potentially be a highly efficient and effective method for capturing small space junk.

Versatility: Frog tongues are capable of capturing objects of various sizes and shapes. This versatility could be advantageous for capturing different types of space debris, including small fragments and particles.

Cost-effective: Developing a frog tongue-inspired mechanism for capturing space junk could potentially be a cost-effective solution compared to other methods, such as using robotic arms or nets.

Self-cleaning: Frog tongues possess self-cleaning properties, where debris adhering to the tongue is automatically removed during retraction. If we can replicate this feature in a modeled system, it would reduce the maintenance requirements and improve the longevity of the capturing mechanism.

6.2 Disadvantages

Complexity: Modeling the frog tongue mechanism accurately is a complex task. The frog tongue consists of a combination of soft tissues, muscles, and specialized adhesion properties, making it challenging to replicate in an artificial system.

Scale and environment: The frog tongue is designed to operate in a specific scale and environment, optimized for capturing prey in terrestrial environments. Adapting this mechanism for the vacuum and extreme conditions of space could pose challenges in terms of design and material selection.

Precision and control: Capturing space debris requires precise control and coordination, especially when dealing with delicate or valuable objects. The frog tongue mechanism may not provide the necessary level of precision required for such operations.

Limited applicability: While the frog tongue-inspired mechanism may be suitable for capturing smaller space debris, it may not be effective for larger objects or more complex space debris, such as defunct satellites or rocket stages. A diverse range of capture methods would still be needed to address the variety of space junk present.

Overall, while modeling the frog tongue for capturing small space junk offers some intriguing advantages, it also presents significant challenges and limitations. Further research and development would be required to determine the feasibility and effectiveness of such an approach.

7 Conclusion

Space activities pose a significant threat to existing space systems, with an increasing number of small space debris posing risks to the orbital environment. This debris can jeopardize the sustainable use of space and the Earth's atmosphere. Effective control and cleanup of space debris is therefore vital for the future use of space. The space industry's interest is addressing this problem through the development of innovative and advanced systems. However, taking inspiration from nature and exploiting the potential offered by biological systems could be an important approach. Usually in the literature, studies on how to capture large space debris have been done, but effective capture methods for small space debris are still lacking. In this paper, we propose a frog tongue-inspired model for capturing small space debris. The fast and strategic movements of frogs while capturing their prey are the basis for the development of this model. The sticky property of the frog's tongue makes it almost impossible for prey to escape, making it a suitable model for the safe capture of small space debris. The proposed model could provide significant benefits to the space industry. This biologically-based approach to the effective capture of small space debris could bring a new perspective to existing technologies and help clean up space junk in the future. This study could also be an important step towards the protection of the space environment and the sustainability of space activities.

However, going forward, a number of future recommendations can be made based on the results of this study. First, it is important to conduct further empirical studies on the practical applicability and effectiveness of the proposed model. These studies will evaluate and optimize how the model will perform in real-world applications. Furthermore, this biologically based approach needs to be integrated and work in harmony with existing space cleanup technologies. This is important to develop an effective and efficient strategy for space debris cleanup.

Future research could take a broader perspective and focus on space debris as well as other space environment issues. This could provide a more comprehensive solution for the sustainability of space activities and the protection of the space environment. Finally, the results and future recommendations of this study can be used to promote collaboration and develop joint solutions among stakeholders in the space industry. This study could be an important step towards protecting the space environment and ensuring sustainable space activities for future generations.

References

1. Schwartz JSJ, Milligan T (2016) Introduction; The scope and content of space ethics. In: Schwartz JSJ, Milligan T (eds) The ethics of space exploration. Springer
2. Campbell JC, Hughes K, Vignjevic R, Djordjevic N, Taylor N, Jardine A (2022) Development of modelling design tool for harpoon for active space debris removal. Int J Impact Eng 166:104236
3. Liou J-C, Johnson NL, Hill NM (2010) Controlling the growth of future LEO debris populations with active debris removal. Acta Astronaut 66:648–653
4. St-Onge D, Sharf I, Sagniéres L, Gosselin C (2018) A deployable mechanism concept for the collection of small-to-medium-size space debris. Adv Space Res 61:1286–1297
5. Takahashi K, Charles C, Boswell RW, Ando A (2018) Demonstrating a new technology for space debris removal using a bi-directional plasma thruster. Sci Rep 8:14417. https://doi.org/10.1038/s41598-018-32697-4
6. Savioli L, Sguotti G, Francesconi A, Branz F, Krahn J, Menon C (2014) Morphing electroadhesive interface to manipulate uncooperative objects. In: Proceedings SPIE 9061, sensors and smart structures technologies for civil, mechanical and aerospace systems, p 906129. https://doi.org/10.1117/12.2045065
7. Jiang P, Liu C, Yang W, Kang Z, Fan C, Li Z (2022) Automatic extraction channel of space debris based on widefield surveillance system. Microgravity 8(14). https://doi.org/10.1038/s41526-022-00200-z
8. Pardini C, Anselmo L (2017) Revisiting the collision risk with cataloged objects for the Iridium and COSMO-SkyMed satellite constellations. Acta Astronaut 134:23–32
9. Li M, Gong Z, Liu G (2018) Frontier technology and system development of space debris surveillance and active removal. Chin Sci Bull 63(25):2570–2591
10. Kurihara M, Higashide M, Takayanagi Y, Arai K, Yano H, Tabat AM (2015) Impact frequency estimate of micron-sized meteoroids and debris on Tanpopo capture panels on the ISS. Proc Eng 103:334–340
11. Gong Z, Gao K, Song G, Li M, Wu Q, Cao Y (2018) Research progress and prospect of hypervelocity impact and its space application. Mod Appl Phys 9(4):11–23
12. Nishida S-I, Kawamoto S, Okawa Y, Terui F, Kitamura S (2009) Space debris removal system using a small satellite. Acta Astronaut 65:95–102
13. Zhang X, Liu J, Feng J, Liu Y, Ju Z (2020) Effective capture of nongraspable objects for space robots using geometric cage Pairs. IEEE/ASME Trans Mech 25(1)

14. Jiang H, Hawkes EW, Fuller C, Estrada MA, Suresh SA, Abcouwer N, Han AK, Wang S, Ploch CJ, Parness A, Cutkosky MR (2017) A robotic device using gecko-inspired adhesives can grasp and manipulate large objects in microgravity. Sci Robot 2(7). https://doi.org/10.1126/scirobotics.aan4545
15. Liou J-C (2011) An active debris removal parametric study for LEO environment remediation. Adv Space Res 47(11):1865–1876
16. Sylvestre H, Parama VRR (2017) Space debris: reasons, types, impacts and management. Indian J Radio Space Phys 46:20–26
17. Shan M, Guo J, Gill E (2016) Review and comparison of active space debris capturing and removal methods. Prog Aerosp Sci 80:18–32
18. Loomis I (2017) NASA sensor to study space junk the size of dust. Science 358(6368). https://doi.org/10.1126/science.358.6368.1234
19. Kessler DJ, Cour-Palais BG (1978) Collision frequency of artificial satellites: the creation of a debris belt. J Geophys Res 83(A6):2637–2646. https://doi.org/10.1029/JA083iA06p02637
20. Hutagalung JM, Tobing CI, Debastri J, Amanda RT (2020) Space debris as environmental threat and the requirement of Indonesia's prevention regulation. In: IOP conference series: earth and environmental science, vol 456, p 012081. https://doi.org/10.1088/1755-1315/456/1/012081
21. Byers M, Boley A (2023) Who owns outer space?. Cambridge University Press
22. Astroscale (2022) Space sustainability safe and sustainable development of space through on-orbit services. https://astroscale.com/space-sustainability/. Accessed 02 Dec 2022
23. Runnels MB (2023) Protecting earth and space industries from orbital debris: implementing the outer space treaty to fill the regulatory vacuum in the FCC's orbital debris guidelines. Am Bus Law J 60(1):175–229
24. Haff PK (2014) Technology as a geological: implications for human well-being. Geol Soc, London, Special Publications 395(1):301–309
25. Nishikawa KC, Roth G (1991) The mechanism of tongue protraction during prey capture in the frog Discoglossus pictus. J Exp Biol 159:217–234
26. Noel AC, Guo H-Y, Mandica M, Hu DL (2017) Frogs use a viscoelastic tongue and non-Newtonian saliva to catch prey. J R Soc Interface 14:20160764. https://doi.org/10.1098/rsif.2016.0764
27. Kleinteich T, Gorb SN (2014) Tongue adhesion in the horned frog Ceratophrys sp. Sci Rep 4:5225. https://doi.org/10.1038/srep05225
28. Keeffe RM, Blob RW, Blackburn DC, Mayerl CJ (2022) XROMM analysis of feeding mechanics in toads: interactions of the tongue, hyoid and pectoral girdle. Integr Organismal Biol 4(1). https://doi.org/10.1093/iob/obac045
29. Ritter D, Nishikawa K (1995) The kinematics and mechanism of prey capture in the African pig-nosed frog (Hemisus Marmoratum): description of a radically divergent anuran tongue. J Exp Biol 198:2025–2040
30. Regal PJ, Gans C (1976) Functional aspects of the evolution of frog tongues. Evolution 30:718–734
31. Schwenk K, Rubega M (2005) Diversity of vertebrate feeding systems. In: Starck JM, Wang T (eds) Physiological and ecological adaptations to feeding in vertebrates. Science Publishers, Enfield, NH
32. Monroy RA, Nishikawa K (2011) Prey capture in frogs: Alternative strategies, biomechanical trade-offs, and hierarchical decision making. J Exp Zool 315:61–71
33. Schaerlaeken V, Meyers JJ, Herrel A (2007) Modulation of prey capture kinematics and the role of lingual sensory feedback in the lizard Pogona vitticeps. Zoology 110:127–138
34. Brau F, Lanterbecq D, Zghikh L-N, Bels V, Damman P (2016) Dynamics of prey prehension by chameleons through viscous adhesion. Nat Phys 12:931–935
35. Rowland K (2012) Hungry plant traps worms underground. Nature. https://doi.org/10.1038/nature.2012.9757
36. Nishikawa KC (1999) Neuromuscular control of prey capture in frogs. Philos Trans R Soc Lond B Biol Sci 354:941–954. https://doi.org/10.1098/rstb.1999.0445

37. Ansdell M (2010) Active space debris removal: needs, implications, and recommendations for today's geopolitical environment. J Public Int Affairs 7–22
38. NASA (2017) Sensor to monitor orbital debris outside space station. https://www.nasa.gov/mission_pages/station/research/news/sensor_to_monitor_orbital_debris_outside_iss. Accessed 02 June 2023
39. Srikrishnan S, Dash PK, Pillai SN, Arunvinthan S (2015) An approach for space debris cleaning using space based robots. Int J Eng Res Manage 02(06)
40. Ellery A (2020) Tutorial review of bio-inspired approaches to robotic manipulation for space debris salvage. Biomimetics 5:19. https://doi.org/10.3390/biomimetics5020019
41. RemoveDEBRIS (2022) Testing technology to clear out space junk. https://www.airbus.com/en/products-services/space/in-space-infrastructure/removedebris. Accessed 02 Dec 2022
42. Piergentili F, Battagliere ML, Candini GP, Piattoni J, Romei F, Spadanuda A, Toschi S, Valdatta M, Santoni F (2010) Redemption: a microgravity experiment to test foam for space debris removal. In: 62nd international astronautical congress, Cape Town, SA
43. Sun C, Wan W, Deng L (2019) Adaptive space debris capture approach based on origami principle. Int J Adv Robot Syst 1–11. https://doi.org/10.1177/1729881419885219
44. Trivailo PM, Kojima H (2016) Dynamics of the net systems, capturing space debris. Trans JSASS Aerosp Tech Japan 14(30):57–66
45. Ru M, Zhang Y, Cheng B, Zhang Y (2022) Capture dynamics and control of a flexible net for space debris removal. Aeorspace 9:299. https://doi.org/10.3390/aerospace9060299
46. Takeichi N, Tachibana N (2021) A tethered plate satellite as a sweeper of small space debris. Acta Astronaut 189:429–436. https://doi.org/10.1016/j.actaastro.2021.08.051

Comparative Analysis of Probabilistic Models for Intermittent Demand Forecasting

Mevlüde Sezer and Ferhan Çebi

Abstract The distinguishing characteristic of intermittent demand compared to non-intermittent demand lies in the occurrence of zero values for certain time periods, defining intervals that transform it into a time series. Typically, a single interval value is chosen for all stock keeping units (SKUs) in software packages. When analyzing demand distribution on a SKU basis, demand is considered fast-moving if its frequency exceeds the selected interval, otherwise, it is classified as slow-moving or intermittent demand. Given the absence of a fixed interval in real-life scenarios, it is reasonable to assume that all demand distributions exhibit intermittent behaviour. This paper aims to apply relatively novel methods to forecast both expected demand and occurrences in the next forecasting period for a sample of automotive spare parts. Intermittent demand forecasting traces back to Croston's work in 1972, which was based on single exponential smoothing, disregarding trend and seasonality and being prone to positive bias. To address these limitations, newer methods such as Holt-Winter, Syntetos and Boylan Approximation (SBA), and Teunter, Syntetos, and Babai (TSB) have been introduced. In this study, Croston's method, SBA, and TSB methods are utilized for forecasting intermittent demand of automotive spare parts, with a comparative evaluation based on forecast error types including mean squared error, root mean squared error, and mean absolute error.

Keywords Intermittent demand · Forecasting · Spare parts

M. Sezer (✉) · F. Çebi
Istanbul Technical University, Istanbul, Turkey
e-mail: sezer19@itu.edu.tr

F. Çebi
e-mail: cebife@itu.edu.tr

F. Çebi et al. (eds.), *Engineering and Technology Management in Challenging Times*, Lecture Notes in Networks and Systems 1137,
https://doi.org/10.1007/978-3-031-70935-7_12

1 Introduction

It is crucial to be able to plan on stock keeping units (SKU) basis by Supply Chain or Planning departments working on forecasting in daily business life. When it comes to calculating order-up-to levels and safety stock levels, especially on SKU basis, planners tend to avoid out-of-stock and keep extra inventory in warehouses, i.e., known as bullwhip effect [1]. Since the cost of out-of-stock status is considered as loss of sales, keeping inventory is preferable. However, it is accepted that the tendency to keep extra inventory leads to increasing in inventory holding costs and scrap rates when there is a risk of obsolescence in real-life examples [2].

Working with many kinds of products with huge number of SKU which have different demand patterns makes it hard to forecast with a single approach in high accuracy. If the demand interval chosen in data collecting cannot prevent the periods of zero demand, it will be difficult to predict future demand size and demand occurrence with conventional forecasting methods, e.g., moving averages or single exponential smoothing (SES), because intermittent demand intervals can increase the replenishment levels and estimated average demand [3].

Although SES method is a common approach for forecasting, it is insufficient in the case of intermittent demand, thus a new probability approach is published by Croston [3], in which demand interval is also calculated. The irregular time intervals or irregularity between demand occurrences begins to be studied [4]. Henceforth, intermittent demand forecasting becomes a new forecasting field to study, and it is expanded with new approaches.

In this paper, a comparative analysis is studied with an application on automotive spare parts. Demand types and common forecasting methods for intermittent demand forecasting will be identified in the literature review and they will be detailed in methodology section. An application will be revealed through the methods: Croston's Method, SBA and TSB. The results will be discussed in the conclusions part.

2 Literature Review

Initially, demand types will be explained and classified in order to understand the characteristics of intermittent demand those distinguish from other types of demand, and then the forecasting methods commonly used for intermittent demand structure are revealed.

2.1 Demand Types

Demand types for forecasting can be described as the distribution of demand values. Identifying demand distribution is as important as selection of forecasting method.

Previous research, whose main concern is the statistical methods for estimating demand requirements, on intermittent demand forecasting has indicated that forecast accuracy is closely related to specific demand characteristics [5].

The categorization schemes, studied by Williams [6] and Eaves [7] with different cutoff points is shown in Fig. 1, are investigated by Syntetos et al. and the demand type is divided into four categories [8]. Syntetos et al., is shown in four categories based on inter-demand interval and the coefficient of variation.

These four demand categories are smooth (regular demand occurrence with low variance in demand values), intermittent (irregular demand occurrence with low variance in demand values), erratic (regular demand occurrence with high variance in demand values), and lumpy (irregular demand occurrence with high variance in demand values). This categorization is shown in Fig. 2 [9].

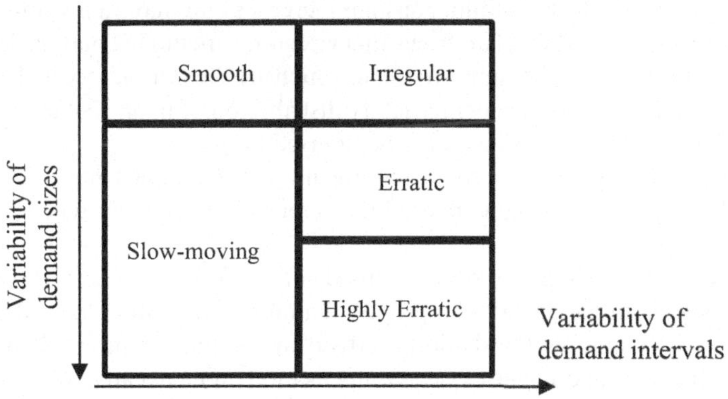

Fig. 1 Demand categorization scheme proposed by Williams and Eaves with different cutoff values

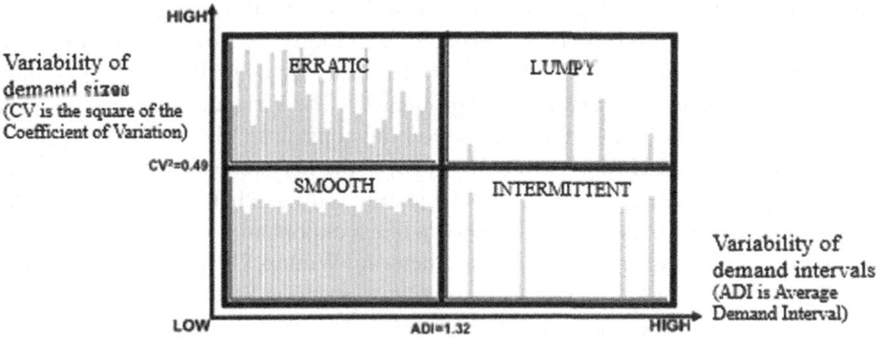

Fig. 2 Demand categorization scheme proposed by Syntetos et al. [8]

2.2 Forecasting Methods for Intermittent Demand

Croston's Method. Croston claims that inventory level is calculated with exponential smoothing is higher than the maximum demand value occurred, especially in cases where demand has intermittent pattern. He proposed a new model to prevent excess stock levels caused by avoiding out-of-stock situations for intermittent demand [3].

Regarding with Croston's method (Cr), he ignores the zero-valued demand intervals and takes non-zero demand values into consideration and calculates the demand size through exponential smoothing with low alpha constant values (between 0.1 and 0.3). In order to include inter-demand intervals with zero-valued into the model, he published an alternative method based on the probability of demand occurrence [10].

Syntetos and Boylan Approximation. Croston's method solves "issue point bias" problem of SES approach but induces "inversion bias", which grows as data becomes sparse and the smoothing constant increases and may causes inflated stock levels [10]. Syntetos and Boylan shows that Croston's method is biased. This bias is not about assumptions of Croston's method, which are the stationarity of demand size and demand interval, and the geometrically distribution of inter-demand intervals, it is due to the smoothing constant value being used [11].

Syntetos and Boylan proposes a modification to Croston's method and claim empirically that with their new method the forecasting accuracy will be improved [12].

Teunter et al. Method. Croston's method suffers from a positive bias as studied by Syntetos and Boylan [8, 9] who proposed a modified method, i.e., SBA. On the other hand, SBA ignores the damping effect on the bias of the probability for a demand occurrence and it causes overcompensation and a negative bias [13].

Teunter et al. exploits the intermittent demand character by modelling lead time demand as a compound of binomial process [14]. The proposed method by Teunter et al. (TSB) updates the demand probability instead of the demand interval unlike Croston based methods. TSB is also suitable to decrease obsolescence risk, which is an important topic for inventory management, especially for slow moving, intermittent demand. TSB method achieves a high flexibility by using different smoothing constants for demand size and demand probability. The variance of the new method is supposed to be useful in setting the smoothing constants [15].

Wright Modified Holt Method. Holt [16] is the researcher who adds the trend factor through exponential smoothing calculation with another smoothing constant for forecasting, which method is also known linear exponential smoothing [10]. Holt method is extended with seasonality parameter by Winters [17], this method is also called as Holt-Winter method, and graphically revealed in Fig. 3 as follows [18].

Croston ignores the seasonality and trend metrics in his study. Wright prefers to use Holt-Winter method for intermittent demand rather than simple exponential smoothing because of the computational easiness of changing smoothing parameters efficiently. Wright Modified Holt (WMH) method is an extension of Holt's two-parameter smoothing method [4].

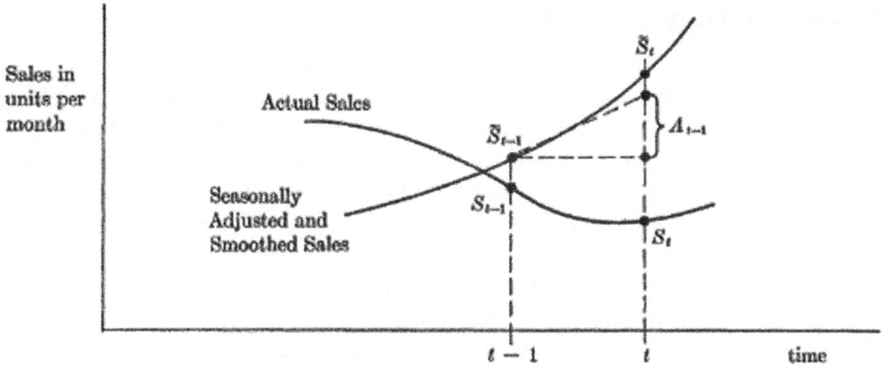

Fig. 3 Sales with seasonal and trend patterns

Altay et al. investigate that WMH is a viable alternative method to the Croston's method for intermittent demand forecasting especially in spare parts industry due to the trend factor of Holt method. Essentially, they use Systetos and Boylan Approximation (SBA) instead of Croston's method in their case study to compare with WMH. Altay et all claim that SBA always carries less inventory than WMH, meanwhile WMH provides superior service levels [19].

Poisson Based Models. Poisson based models for intermittent demand are derived from some geometric distributions to forecast of the next non-zero demand, i.e., Poisson, Negative Binomial, and Hurdle Shifted Poisson (HSP) [10, 20].

The study of Hyndman et al. [21] accepts that inter-demand intervals between non-zero demand values are assumed to follow a geometric distribution in their HSP method. Snyder et al. compares the predictions based on three distributions, namely Poisson, Negative Binomial, and HSP and concludes that HSP performs better than Poisson but not as well as Negative Binomial method [20]. These kinds of approaches to modelling generate one-step-ahead forecasts but do not give equivalent forecasts for longer horizons [10].

Bootstrap Method. Snyder [22] introduces more complex variations of the Croston method, which involve bootstrapping. Bootstrapping is also proposed by Porras Musalem [23] and Willemain et al. [24]. The main advantage of bootstrapping is that (the mean and variance of) the lead time demand distribution is forecasted directly by repeated sampling from demand occurrences. Willemain et al. uses estimating transition probabilities in a Markov model and create a sequence of zero/non-zero demand values through this method. The bootstrapping method proposed by Porras Musalem is simpler which can capture demand autocorrelation by restrictive sampling [25].

3 Methodology

The problem of the case study is forecasting future demand for the next period of range of intermittent demands in different distributions. The methods declared in the study generally focus on demand forecast as of the end of the range and calculates the prediction for future demand. Especially using software makes it easy to generate information for each dedicated time and it is common way of working with planning horizons.

In this study, the predictions are calculated iteratively for each time and the forecast errors are generated for all demand range. In this way, it can be observed how much the range of time may affect the calculations in short term.

The methodologies used for the application in the case study are Croston's method, SBA and TSB. These methods will be used for comparative analysis of intermittent demand in different demand distributions. Alpha constant is taken as 0.2 and Beta contact is taken as 0.3 for all calculations.

The accuracy measures will be examined to choose the best method for each demand distribution. The criterion for this selection will be based on average forecast errors which are determined as Mean Squared Error (MSE), Root Mean Squared Error (RMSE) and Mean Absolute Error (MAE). Mean Absolute Percentage Error (MAPE) is not suitable because of the zero-valued demand intervals, it gives the "division by zero" error. Thus, MAPE is out-of-scope of this study. Although it is aimed that the proper forecasting method is the main target of this study, the accuracy measures are also evaluated for comparison.

All calculations are generated in tables by means of Microsoft Excel for this application.

3.1 *Croston's Method*

Croston's method proposes two separate exponential smoothing calculations those estimates the average size of demand (when demand occurs) and the average interval demand occurrences. This method updates the calculations when non-zero demand occurs.

The method has two dimensions for the calculation. One is the mean demand size, and the other one is mean demand interval. Equation 1 is the formula of the forecast of the mean demand size for period t + 1 and Eq. 2 shows the mean interval of demand forecast. The result function as the mean demand forecast for the next period is formulised in Eq. 3:

$$\widehat{\mu}_{S,t+1} = \alpha S_t + (1-\alpha)\widehat{\mu}_{S,t}, \tag{1}$$

$$\widehat{\mu}_{I,t+1} = \beta I_t + (1-\beta)\widehat{\mu}_{I,t}, \tag{2}$$

$$\hat{\mu}_{t+1} = \frac{\hat{\mu}_{S,t+1}}{\hat{\mu}_{I,t+1}},\ 0 < \alpha, \beta < 1. \tag{3}$$

In Eq. 1 S_t denotes the demand size in period t. I_t in Eq. 2 denotes the most recent demand interval. Alpha and Beta are the constant of exponential smoothing (it is supposed to be the same both for demand size and demand interval for this method, $\alpha = \beta$).

3.2 SBA Method

SBA method is derived from Croston's method with a correction constant shown in Eq. 4 as the forecast of the mean demand for period t + 1:

$$\widehat{\mu}_{t+1} = \left(1 - \frac{\alpha}{2}\right)\frac{\widehat{\mu}_{S,t+1}}{\widehat{\mu}_{I,t+1}}. \tag{4}$$

where alpha is the smoothing constant used for the inter-demand intervals.

3.3 TSB Method

TSB method has a probability function, identified as *p*, is different than Croston's method which is described with the notations below. There are two conditions for the methodology: If demand occurs or not.

If demand occurs,

$$\widehat{\mu}_{S,t+1} = \alpha S_t + (1-\alpha)\widehat{\mu}_{S,t}, \tag{5}$$

$$\widehat{p}_{S,t+1} = \widehat{p}_t + \beta\left(1 - \widehat{p}_t\right), \tag{6}$$

$$\widehat{\mu}_{t+1} = \widehat{p}_{S,t+1}\ \widehat{\mu}_{S,t+1} \tag{7}$$

Else

$$\widehat{\mu}_{S,t+1} = S_t, \tag{8}$$

$$\widehat{p}_{S,t+1} = \widehat{p}_t + \beta\left(0 - \widehat{p}_t\right), \tag{9}$$

$$\widehat{\mu}_{t+1} = \widehat{p}_{S,t+1}. \tag{10}$$

where p denotes the probability of demand occurrence and alpha and beta constants are between 0 and 1.

4 Application and Results

4.1 Case Study Background

An automotive spare parts company, the producer of public transport vehicles and located in Turkey, is selected for the case study. Their main concern about inventory management is using stock classification according to demand distributions not to be in out-of-stock condition due the customer expectations.

Although different demand patterns are observed in the spare parts industry, intermittent demand is accepted as the main demand type which has zero values for some demand periods. The datasets are chosen among intermittent demand samples. Data samples have 20 weeks of time series and different kinds of demand distributions.

4.2 Analysis of Data

There are four different datasets in the application, and they are numbered in order. Scatter plots for demand values and histograms for demand distributions are revealed in Fig. 4 as descriptive statistics.

4.3 Results

In this study MSE, RMSE and MAE are used for forecasting error calculations. Figure 5 shows the graphical results of forecast errors with iterative calculations for each demand week.

Average forecast error values are compared in Table 1 and the results are shown for each forecasting method and dataset. As it is revealed from the table that dataset2 has the lowest forecast errors whose demand distribution is likely to normal distribution.

Regarding with forecasting method, some of them are seen as appropriate ones according to forecasting errors and it is shown in Table 2.

TSB is the most accurate method for dataset1 and dataset4 for each error type. SBA is the more accurate method for dataset2 and dataset3, although the results are very close to TSB method. It can be inferred that the distribution of dataset3 is likely to Poisson distribution.

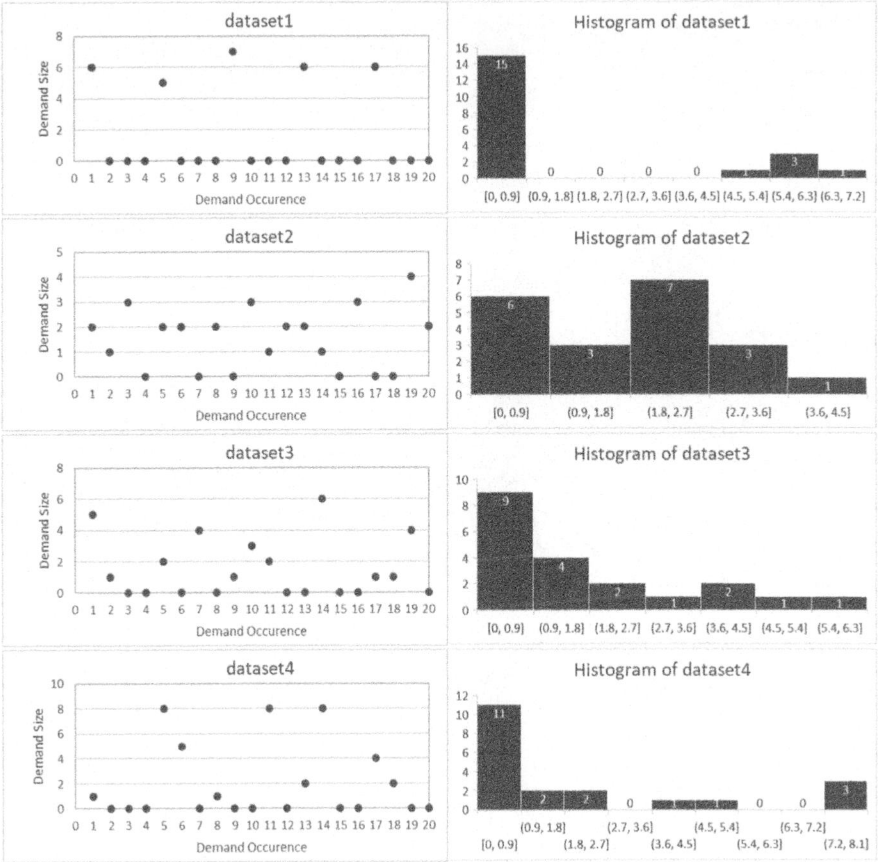

Fig. 4 Descriptive statistics for the dataset

5 Conclusions

In this study, basic probabilistic models for intermittent demand forecasting are examined. It is studied which method is much more appropriate for which type of demand distribution.

For lumpy demand pattern TSB is the selected method. For smooth or erratic demand with normal distribution, all methods can be used. For demand with Poisson distribution, SBA or TSB is recommended to be used. If the demand distribution is known, a suitable method can be selected. In this study, this comprehensive method is TSB. Nevertheless, material planning software packages are not able to give decisions about proper forecasting methods but are set-up with dedicated rules. For this reason, a method that may reach the target with high forecasting accuracy should be selected before running business. So, Croston's method cannot be selected because of the relatively poor performance for some demand types regarding this study.

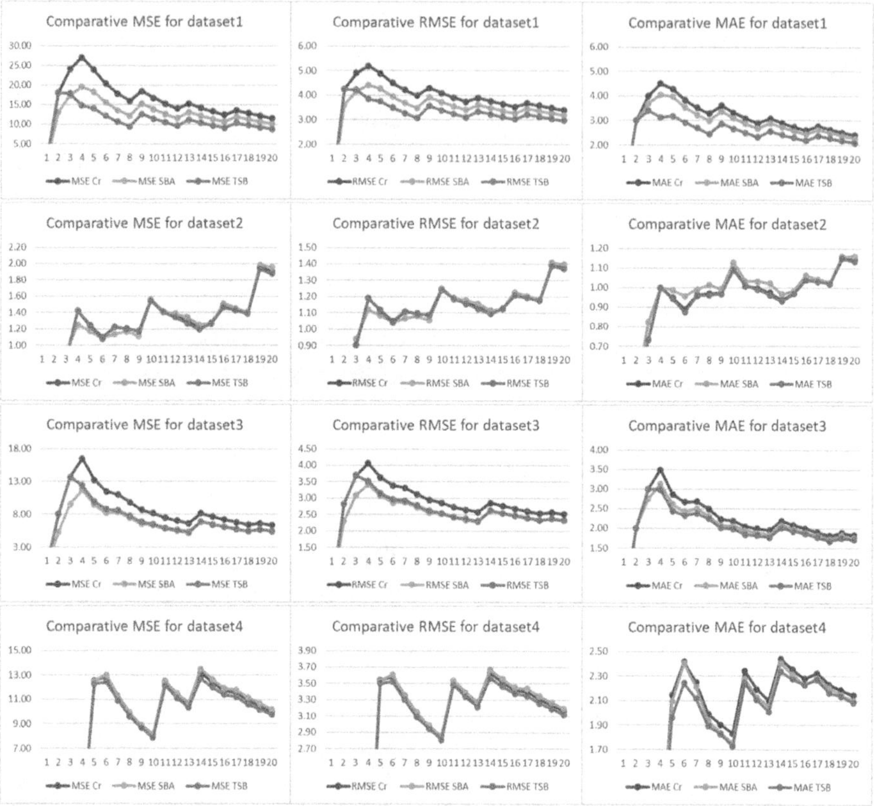

Fig. 5 Inferential statistics for the datasets

Consequently, there is a need to be aware of the reason for zero values of demand. Out-of-stock or other reasons are unknown, and they are out-of-scope of these kinds of research. When there is an out-of-stock status, it is not known whether this demand will be postponed or made by new customers. Probabilistic models are generally used with historical data as if there is always inventory in stock.

If the historical data is not available or not suitable to use for the current conditions, there will be a need for possibility of demand occurrence. This study claims that there is a lack of information on the possibility of demand occurrence and new approaches for further research.

Table 1 Forecast errors for the datasets

Average forecast errors	Croston				SBA				TSB			
	dataset1	dataset2	dataset3	dataset4	dataset1	dataset2	dataset3	dataset4	dataset1	dataset2	dataset3	dataset4
MSE	15.85	1.25	8.54	9.01	12.70	1.23	6.53	9.10	11.01	1.24	7.05	8.73
RMSE	3.85	1.08	2.82	2.78	3.46	1.07	2.47	2.78	3.21	1.08	2.56	2.73
MAE	3.05	0.92	2.17	1.85	2.84	0.94	2.03	1.81	2.48	0.91	1.98	1.76

Table 2 Chosen methods for the datasets

Method selection	Croston			SBA			TSB		
	MSE	RMSE	MAE	MSE	RMSE	MAE	MSE	RMSE	MAE
dataset1	15.85	3.85	3.05	12.70	3.46	2.84	11.01	3.21	2.48
dataset2	1.25	1.08	0.92	1.23	1.07	0.94	1.24	1.08	0.91
dataset3	8.54	2.82	2.17	6.53	2.47	2.03	7.05	2.56	1.98
dataset4	9.01	2.78	1.85	9.10	2.78	1.81	8.73	2.73	1.76

Acknowledgements The company, is in automotive spare parts industry and located in Turkey, is appreciated for their support in finding real-life examples. Systematic integration into planning software for possible solutions is on the agenda.

References

1. Zottero G (2013) An empirical investigation on causes and effects of the Bullwhip-effect: evidence from the personal care sector. Int J Prod Econ 143:489–498
2. Johnston FR, Boylan JE (1996) Forecasting for items with intermittent demand. J Oper Res Soc 47:113–121
3. Croston JD (1972) Forecasting and stock control for intermittent demands. J Oper Res Soc 23:274–293
4. Wright DJ (1986) Forecasting data published at irregular time intervals using an extension of Holt's method. Manage Sci 32:499–510
5. Syntetos AA, Nikolopoulos K, Boylan JE, Fildes R, Goodwin P (2009) The effects of integrating management judgement into intermittent demand forecasts. Int J Prod Econ 118:72–81
6. Williams TM (1984) Stock control with sporadic and slow moving demand. J Opl Res Soc 35:939–948
7. Eaves A (2002) Forecasting for the ordering and stock-holding of consumable spare parts. Thesis (Ph.D.), Lancaster University
8. Syntetos AA, Boylan JE, Croston JD (2005) On the categorization of demand patterns. J Oper Res Soc 56:495–503
9. Rozanec JM, Fortuna B, Mladenic D (2022) Reframing demand forecasting: a two-fold approach for lumpy and intermittent demand. Sustainability 14:92–95
10. Syntetos AA, Boylan JE (2021) intermittent demand forecasting: context, methods and applications. Wiley & Sons
11. Syntetos AA, Boylan JE (2005) The accuracy of intermittent demand estimates. Int J Forecast 21:303–314
12. Syntetos AA, Boylan JE (2006) On the stock control performance of intermittent demand estimators. Int J Prod Econ 103:36–47
13. Teunter RH, Sani B (2009) On the bias of Croston's forecasting method. Eur J Oper Res 194:177–183
14. Teunter RH, Syntetos AA, Babai MZ (2010) Determining order-up-to levels under periodic review for compound binomial (intermittent) demand. Eur J Oper Res 203:619–624
15. Teunter RH, Syntetos AA, Babai MZ (2011) Intermittent demand: linking forecasting to inventory obsolescence. Eur J Oper Res 214:606–615
16. Holt CC (1957) Forecasting seasonals and trend by exponentially weighted moving averages. ONR Memorandum 52

17. Winters PR (1960) Forecasting sales by exponentially weighted moving averages. Manage Sci 6(3):324–342. https://doi.org/10.1287/mnsc.6.3.324
18. Winters PR (1960) Forecasting sales by exponentially weighted moving averages. Manage Sci 6:324–342
19. Altay N, Rudisill F, Litteral LA (2008) Adapting Wright's modification of Holt's method to forecasting intermittent demand. Int J Prod Econ 111:389–408
20. Snyder RD, Ord JK, Beaumont A (2012) Forecasting the intermittent demand for slow-moving inventories: a modelling approach. Int J Forecast 28:485–496
21. Rob J, Hyndman Anne B, Koehler (2006) Another look at measures of forecast accuracy. Int J Forecast 22(4):679–688. https://doi.org/10.1016/j.ijforecast.2006.03.001
22. Ralph S (2002) Forecasting sales of slow and fast moving inventories. Eur J Oper Res 140(3):684–699. https://doi.org/10.1016/S0377-2217(01)00231-4
23. Porras ME (2005) Inventory theory in practice: joint replenishments and spare parts control. Erasmus University Rotterdam, Amsterdam
24. Thomas R, Willemain Charles N, Smart Henry F, Schwarz (2004) A new approach to forecasting intermittent demand for service parts inventories. Int J Forecast 20(3):375–387. https://doi.org/10.1016/S0169-2070(03)00013-X
25. Teunter RH, Duncan L (2009) Forecasting intermittent demand: a comparative study. J Oper Res Soc 60:321–329

Culture of Change and Digitalization

Ibrahim Sapaloglu and Işık Çiçek

Abstract Digitalization stands out as a significant catalyst for organizational evolution. Even when companies successfully integrate new technologies and implement changes, digital transformation initiatives often face a high risk of failure. This is primarily because not all employees readily embrace all planned changes. Consequently, organizations encounter substantial challenges in achieving successful change implementation. The corporate culture plays a pivotal role in the outcome of digital technology initiatives, with many such initiatives faltering due to cultural misalignment. Organizations are reshaping their cultures to navigate this complex terrain to align with attributes like interactivity, agility, speed, and connectivity. The rapid acceleration of digital transformation in organizations, especially during the Covid-19 pandemic, emphasizes the critical role of organizational culture in this process. Despite advancements in technology, the success of digital initiatives often stalls due to cultural misalignments stemming from the "culture of change". A fundamental question emerges: How the content of "culture of change" affects digital transformation? This paper explores how the adoption of digital technologies influenced by the culture of change within organizations, highlighting the necessity of fostering a culture that supports continuous adaptation and learning. For this aim, systematic literature review is carried out through Scopus databases. We discuss the essential digital competencies that facilitate this cultural shift and address the impact of uncertainties that digital transformations frequently introduce. By analyzing how organizations can develop a resilient change culture, this study contributes to understanding the dynamics between digital transformation and change of culture in organization. Readers will find insights into managing culture of change in digital transformation effectively, which is crucial for navigating the uncertainties of digitalization.

Keywords Digitalization · Digital culture · Culture of change · Resistance to change · Uncertainty

---

I. Sapaloglu (✉)
Management Engineering Department, Istanbul Technical University, Istanbul, Turkey
e-mail: sapaloglu19@itu.edu.tr

I. Çiçek
Aviation Management Department, Necmettin Erbakan University, Konya, Turkey

F. Çebi et al. (eds.), *Engineering and Technology Management in Challenging Times*, Lecture Notes in Networks and Systems 1137,
https://doi.org/10.1007/978-3-031-70935-7_13

1 Introduction

Organizational transformation is now being driven by digitalization, which has ushered in a period of unprecedented technological innovation and adaptation. While businesses work to implement organizational changes and fully utilize the potential of new technologies, the road to successful digital transformation is not without its difficulties. Employee resistance to change is one of the main challenges, and it frequently impedes the advancement of initiatives for digital transformation [1]. The influence of corporate culture on the success of digital technology initiatives cannot be overstated in this dynamic environment, as many such initiatives fail as a result of a mismatch between cultural values and digitalization goals [2].

Organizations are becoming more aware of the need to transform their cultures in order to successfully negotiate the challenging landscape of digital transformation. Aligning the corporate culture with the attributes that the contemporary digital landscape demands, such as interactivity, agility, speed, and connectivity, becomes crucial. A lot of organizations were motivated to accelerate their pursuit of digital transformation as a result of the Covid-19 pandemic's onset. That is to say, the onset of the Covid-19 pandemic acted as a catalyst for numerous organizations to intensify their pursuit of digital transformation. In order to support efficient change management in today's changing business environment, organizations accelerated their technological initiatives, implemented significant changes in their processes, and re-evaluated their cultural paradigms [3].

The cultural shift is essential for triumphing in the demanding journey of digital transformation. Companies swiftly advanced their technological endeavors, introduced sweeping process alterations, and revisited their cultural paradigms to facilitate change management in today's evolving business landscape. A fundamental query arises in the midst of this transformative environment: Can the adoption of digital technologies foster a culture that actively supports and encourages digital transformation? Here, the idea of the "culture of change" assumes a prominent role. The degree to which a company is dedicated to improving its processes, technologies, and practices is indicated by its culture of change. A change in culture in the digital sphere makes adopting a new culture with few barriers to producing shared knowledge increasingly possible. However, despite its growing significance in the process of digitalization, organizations frequently struggle to improve their change culture. A report by McKinsey showed that the biggest obstacle to digital transformation is culture. 33% of respondents in a survey of 2135 people said that they had the most cultural and behavioral challenges with digital transformations [4].

Building a strong "culture of change" turns out to be essential to the process of digital transformation. Maintaining core values and fostering employee engagement with little resistance depends heavily on culture of change. In order to survive and remain competitive in this rapidly changing business environment, it is crucial that organizations adopt a culture of change that makes them more willing to implement necessary changes and initiatives effectively and efficiently [5, 6].

Although organizational culture and digital culture are important factors in the digitalization process, there is a significant gap in the study of the components and correlates of "change culture" in the context of digitalization.

Through an examination of the dynamics of "culture of change" in the context of digital transformation, this study seeks to close the gap in the literature. We examine the ways in which digitalization affected by organizational culture as well as the particular channels via which a strong "culture of change" can reduce resistance to change and manage uncertainty.

The primary objectives of this study are to:

- Analyze how the culture of change affects digital transformation
- Determine the best ways to cultivate a culture of change that will enable a successful digital transformation
- Examine the elements and correlates of "change culture" in the context of digitalization to close the gap in the literature.

This study aims to contribute to a broader understanding of how digitalization affects organizational behavior and culture by focusing on these areas. Through this exploration, we aim to provide valuable insights into fostering the culture of change enabling digitalization process that is necessary to maintain competitive advantage in the rapidly evolving business environment.

In the study, after the details of method utilized in this study is clarified, the digital transformation process was explained and the role of the organizational culture in the digitalization process was emphasized. Then, the aspect of cultural transformation in digitalization context was highlighted. Finally, the role of "the culture of change" in the digitalization process was disclosed.

2 Method

Systematic literature review is carried out in this study.

To further explore the relationship between cultural transformation and digitalization, we conducted a bibliometric analysis of the existing literature. A comprehensive review of the Scopus database yielded 382 studies related to the keywords 'culture' and 'digitalization.' This review focused on works published between 2019 and 2024, including articles, books, book chapters, conference papers, conference reviews, and review papers for the evaluation. R program was used in the analysis part. The findings from this analysis are presented in Figs. 1 and 2. Detailed explanations for each figure are provided below the corresponding figures.

The Sankey diagram in Fig. 1, visualizes the evolution of topics in the literature on digital transformation from 2019 to 2024. The flow of topics suggests a consistent focus on digital transformation, with emerging connections to broader societal and economic issues such as the digital economy and cultural heritage. The varying thickness of the connections indicates the changing emphasis and interrelationships between topics over time. The figure indicates that the COVID-19 pandemic's

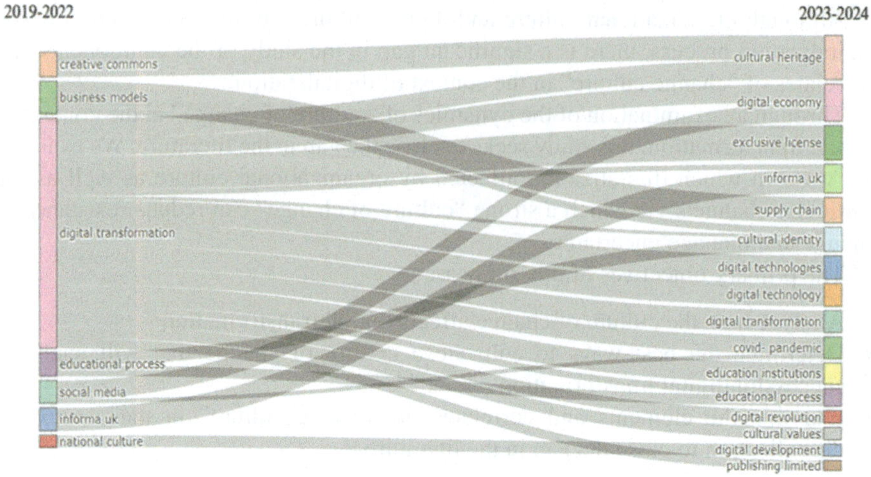

Fig. 1 Thematic evaluation change of literature based on Scopus database

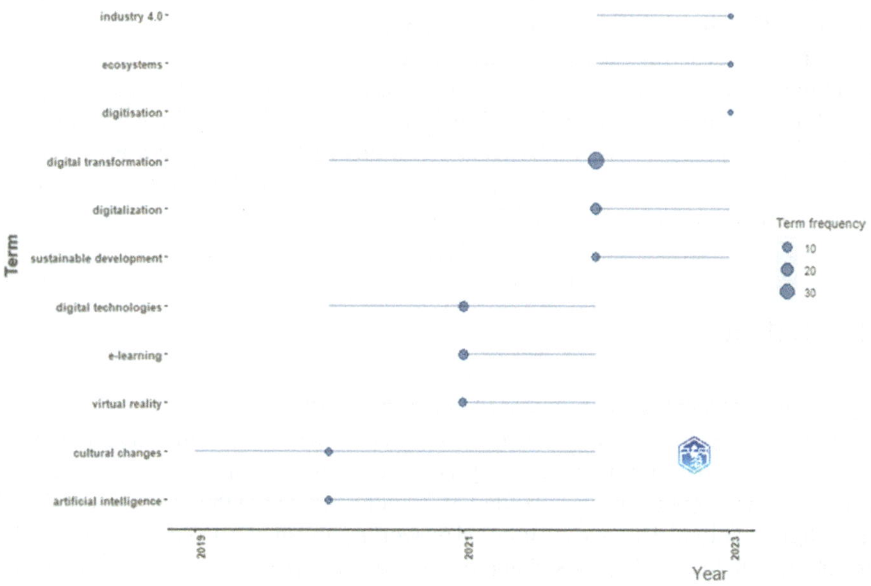

Fig. 2 Trend topics of literature based on Scopus database

continued effects on digital transformation and related fields are reflected in the themes that are present. In the meantime, new concepts like "digital economy" and "cultural heritage" point to a developing conversation that combines digital transformation with more general social and economic issues. Cultural identity is also seen as a very niche area.

Figure 2's chart offers a visual depiction of how, within the context of digital transformation, discussions and emphasis on particular subjects have changed over time, reflecting changes in innovation, focus, and the incorporation of technology into a range of societal and industrial domains. The topic of "digital transformation" appears to have been discussed frequently during the period, peaking in 2022, as illustrated in Fig. 2. A more recent focus on the effects of digital transformation on organizational and societal culture may be indicated by the later years where the phrase "cultural changes" appears. In this context, this study adds the role of culture of change" as a dimension to the digitalization process.

3 Digital Transformation

Digital transformation, which is the use of technology to significantly increase an enterprise's performance or reach, is a hot topic for businesses all over the world. Executives across all sectors are changing customer relationships, internal processes, and value propositions by leveraging digital innovations like analytics, mobility, social media, and smart embedded devices, as well as by enhancing their use of more established technologies [7]. Executives are pushing improvements in customer connections, internal procedures, and value propositions through digital breakthroughs like analytics, mobility, social media, and smart devices, as well as by optimizing existing technology, and this is crucial for businesses all over the world. This transition involves updates to organizational culture, management paradigms, and business procedures in addition to technology and software advancements. To transform data into useful knowledge that is then applied across all aspects of production, operations, and management, digital transformation involves improving technology and software while also updating organizational culture, management models, and business operations. People with the necessary skills, an in-depth knowledge of the company, and the ability to apply information and data effectively are required for this process [8].

The implementation of a flexible management framework to stay competitive, quick adaptation to shifting demand patterns, and a complete reimagining of business operations to embrace digitalization and expand supply chain networks are all included in the multifaceted approach that makes up digital transformation. Changes to job creation and income generation strategies, the implementation of an adaptable management framework to withstand competition, quick response to changing demand patterns, and reimagining a business to digitalize operations and establish wider supply chain connections are all part of the process of digital transformation. It embraces a data-driven management approach to design, production, marketing, and sales [9]. Organizational change brought about and shaped by the adoption of digital technology is known as digital transformation. A shift toward malleable organizational designs that are integrated into and driven by digital business ecosystems is the context for this change [10].

Fig. 3 Lewin's ice cube example for digital-change management (adapted from Pacolli [13])

Additionally, using digital technology has advantages such as lower transaction costs, less information asymmetry, increased information transparency, and increased support for broader social responsibility and governance activities [11]. Investing in digital technologies, developing employee digital skills, and developing digital transformation strategies are three essential elements for companies that will allow them to undergo digital transformation and continue their sustainable development [12].

Lewin's Ice Cube example for digital-change management can be mentioned in this respect (see Fig. 3). His model is known as Unfreeze—Change—Refreeze, which refers to the three-stage process of change that he describes. A new ground is emerging where the elements of the environment in which digital transformation will take place encounter new factors that will enable this transformation.

4 The Role of Organizational Culture in Digital Transformation

Culture represents the foremost factor in upholding core values and ensuring employee engagement with minimal resistance (Fig. 4). Values, beliefs, and underlying assumptions (i.e., required competencies) in the organization can affect the individual resistance for digital transformation. Suppose the presence of certain digital competencies for employees becomes an accepted requirement as organizational or departmental artifacts. Digital skills developed through collaboration and networking are digital knowledge, information management, digital communication, networking, continuous learning, strategic vision, network leadership, customer orientation. In that case, it will be easy to build an organizational culture in the context of digitalization.

The change created by digital transformation also requires a qualified workforce that is open to change. Digital technology requires qualified employees who are adept at using information and data and have a full understanding of the business in order

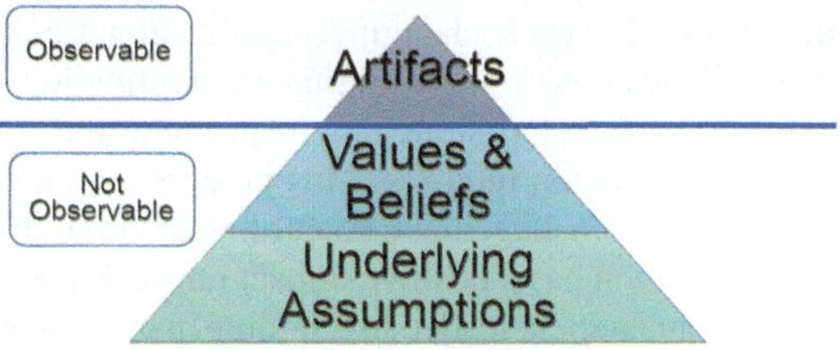

Fig. 4 The iceberg model of culture (Schein 1985) (adapted, Hammerich [14])

Fig. 5 Culture a starting point in digitalization process (adapted from Rambabu and Pilli [16])

to collect and amass knowledge [8]. So that; one of the key barriers to digitalization in services is a lack of qualified employees to develop and provide such services [15]. Creating a culture embarked on by qualified employees and open to change is essential for digital transformation (Fig. 5).

There are many studies in the literature, that briefly emphasize the role of culture in successful digital transformation [17–20]. These studies highlight the significance of creating an environment where staff members are excited about accepting change and are not merely comfortable with it. Such a culture encourages employee ownership and engagement, which eventually results in more successful and long-lasting digital transformation initiatives. A company's organizational culture may need to drastically change in light of digital transformation, which involves implementing new technologies, and developing new skills [15, 21]. After conducting an extensive review of published articles from 1970 to 2016, Morente et al. [22] found that two essential components of a successful organization are culture and innovation. Organizational culture has been found by Odagiu [23] to be a significant factor in maximizing employee performance during transformation processes.

For digital transformation to be successful, organizational culture is crucial. A business must adopt a culture that responds to the adjustments necessitated by digital transformation, fosters innovation, and offers its staff members opportunities for continuous learning. Projects for digital transformation can be more successful and sustainable in this way.

Digital transformation often changes traditional business processes and requires a faster, more flexible, and more data-driven approach. However, these changes may encounter resistance within the organization [24]. This is where organizational culture can cause many employees to embrace or reject digital transformation. If a company's culture is open to change and encourages innovation, employees will be more willing and have an easier time adopting these new digital tools and processes. However, changing organizational culture is not an easy task. Companies that want to initiate digital transformation must first evaluate their current culture and develop strategies to promote the values, behaviors, and beliefs that the transformation requires. This process is more effective with the leadership and support of top management. According to the study of Alieva and Powell [25], digital transformation has a positive impact on work-family balance, psychological capital, senior management leadership, middle management engagement, employee training, corporate social responsibility focus, innovation, knowledge sharing, and employee commitment.

The COVID-19 pandemic has rapidly pushed organizations towards digitalization, and as already mentioned transformation may provide organizations with several advantages. However, the digitalization process needs to be made a long-term strategy rather than just an immediate response. Organizations should use this process to gain a sustainable competitive advantage and be prepared for future uncertainties.

4.1 Cultural Transformation in Digitalization Context

It is commonly assumed that a stronger culture of change within organizations leads to a higher likelihood of continued usage of new digital technologies. Internal and external collaboration, customer centricity, seamless information flow, decision-making at lower organizational levels, responsiveness and quick action, innovation and an entrepreneurial mindset, risk aversion, flexibility, agility, and adaptability, learning and growth orientation, data-driven decisions, a digital-first mindset, openness to unusual career developments, and change openness and readiness are all characteristics of digital culture. Technological trends, in conjunction with the globalization of information and economies, have sparked a transformation aimed at improving organizational effectiveness and utilizing the digital resources at hand in order to change organizational cultures so that they embrace and make use of digital instruments and tools. New paradigms that are developing in the digital world are leveraged as a turning point for digital transformation. It has caused experts to reevaluate how an institution can maintain its validity and long-term sustainability while undergoing social transformation [26, 27].

Considering the observable and not observable layers of organizational culture, cultural components that will provide digital transformation for each layer become important (Fig. 6). For artifacts, visibility of digital process, sharing of best digital practices, available digital equipment gain importance. If open communication, willingness to help, openness to change, trust, cooperation, willingness to take risk, willingness to learn are adopted as values and developed as belief, the organizational culture that will enable digital transformation will emerge more effectively. Underlying assumptions related to digitalization of organizational culture consisting of internal or learned judgments in individual level and departmental level will make the process easier.

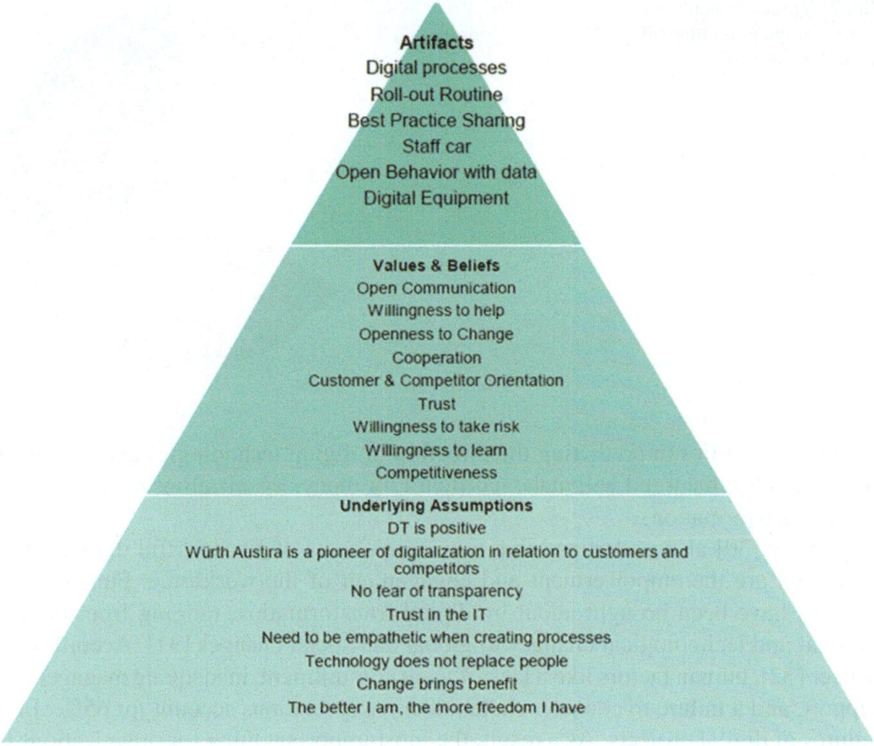

Fig. 6 Digital transformation culture (adapted from Hammerich [14])

4.2 *The Role of "Culture of Change" in the Digitalization Process*

Culture of change signifies the extent to which an organization is dedicated to the continual enhancement of methods, technologies, and practices. Embracing a new culture of participation marked by minimal barriers to generating shared knowledge is attainable through a "culture of change" in the digital sphere. Nevertheless, despite its growing significance in the digitalization journey, organizations seem to face difficulties in enhancing their change culture. Developing a "change culture" is a critical success factor in the digital transformation process.

Creating a culture of change in organizations varies depending on the organization itself. It is easier to establish this change culture in organizations that are open to new approaches, new ideas, and new perspectives, and that understand the importance of change [28]. In general, building a culture of change consists of three elements: leadership, employee inclusion, and communication (Fig. 7).

However, Schwarzmüller et al. [29] concluded that one of the key elements in digital transformation is leadership as a vital duty for transformation. The lack of

Fig. 7 Three elements for culture of change (adapted from Pacolli [13])

leadership abilities in predicting the direction of digital technologies and preparing employees for that trend will make it difficult for many organizations to sustain their digital transformation.

Sainger [30] also underlined that two components of a successful digital transformation are the empowerment and engagement of the workforce. Fundamental changes have been brought about by digital transformation, ranging from organizational and technological changes to global and social changes [31]. According to Scheer [32], human factors like a lack of team commitment, inadequate management support, and a failure to consider stakeholders' expectations account for 65% of the failures of digital projects. As a result, the most important thing for organizations to do is to establish a culture of change.

However, culture of change in organizations varies from organization to organization. When we look at it in general terms; Creating and changing a culture of change in organizations is difficult and time-consuming [33].

There will always be some resistance in any organization where change is occurring; however, when organizational change is imminent, individuals who do not embrace change and the opportunities it brings about will create resistance to upholding the status quo [28, 34]. It is crucial to examine the difficulties involved in promoting such a culture after highlighting the differences in change cultures among organizations and the potential ease with which they can be established in those open to innovation. These difficulties include maintaining the culture over an extended period of time, managing internal resistance, and ensuring leadership commitment.

In order for digitalization to occur successfully, organizations must embrace and adapt to this change. It is important for employees and management not to resist change but to be a part of this process. It is not possible to make digital technologies usable without creating an effective change culture. Digital technologies should be integrated with the culture of change [35].

Companies that don't have a "culture of change" may find it difficult to stay up to date with new technology and eventually lose their competitive advantage. On the other hand, workers who are part of a culture of change are better equipped to accept, understand, and use new technologies, improving the organization's chances of long-term success. Competencies like programming, application management, hardware and software handling, information processing, data organization, communication, teamwork, networking, netiquette, data sharing, cultural awareness, security, legal obligations, and ethical considerations are all challenged by digital technologies [36].

One of the factors limiting the facilitating effect of change culture in the digital change process is national culture. Strong organizational cultures, ample resources for digital projects, the application of strong digital competencies, and information systems management are some of the ways that effective leadership may mitigate the complex effects of national cultures on digital transformation [37].

Non-Information sharing culture and the value of uncertainty avoidance will be the basic building blocks of the culture of change that will support the digitalization process. Employees' insecurity about the future to handle extreme uncertainty can be diminished through designing a change culture. Given that digital technologies are best described as a continuous state of change, influenced by occasional bursts of transformation they can introduce uncertainties into organizations. Therefore, it becomes crucial to gain insights into the nature of these uncertainties and how best to manage them. Organizations should redesign the change culture to handle extreme uncertainty in the context of digitalization.

These low-level of cultural layers, which can change even among departments, can limit the facilitating role of digital transformation. If there is a corporate culture that develops with the understanding that digital skills are necessary for employees, it will be easier to build a culture of change in the context of digitalization.

Critical success factors can be shown at Fig. 8. They should be considered for building an efficient culture of change enabling digitalization process.

5 Conclusion

Considering the literature on digitalization, it can be seen that the importance of organizational culture and digital culture in the digitalization process are examined. The explanation of the elements and correlates of "change culture" in the digitalization process are missing in the literature. In this study, what affects the culture of change in the context of digitalization and the effective elements in building the culture of change to handle extreme uncertainty are explained in detailed.

To sum up, a culture of change plays a key role in coping with the uncertainty that will arise during the adaptation process of new technologies. A culture like this promotes adaptability, innovation, and resilience in the face of unanticipated challenges. Accepting change not only makes it easier to integrate new technologies, but it also puts organizations in a position to succeed in a digital environment that is constantly changing, promoting growth and competitive advantage. The change culture greatly affects the ability of organizations to adapt to technological changes. The advantages of a culture of change for organizations arise when

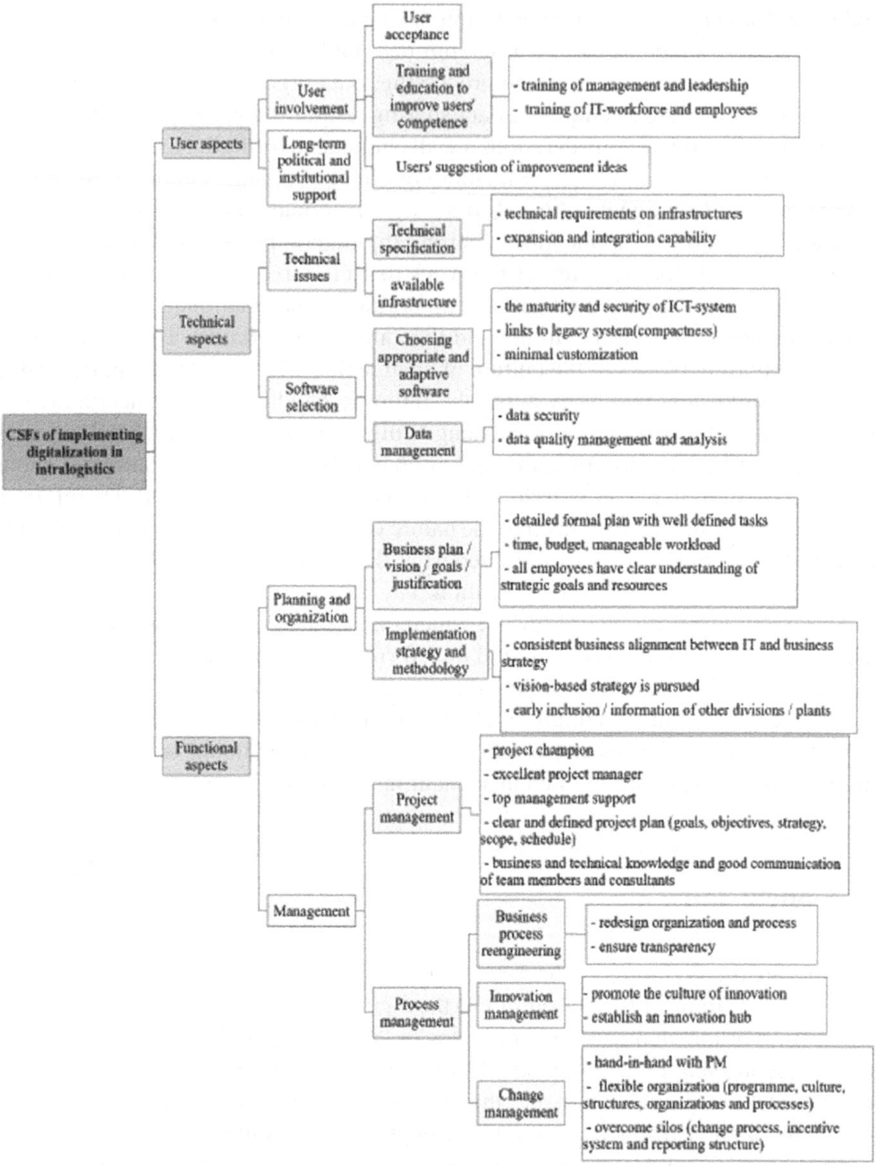

Fig. 8 Identified critical factor that enables culture of change enabling digitalization process [38]

employees increase their openness to changes, learning abilities, flexibility, and risk-taking abilities. Additionally, a culture of change supports the more effective use of new technologies by encouraging open communication and collaboration.

As a result, organizations must embrace a change culture to keep up with technological developments and achieve sustainable success. Leaders need to foster this culture, provide employees with training and development opportunities, and encourage open communication. However, it may take time and careful planning for a culture of change to seep into within the organization. However, this effort helps organizations adapt to technological changes more quickly and effectively and contributes to maintaining their competitive advantage in the long term.

References

1. Soluk J, Kammerlander N (2021) Digital transformation in family-owned Mittelstand firms: a dynamic capabilities perspective. Eur J Inf Syst 30(6):676–711. https://doi.org/10.1080/0960085X.2020.1857666
2. Cracking the culture code for successful digital transformation. Accessed 4 Oct 2023. [Online]. https://sloanreview.mit.edu/article/cracking-the-culture-code-for-successful-digital-transformation/
3. Cordes C, Richerson PJ, Schwesinger G (2010) How corporate cultures coevolve with the business environment: the case of firm growth crises and industry evolution. J Econ Behav Organ 76(3):465–480. https://doi.org/10.1016/J.JEBO.2010.09.010
4. Manyika J, Lund S, Bughin J (2016) Digital globalization: the new era global flows. McKinsey Global Institute
5. DC D, Griffiths A, Benn S (2007) Organizational change for corporate sustainability. A guide for leaders and change agents of the future
6. Omar A, Almaghthawi A (2020) Towards an integrated model of data governance and integration for the implementation of digital transformation processes in the Saudi Universities. Int J Adv Comput Sci Appl 11(8):588–593. https://doi.org/10.14569/IJACSA.2020.0110873
7. Westerman G, Calméjane C, Bonnet D, Ferraris P, McAfee A (2011) Digital transformation: a roadmap for billion-dollar organizations. MIT Center for digital business and Capgemini consulting, vol 1, pp 1–68
8. Dou B, Guo S, Chang X, Wang Y (2023) Corporate digital transformation and labor structure upgrading. Int Rev Financ Anal 90:102904. https://doi.org/10.1016/j.irfa.2023.102904
9. Schallmo D, Williams C, Boardman L (2017) Digital transformation of business models—best practice, enablers, and roadmap. Int J Innov Manag 21:1740014. https://doi.org/10.1142/S136391961740014X
10. Hanelt A, Bohnsack R, Marz D, Antunes Marante C (2021) A systematic review of the literature on digital transformation: insights and implications for strategy and organizational change. J Manage Stud 58(5).1159–1197. https://doi.org/10.1111/joms.12639
11. Zhao X, Cai L (2023) Digital transformation and corporate ESG: evidence from China. Financ Res Lett 58:104310. https://doi.org/10.1016/j.frl.2023.104310
12. Teng X, Wu Z, Yang F (2022) Research on the relationship between digital transformation and performance of SMEs. Sustainability (Switzerland) 14(10). https://doi.org/10.3390/su14106012
13. Pacolli M (2022) Importance of change management in digital transformation sustainability. IFAC-PapersOnLine 55(39):276–280
14. Hammerich L (2021) The digital transformation: the role of culture in the DT change process
15. Peillon S, Dubruc N (2019) Barriers to digital servitization in French manufacturing SMEs. Proc CIRP 146–150. Elsevier B.V. https://doi.org/10.1016/j.procir.2019.04.008
16. Rambabu A, Pilli K (2018) "Revolutionizing organisational culture for the digital economy," in two day national seminar GST and digital economy-implications on trade and commerce, p 124
17. Duerr German S, Holotiuk F, Beimborn D, Wagner German H-T, Weitzel T, What is digital organizational culture? Insights from exploratory Case studies. [Online]. http://hdl.handle.net/10125/50529

18. Yu J, Moon T (2021) Impact of digital strategic orientation on organizational performance through digital competence. Sustainability (Switzerland) 13(17). https://doi.org/10.3390/su13179766
19. Busca L, Bertrandias L (2020) A framework for digital marketing research: investigating the four cultural eras of digital marketing. J Interact Mark 49(1):1–19. https://doi.org/10.1016/j.intmar.2019.08.002
20. Robinson K (2008) The other climate crisis: digital culture, demography and education. Media Int Aust 128(1):49–58. https://doi.org/10.1177/1329878X0812800107
21. Mittal S, Khan MA, Romero D, Wuest T (2018) A critical review of smart manufacturing & Industry 4.0 maturity models: implications for small and medium-sized enterprises (SMEs). J Manuf Syst 49:194–214. Elsevier B.V. https://doi.org/10.1016/j.jmsy.2018.10.005
22. Morente F, Ferràs X, Zizlavsky O (2018) Innovation cultural models: review and proposal for next steps. Revista Universidad y Empresa 20(34):53–81
23. Odagiu C (2013) Organizational culture, change management and performance. In: Managerial challenges of the contemporary society, no. 6
24. Rusu L, Balasuriya PBL, Bah O (2020) Cultural barriers in digital transformation in a public organization: a case study of a Sri-Lankan organization. In: Information systems: 17th European, Mediterranean, and Middle Eastern conference, EMCIS 2020, Dubai, United Arab Emirates, Nov 25–26, 2020, Proceedings 17. Springer, pp 640–656
25. Alieva J, Powell DJ (2023) The significance of employee behaviours and soft management practices to avoid digital waste during a digital transformation. Int J Lean Six Sigma 14(1):1–32. https://doi.org/10.1108/IJLSS-07-2021-0127
26. Branch J, Burgos D, Arango-Serna M, Ortega G (2020) Digital transformation in higher education institutions: between myth and reality, pp 41–50. https://doi.org/10.1007/978-981-15-4952-6_3
27. Firican DA (2023) Digital transformation and digital culture: a literature review of the digital cultural attributes to enable digital transformation. Proc Int Conf Bus Excellence 17(1):791–799. https://doi.org/10.2478/picbe-2023-0073
28. Aktan CC, Yay S (2016) Organizasyonel değişimde iki farkli kültür: değişim kültürü ve değişime direnç kültürü", Organizasyon ve Yönetim Bilimleri Dergisi, c 8, sy 2
29. Schwarzmüller T, Brosi P, Duman D, Welpe IM (2018) How does the digital transformation affect organizations? Key themes of change in work design and leadership. Manage Revue 29(2):114–138 [Online]. https://www.jstor.org/stable/26491473
30. Sainger G (2018) Leadership in digital age: a study on the role of leader in this era of digital transformation. Int J Leadersh 6(1):1
31. Reis J, Amorim M, Melão N, Matos P (2018) Digital transformation: a literature review and guidelines for future research. In: Rocha Á, Adeli H, Reis LP, Costanzo S (eds) Trends and advances in information systems and technologies. Springer International Publishing, Cham, pp 411–421
32. Scheer V (2016) Process improvement for implementing and sustaining change in healthcare. Diss. The College of St. Scholastica
33. Jauch LR, Glueck WF (1988) Business policy and strategic management. New York McGraw-Hill Book Company, (5th. Edition). - References - Scientific Research Publishing. Accessed 16 Oct 2023. [Online]. https://www.scirp.org/(S(i43dyn45te-exjx455qlt3d2q))/reference/referencespapers.aspx?referenceid=136373
34. Pearce JA, Robinson RB (1989) Management. McGraw-Hill Ryerson, Limited [Online]. https://books.google.com.tr/books?id=S6GsAAAACAAJ
35. Terzioğlu F (2020) COVID-19 sürecinde sağlik sistemlerinin zorluklarin üstesinden gelebilme kapasitesinin geliştirilmesi ve hemşireler ile diğer sağlik profesyonellerinin güçlendirilmesi. Turkish J Pediatr Dis 76–83. https://doi.org/10.12956/tchd.733123
36. Oberländer M, Beinicke A, Bipp T (2020) Digital competencies: a review of the literature and applications in the workplace. Comput Educ 146. [Online]. https://api.semanticscholar.org/CorpusID:209063012

37. Alshammari K, Alshammari A (2023) Managing digital transformation in a global environment: the role of national culture. Revista Amazonia Investiga 12:112–125. https://doi.org/10.34069/AI/2023.68.08.11
38. Winkler H, Zinsmeister L (2019) Trends in digitalization of intralogistics and the critical success factors of its implementation. Braz J Oper Prod Manage 16(3):537–549. https://doi.org/10.14488/BJOPM.2019.v16.n3.a15

Detecting Gait Disorders Using Machine Learning Analysis Based on Kinect and Smartwatch Data

Süheyla Uygur, Zeynep Öztürk, Murat Emeç, and Mehmet Hilal Özcanhan

Abstract Gait disorders are significant health issues arising from neurological conditions or accidents. Evaluating and monitoring these disorders objectively using traditional methods is challenging. Therefore, advanced technological tools such as the Kinect v2 Xbox One camera sensor and a smartwatch, supported by Machine Learning methods, can provide doctors with more in-depth gait analyses. This study recorded subjects' lower and upper extremity angles while walking for a specific duration using two Kinect camera sensors. Data from a smartwatch complement the extremity angular data. The data is processed using supervised Machine Learning techniques. The models obtained by using different Machine Learning algorithms are compared. The final two models provide objective insight into gait disorders, replacing the traditional subjective observational therapies. As such, our research presents a novel tool for detecting gait disorders. Further, clinical tests are needed for a final model suitable for the real world.

Keywords Kinect sensor · Gait disorder · Machine learning

S. Uygur · Z. Öztürk · M. H. Özcanhan
Computer Engineering Department, Dokuz Eylül University, Izmir, Turkey
e-mail: suheyla.uygur@deu.edu.tr

Z. Öztürk
e-mail: zeynep.ozturk@ceng.deu.edu.tr

M. H. Özcanhan
e-mail: mehmet.ozcanhan@deu.edu.tr

M. Emeç (✉)
Informatics Department, Istanbul University, Istanbul, Turkey
e-mail: murat.emec@istanbul.edu.tr

F. Çebi et al. (eds.), *Engineering and Technology Management in Challenging Times*, Lecture Notes in Networks and Systems 1137,
https://doi.org/10.1007/978-3-031-70935-7_14

1 Introduction

In medicine, abnormally performed or out-of-control walking is defined as a gait disorder. Gait disorders (GD) occur because of genetic factors, diseases, or injuries. Patients with GD require regular physical therapy and medical treatment because they can cause problems in the muscle, bone, and nerve structures of the legs over time [1]. Gait disorders are among the most common issues in injured patients or those with neurological disorders. Inadequacies such as lack of coordination and muscle weakness cause gait with a typically shorter stride length and an increased double support phase. Fatigue caused by GD adversely affects a patient's maximum walking distance and time. Therefore, physicians evaluate patients walking performance using gait assessment tools or function tests. The reliability of such evaluations is subjective and depends on examiners' experience. In recent years, computer-assisted gait analysis systems have helped assess functional alterations in gait patterns [2].

Our work aimed to develop Artificial Intelligence (AI) based software to detect gait disturbances. Our work includes all patient groups with dysfunction in the lower extremities, that is, individuals with GD. In our tests, the 16 extremity angles, except the finger and toe angles, were recorded using two cameras. The recorded images were inputted into an AI model for evaluation. The outcome of our AI model is an assistive tool for physiotherapists to evaluate gait disturbances in patients.

In some cases, our tool has the potential to detect minor defects or minor improvements that cannot be observed. The main hardware tool used to obtain the extremity joint angles was the Kinect V2 Xbox One (TM Microsoft) camera sensor. Complementary data were obtained from a smartwatch.

2 Related Works

Various studies have been conducted on GD using Kinect. Below is an account of GD studies using the Kinect and early AI methods. One of the first alternatives to physical therapy was proposed in the work [3]. A total of 50 patients were tested for improvement using telerehabilitation. The results of this study showed that a telerehabilitation program based on a virtual reality system provided therapeutic function in individuals with Multiple Sclerosis (MS). In another tele-rehabilitation study, Kinect was also used to measure the physical performance of people with MS [4].

Other researchers performed gait analysis in patients with paralysis. The spatial and kinematic walking parameters were estimated using a Kinect-based system, a low-cost alternative to full-hospital gait assessments [5]. Another low-cost design using the Kinect was proposed in work [6], where the positions of the body joints were calculated.

A real-time 3-D analysis of anatomical position was performed in the work [7]. Forward reach, lateral reach and one-legged standing eyes-closed balance were tested. The test results of 20 healthy subjects revealed that Kinect can be useful in validating kinematic strategies. The concurrent validity of Kinect in measuring gait parameters was tested favorably in the work [8]. Another study investigated the validity of Kinect's performance in gait analysis [9]. The work showed that the pelvis displacement and other gait parameters can be evaluated using Kinect, while ankle and knee joint analyses are limited.

Studies have used Kinect on step length, duration, cadence, and peak foot-swing velocity parameters. The effect of Kinect's viewing angle was reduced in work [10]. The distribution of data obtained from an expert system called OptiTrack was matched to that obtained from Kinect. The analysis results of the two systems agreed on the significant difference between the parameters of the young and older subjects. In another similar study, the measurements of a marker-based 3-D Motion Analysis system were compared to those of a Kinect-based system. The study concluded that the low-cost Kinect can be used in clinics [11].

The reliability of Kinect in evaluating a subject's performance after a stroke was also studied [12]. The results showed that the variables obtained by Kinect were reliable. Having observed the positive reports about the reliability of Kinect in gait evaluation, researchers started merging Kinect-based work with AI. In work [13], a human gait recognition method was proposed. The study used static (joint distances) and dynamic (joint angles) gait features. Two AI methods, k-Nearest Neighbor (KNN) and Long Short-Term Memory (LSTM), were used to classify subject walking parameters. The depth data from Kinect analyzed the gait cycle for both feet. The KNN and LSTM models identified the gaits of 20 subjects with a success rate of 100% and 98.33%, respectively. In another study, the gait cycle data from Kinect was used to train a Random Forest (RF) and a Multi-Layer Perceptron (MLP) model [14]. The RF and MLP models' accuracies were 81.8% and 87.8%, respectively. Model training has been reported to be faster in RF.

Another interesting external tool has been used together with ML for gait disorder detection [15]. A walking aid (elbow crutch) customized with a sensor tip sends data to an ML model. Unfortunately, since low-gait disorder patients do not use walking aids, using the crutch while testing may cause misconceptions. The disadvantage of another Kinect-based ML sensor merged work is the repeated calibration of the test set every time a new subject is tested [16]. Frequently repeated calibrations can be time-consuming and cumbersome.

3 Material and Methods

3.1 Materials

Two Kinect V2 Xbox One camera sensors are the primary tools for obtaining the subject' walking angles (Anterior_Kinect and Lateral_Kinect). The second source of complementary personal data comes from a Smartwatch. The choice of input devices follows previous scientific studies, where, although originally designed for the entertainment industry, Kinect can also monitor human movements. Microsoft Visual Studio and C# were used to record joint angles .Net technology. The tool used for storing data was MSSQL, which forms the input block for the AI engine. The Windows Presentation Foundation (WPF) was designed. WPF is an Extensible Application Markup Language (XAML)-based system that creates a graphics 'infrastructure .Net framework'. Two of the indispensable libraries used were as follows

(1) LightBuzz Vitruvius: A set of easy-to-use Kinect utilities. WPF and Windows stores
(2) System.Data.SqlClient: Provides data to the SQL Server.

Our work's main material is raw data coming from three different sources. Sample manually-entered data before feature selection can be seen in Table 1a. The manually-entered data contain the basic profile information of the subjects. Each data row continues with Kinect-recorded extremity angles. The extremity angles are followed by labeling the subject in the furthest right column of the table. A '1' label means a subject with GD, and a '0' implies a subject without GD. The table continues further with the raw data of subjects' distance walked, calories spent, and heartbeat.

Table 2 shows sample Kinect data after feature creation and selection. Feature selection is further detailed in the next Section.

3.2 Methods

An application software was developed to manually input the profile data and determine the new subjects' gait anomalies. On the first screen page of the new subject software, the user is welcomed, and the subject is registered. The second screen of post-AI analysis is shown in Fig. 1. The numerical information of the registered subjects was entered manually.

In our analysis, the Python programming language and its libraries were used (detailed in the previous Section). First, data preconditioning was performed as the first step of AI analysis. Data was preconditioned by eliminating null and outlier data. Then, the data was normalized. In the next step, the correlation of the original data to the output (GD prediction) was analyzed. The resulting correlation matrix is presented in the next Section. However, the two most appropriate feature extraction methods, Principal Component Analysis (PCA) and Linear Discriminant Analysis (LCA), were compared. The K-fold cross-validation technique was used because feature extraction methods depend on the specific characteristics of the data set used. Another analysis, SHapley Additive ExPlanations (SHAP), was performed to determine the features that greatly impacted the prediction outcome. SHAP shows the degree of importance of each feature on the model's predictions after considering all possible combinations of features [17]. The SHAP results are also in the next Section. The main features extracted from the raw data are the Minimum, Maximum, Mean,

Table 1 Sample raw data before feature selection

Gender	Weight	Height	Age	Heart_rate	Distance	Steps	Calories
1	65	160	27	90.250000	991.250000	1419.000000	28.000000
0	55	166	23	88.666667	1081.666667	1570.333333	30.000000
1	50	160	24	89.333333	1148.333333	1679.000000	32.000000
1	68	175	59	94.000000	1200.000000	1732.000000	33.000000
0	77	172	48	94.333333	1226.000000	1772.333333	34.333333

(a) *Manually-entered raw data*

spinBaseAnteriorMeanDiff	spinBaseAnteriorMinDiff	…	leftPelvisAnteriorMaxDiff	leftPelvisAnteriorP2PDiff	leftPelvisAnteriorRmsDiff	rightPelvisAnteriorMeanDiff
−0.002086	0.609786	…	−0.367278	−0.309586	−0.110199	−0.486200
0.169383	0.677636	…	−0.225870	−0.206176	0.054964	0.200558
−0.204695	0.210286	…	−0.358215	−0.194654	0.000000	−0.279622
−0.025274	0.596284	…	−0.027907	−0.013672	−0.082343	−0.215038
0.790668	0.805727	…	0.039283	−0.225500	−0.223147	−0.215617

rightPelvisAnteriorMinDiff	rightPelvisAnteriorStdDiff	rightPelvisAnteriorMaxDiff	rightPelvisAnteriorP2PDiff	rightPelvisAnteriorRmsDiff	Disorder
−0.225369	−0.631567	−0.605988	−0.376190	−0.460480	1.0
0.259462	−0.199856	−0.090082	−0.278185	0.149534	0.0
−0.026973	−0.261879	−0.251853	−0.248148	−0.109850	0.0
−0.161710	0.037646	−0.033373	−0.011283	−0.155294	0.0
0.264294	−0.546866	−0.500368	−0.603027	−0.435961	0.0

(b) *Measured angle data from Kinect*

Table 2 Sample data after feature selection

leftPelvisAnteriorP2PDiff	leftPelvisAnteriorStdDiff	leftPelvisAnteriorRmsDiff	rightHipAnteriorP2PDiff
−0.309586	−0.211654	−0.110199	−0.172923
−0.206176	−0.104742	0.054964	−0.324704
−0.194654	−0.099231	0.000000	0.182944
−0.013672	0.008134	−0.082343	−0.430386
−0.225500	−0.261583	−0.223147	−0.081483
leftPelvisAnteriorMeanDiff	leftPelvisAnteriorMinDiff	rightHipAnteriorMinDiff	Disorder
−0.029632	0.091075	0.064236	1
0.154563	0.031130	−0.280941	0
−0.163593	−0.042620	−0.574115	0
0.015910	−0.064253	−0.069622	0
0.159604	0.272777	−0.061639	0

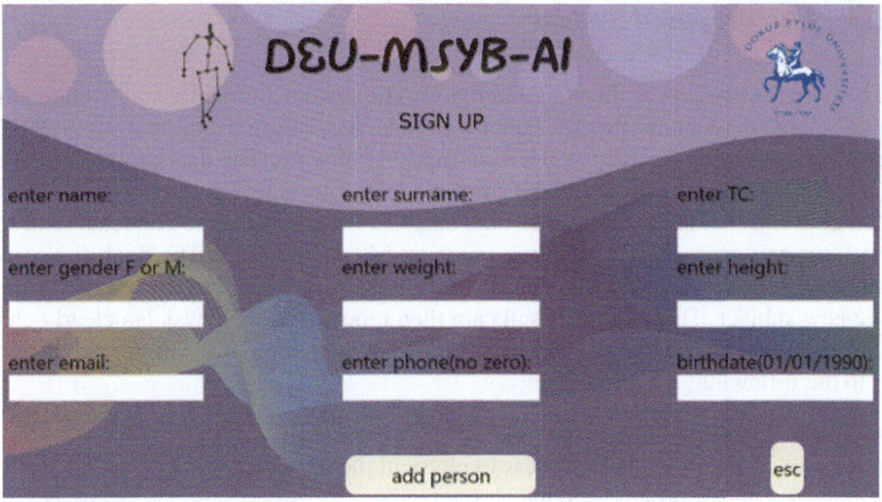

Fig. 1 The main user interface screen

Standard Deviation, Root Mean Square (RMS), and Peak-to-Peak statistical values of each angle and the smartwatch data. In total, 201 features were selected for AI analysis.

The AI analysis constitutes the major bulk of our work. In our AI analysis, supervised ML techniques were used due to the structure of our collected data. ML is the most mature area of AI analysis. Supervised learning is a subcategory of ML. Supervised ML problems can be divided into two groups: regression and classification. In regression, the outcome is predicted. Some supervised ML algorithms are the Random Forest classifier, Support Vector Machine, Decision Tree (DT), XGBoost, Logistic Regression, and K-Nearest Neighbors. In our analysis, the performance of the above algorithms was evaluated according to their accuracy, precision and recall scores. The RF and DT algorithms were the top scorer algorithms.

A total of 21 subjects participated in providing data to our model in non-medical, digital laboratory conditions. Due to the limited number of volunteering subjects, some subjects were asked to repeat the first phase. The most widespread Synthetic Minority Oversampling Technique (SMOTE) was applied to increase the dataset size beyond 5000 instances and improve the success rate of our model. The preconditioned and labeled data set was used to train the target ML algorithms to predict the GD outcomes accurately. As a rule of ML, 80% of the data were defined as the training set and the remaining 20% as the test dataset.

3.3 The Proposed System

The proposed system is illustrated in Fig. 2. The system's modular architecture has three main components: the data capture module, the collected data analysis module, and the new subject diagnosis decision support software. The data capture module collects, organizes, and stores the subject's extremity angle data from the Kinect cameras and the smartwatch. The data analysis module uses the collected data to perform gait analysis. The analysis trains an ML Gait Model. The final diagnosis software registers new subjects. Next, the new gait data is analyzed to detect GD in the new subject. The analysis results are then reported to the physician clearly and intuitively through the user interface. The results help the physician to gain insight into the following:

- Walking abnormalities of new subjects,
- Observation of ongoing progress in physical therapy,
- Sources of gait problems in the new subjects.

The angle measurements were carried out using the connected Kinect cameras. The angles were calculated using the Vitruvius library, which uses the body nodes' x, y, and z coordinates. The computed angles were saved in a database. The stored angles were the Spin Base, Left Hip, Left Knee, Left Ankle, Right Hip, Right Knee, Right Ankle, Neck, Spine Shoulder, Spine Mid, Left Shoulder, Left Elbow, Right

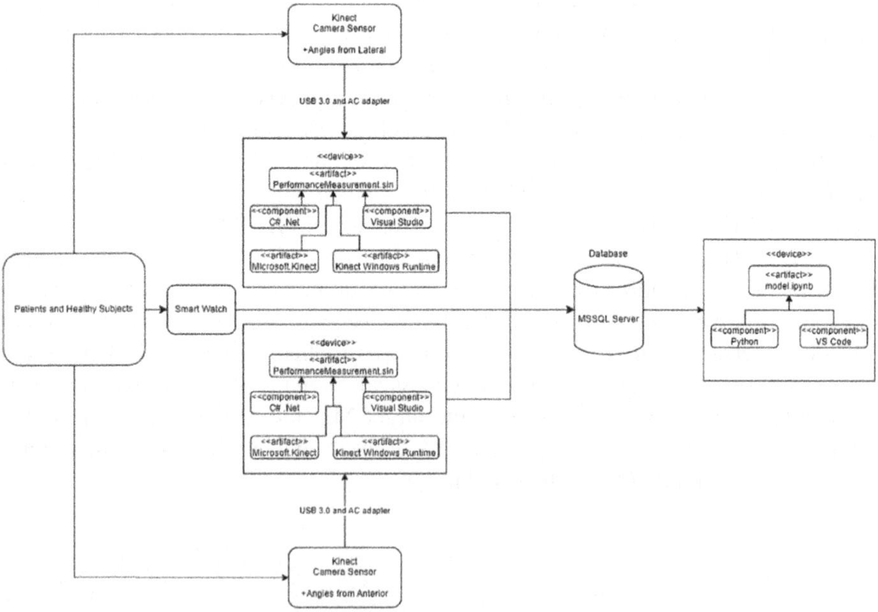

Fig. 2 Deployment diagram of the system

(a) Detecting extremity angles of subjects.

(b) Subjects imitating gait disorder with smartwatch on.

Fig. 3 The angle views with Kinect Xbox One v2 camera sensor

Shoulder, Right Elbow, Left Pelvis, and Right Pelvis. Two instances of our tests are shown in Fig. 3.

Figure 3a shows the detection of extremity angles using Kinect and our custom software. Figure 3b shows a subject wearing a smartwatch and imitating a gait disorder. The smartwatch data is collected using Bluetooth communication between the smartwatch and our notebook. Later, the Kinect and smartwatch data are merged using the computer system time to obtain synchronized subject data. The system is ready to accept new subject data after the best-performing ML model is trained and tested.

4 Experimentations and Results

A new subject is registered, and personal data is entered manually using the developed software. The subject puts on the smartwatch on the non-preferred wrist. Both the anterior and lateral Kinects are turned on. Next, the subject is asked to walk a short distance. The extremity angles and smartwatch data are captured and stored in the database. Then, the collected data is entered automatically into the developed ML model. The predicted result is reported on the screen and stored in the database. The reported result is now ready for the physician's opinion.

4.1 Results

The correlation matrix of the angular differences was examined to determine the features highly correlated to the disorder output. The matrix is shown in Fig. 4. As a rule of statistics, Pearson Correlation values above 0.70 indicated a high correlation

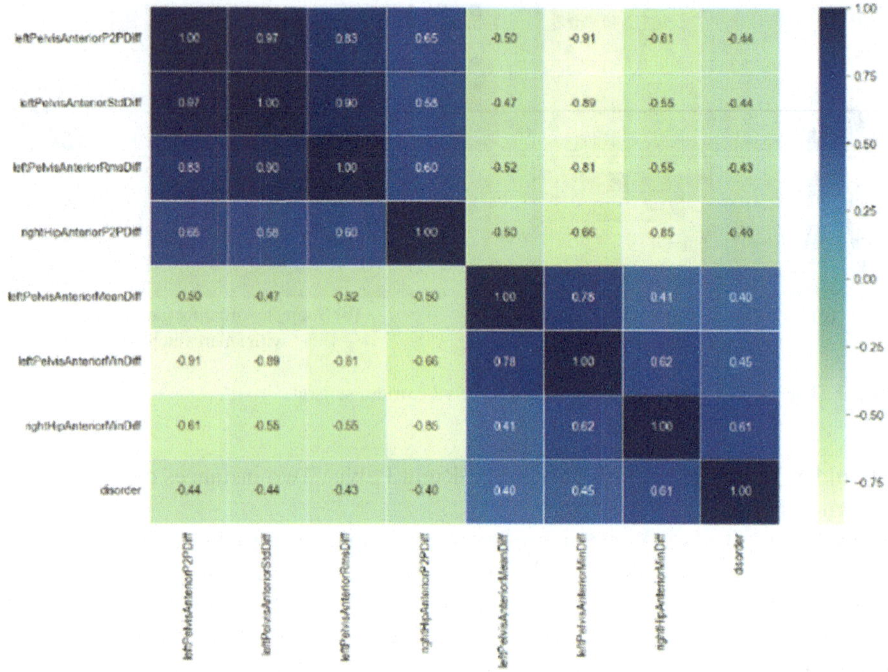

Fig. 4 Correlation matrix of best features

between the two parameters of the matrix. In our work, we accepted values greater than 0.85 as very highly correlated features.

The matrix shows a medium correlation between the Left Pelvis Anterior P2P Difference feature. The rest of the correlations are low. However, some features have been observed to have very high correlations with each other.

With the SHAP diagrams shown in Fig. 5, it can be seen that the two most important features are the rightHipMin and leftPelvisStd angles. However, the effect of manually entered personal and smartwatch data has a very low impact on the model output. The low impact factor of the smartwatch data is discussed in the next Section.

The accuracy, F1 score, recall, and precision performances are presented in Table 3. The performances of the merged Kinect-smartwatch and Kinect-only data are the same. This result is discussed in the next Section. The top performers are the RF and DT algorithms. The RF model achieves 95.00% accuracy, 94.90% F1 score, 95.00% recall, and 95.36% precision. Meanwhile, the DT model achieves 90.00% accuracy, 89.52% F1 score, 90.00% recall, and 91.33% precision. The "Random Forest" model has higher accuracy, F1 score, recall, and precision values, making it a more reliable classification model. These results show that the RF algorithm outperforms the DT model, suggesting it is the preferable algorithm on this dataset. However, it has to be pointed out that the data was balanced using an oversampling technique, and the subjects were not actual patients.

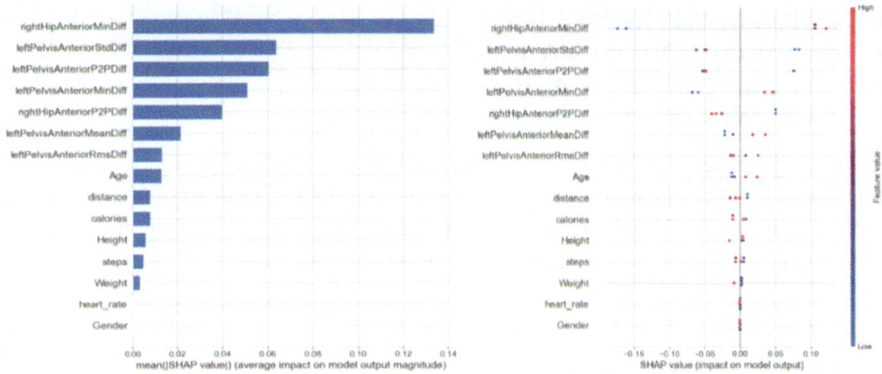

Fig. 5 The SHAP diagrams

Table 3 Performance results of the ML algorithms

Machine learning algorithm(s)	Accuracy	F1 score	Recall	Precision
Random forest	95.00	94.90	95.00	95.36
Decision tree	90.00	89.52	90.00	91.33

The classification report of the RF models is shown in Fig. 6. In this case, Class 0 represents the class with no GD, and Class 1 represents the class with a GD. It can be observed that there are no instances of mislabeling.

The statistical results obtained from the above confusion matrix are as follows:

- True Positive (TP) for Class 0: 300

Fig. 6 Confusion matrix of Random Forest models for smote oversampled dataset

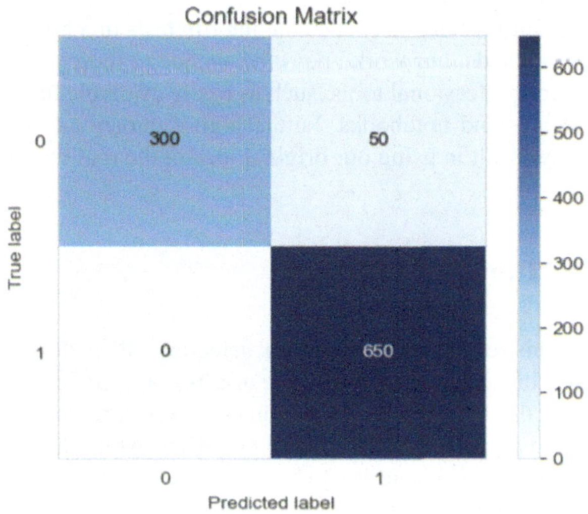

- True Positive (TP) for Class 1: 650
- True Negative (TN) for Class 0: 650
- True Negative (TN) for Class 1: 300
- False Negative (FN) for both classes: 50.

The high True Positive (TP) value indicates that our model correctly identifies a significant number of Class 0 and Class 1 instances. The low number of false negatives (FN) suggests some false alarms. The outcome also indicates that the model could accurately detect and classify mostly GD (Class 1) instances.

5 Future and Discussion

The presented research focuses on detecting GD in patients using two Kinect camera sensors and a smartwatch, supported by ML techniques. The model testing results demonstrate satisfactory performances. The RF algorithm could identify GD with high accuracy. The high performance indicates that ML models can learn the patterns and features of GD and prove to be a reliable predictor for physicians.

However, it is important to acknowledge the limitations of this study. Our work lacks real tests conducted under physician supervision. The research was conducted using a relatively limited-size dataset, and the application of the SMOTE technique addressed the imbalanced data issue. Further validation on larger and more diverse datasets would be beneficial. Additionally, the subjects were not tested in real clinical tests. The subjects consisted of normal people and non-patient gait disorder imitators. Secondly, the subject testing time was very short. Although the high-impact extremity angles can be obtained in a very short time, the impact of the smartwatch data can surface only after a very long tracking period. In other words, there is a skewness in the two merged data sets.

Our present study runs the interim tests in a non-clinical environment. However, as it stands, our work opens the avenue to more sophisticated future research using more professional tools, such as easily available high-resolution cameras on mobile phones and notebooks. Future high-accuracy and low-error clinically tested work may result in using our original idea in the real world.

6 Conclusion

In this research, a method for detecting GD is developed using two Kinect camera sensors and ML methods. The results obtained from the model testing phase showed that the Random Forest model achieved 95.00% accuracy and 94.90% F1 score, indicating high accuracy and precision, while the Decision Tree model performed slightly lower with 90.00% accuracy and 89.52% F1 score. However, both models achieved an accuracy and F1 score above 90%, indicating that both algorithms can effectively detect GD. The RF model has higher performance, suggesting a more reliable prediction. In conclusion, the results indicate that ML can be used to detect GD.

The outcome of this research has significant implications for GD detection and tracking. The high performance achieved by the trained models indicates their potential as valuable tools for healthcare professionals. However, our study has limitations. Therefore, further validation on larger and more diverse datasets is needed before clinical settings.

References

1. Baker JM (2018) Gait disorders. Am J Med 131(6). https://doi.org/10.1016/j.amjmed.2017.11.051
2. Güner S, Alsancak S (2011) Multiple Skleroz ve Yürüyüş. Ankara Sağlık Hizmetleri Dergisi
3. Ortiz-Gutiérrez R, Cano-de-la-Cuerda R, Galán-del-Río F, Alguacil-Diego IM, Palacios-Ceña D, Miangolarra-Page JC (2013) A telerehabilitation program improves postural control in multiple sclerosis patients. A Spanish preliminary study. Int J Environ Res Public Health
4. Naser I (2021) Designing virtual-augmented reality-based testing and rehabilitation software for people with multiple sclerosis. Thesis
5. Latorre J, Colomer C, Alcañiz M (2019) Gait analysis with the Kinect v2: a normative study with healthy individuals and comprehensive study of its sensitivity, validity, and reliability in individuals with stroke. J NeuroEng Rehabil 16. https://doi.org/10.1186/s12984-019-0568-y
6. Sosa GD, Sanchez J, Franco H (2015) Improved front-view tracking of the human skeleton from Kinect data for rehabilitation support in Multiple Sclerosis. In: IEEE conferences
7. Clark RA, Pua Y, Fortin K, Ritchie C, Webster KE, Denehy L, Bryant AL (2012) Validity of the Microsoft Kinect for assessment of postural control. Gait Posture. SciVerse ScienceDirect
8. Dolatabadi E, Taati B, Mihailidis A (2016) Concurrent validity of the Microsoft Kinect for windows v2 for measuring spatiotemporal gait parameters. Med Eng Phys 38(2016):952–958
9. Eltoukhy M, Oh J, Kuenze C, Signorile J (2017) Improved kinect-based spatiotemporal and kinematic treadmill gait assessment. Gait Posture 51(2017):77–83
10. Dubois A, Bresciani JP (2018) Validation of an ambient system for the measurement of gait parameters. J Biomech 69(2018):175–180
11. Clark RA, Bower KJ, Mentiplay BF, Paterson K, Pua YH (2013) Concurrent validity of the Microsoft Kinect for assessment of spatiotemporal gait variables. J Biomech 46(2013):2722–2725
12. Clark RA, Vernon S, Mentiplay BF, Miller KJ, McGinley JL, Pua YH, Paterson K, Bower KJ (2015) Instrumenting gait assessment using the Kinect in people living with stroke: reliability and association with balance tests. Open Access Article, licensee BioMed Central
13. Kim W, Kim Y, Lee KY (2020) Human gait recognition based on integrated gait features using Kinect depth cameras. In: 2020 IEEE 44th annual computers, software, and applications conference (COMPSAC), pp 328–333. https://doi.org/10.1109/COMPSAC48688.2020.0-225
14. Ince OF, Ince IF, Park JS, Song JK (2017) Gait analysis and identification based on joint information using an RGB-depth camera. In: 2017 14th international conference on electrical engineering/electronics, computer, telecommunications and information technology (ECTI-CON), pp 561–563. https://doi.org/10.1109/ECTICon.2017.8096299
15. Asier Zubizarreta A, Portillo E (2023) Machine learning-based gait anomaly detection using a sensorized tip: an individualized approach. In: Neural computing and applications, vol 35. Springer, pp 17443–17459. https://link.springer.com/article/10.1007/s00521-023-08601-1

16. Kim W, Kim Y (2021) Abnormal gait recognition based on integrated gait features in machine learning. In: 2021 IEEE 45th annual computers, software, and applications conference (COMPSAC), pp 1683–1688. https://ieeexplore.ieee.org/stamp/stamp.jsp?tp=&arnumber=9529558
17. Demirci F, Emec M, Doruk GO, Ormen M, Akan P, Ozcanhan HM (2023) Prediction of LDL in hypertriglyceridemic subjects using an innovative ensemble machine learning technique. Turkish J Biochem 48(6):641–652, Published by De Gruyter. https://doi.org/10.1515/tjb-2023-0154

Earthquake Prediction for Turkey: Ensemble Learning Approach

Ahmet Tezcan Tekin and Cem Sari

Abstract The prediction of earthquakes is a critical field of study in seismology and geophysics due to their substantial risks to human lives and infrastructure. The conventional approaches to earthquake prediction frequently depend on historical data and empirical models, which exhibit restricted accuracy and lead times. In recent times, the use of machine learning methodologies has surfaced as a promising approach for enhancing earthquake prediction capabilities. One of the key benefits of machine learning is its capacity to effectively evaluate extensive and heterogeneous datasets, encompassing seismic waveforms, geographical data, and historical records about earthquakes. Machine learning models can use these data sources to discern concealed patterns and connections, offering significant insights into the anticipation of earthquakes and the spatial and temporal distribution of seismic occurrences. In addition, machine learning algorithms have demonstrated potential in real-time seismic monitoring systems, facilitating the implementation of early warning mechanisms and expediting reaction techniques. These systems can offer crucial seconds to minutes of advanced warning, potentially preventing loss of life and reducing damage in locations susceptible to earthquakes. The dataset used in this study consists of earthquake data that occurred in Turkey between 1915 and 2023, taken from the Kandilli Observatory of Boğaziçi University. Extra Trees Regressor, Random Forest Regressor, Catboost Regressor, and Extreme Gradient Boosting algorithms were applied to the dataset, and the Extra Trees Regressor was given the best prediction score. The results of these algorithms were compared briefly.

Keywords Earthquake prediction · Machine learning · Ensemble learning · Cross-validation

---

A. T. Tekin (✉)
Department of Management Engineering, Istanbul Technical University, Istanbul, Turkey
e-mail: tekina@itu.edu.tr

C. Sari
Department of Industrial Engineering, Istanbul Technical University, Istanbul, Turkey
e-mail: saric21@itu.edu.tr

F. Çebi et al. (eds.), *Engineering and Technology Management in Challenging Times*, Lecture Notes in Networks and Systems 1137,
https://doi.org/10.1007/978-3-031-70935-7_15

1 Introduction

Earthquakes are forceful and unpredictable natural calamities that can induce catastrophic repercussions, such as loss of human life and extensive destruction of infrastructure. Despite extensive scientific research and significant improvements in seismology over several decades, correctly predicting earthquakes' precise timing and location remains an elusive objective. Nevertheless, incorporating machine learning (ML) methods in seismic research has emerged as a promising avenue for advancing our comprehension of seismic events and raising the efficiency and precision of earthquake forecasting in recent times.

The predominant methods for earthquake prediction in traditional approaches are generally based on historical data, empirical models, and expert judgment. Although these methodologies have yielded significant findings on fundamental geological mechanisms, they frequently lack the ability to provide timely and effective alerts that can minimize the consequences of seismic occurrences. The application of machine learning, owing to its capacity to analyze vast and complex datasets efficiently, presents a novel methodology for addressing the persistent challenges associated with earthquake prediction [1, 2].

Nevertheless, it is imperative to recognize that although machine learning exhibits significant potential, it is not devoid of inherent difficulties and constraints when applied to earthquake prediction. It is crucial to carefully navigate issues about data quality, model interpretability, and ethical considerations surrounding early warning systems.

The objective of utilizing machine learning for earthquake prediction is to provide insights into the ongoing research and advancements in this crucial domain. The statement highlights the significant influence that machine learning (ML) may have in improving our capacity to predict seismic occurrences, mitigate their destructive consequences, and ultimately protect human lives and infrastructure in areas susceptible to earthquakes.

The objective of this study was to conduct earthquake predictions specifically for the region of Turkey. One of the main contributions of this study is incorporating a dataset encompassing time series attributes. However, throughout feature engineering, the dataset transformed into a regression dataset. The dataset was preprocessed and underwent feature engineering before being utilized in ensemble machine learning techniques, notably bagging and boosting methods, along with classical algorithms. Following that, hyperparameter tuning was applied to the three most promising algorithms to improve the accuracy of forecasts. Additionally, the stability of the models was assessed using the K-Fold cross-validation approach.

The next section of the paper thoroughly examines the extant body of literature on utilizing machine learning techniques in the present study and their many applications in earthquake prediction. The final component of this study describes the dataset used, an overview of the data preparation and feature engineering methods employed, and a comparison of the outcomes achieved. In conclusion, the study concisely summarizes the findings and suggests potential avenues for further research in the concluding section.

2 Literature Review

Predicting earthquakes has historically been a significant difficulty within geophysics, primarily due to seismic events' complex and unexpected characteristics. Historical data, empirical models, and expert intuition have traditionally been employed as tools for earthquake prediction but with restricted efficacy. Recently, the incorporation of machine learning (ML) methodologies has presented novel opportunities for enhancing the precision of forecasts and comprehending seismic phenomena.

The availability of reliable seismic data is of utmost importance for the efficacy of machine learning-based earthquake forecasting. Researchers have utilized diverse data sources, including seismic waveform data, geospatial information, and historic earthquake catalogs. Seismic Waveform Data is the collection of ground motion data through seismometers, which is of paramount importance in earthquake prediction. Jozinović et al. [3], Sipkin [4], Funning et al. [5], and Kim et al. [6] used seismic waveform data for predicting earthquakes. Jena et al. [7] used Geospatial information for earthquake risk assessment, and Shafapourtehrany et al. [8] used Geospatial technology for earthquake preparedness.

Using machine learning techniques in earthquake prediction offers a robust set of tools, facilitating the examination of extensive and heterogeneous datasets to unveil patterns and connections that were previously difficult to discern. Numerous machine-learning algorithms have been employed in the field of earthquake prediction.

Initially developed for image analysis, convolutional Neural Networks (CNNs) have been modified to examine seismic data, demonstrating its efficacy in detecting spatial patterns and discerning seismic precursors. Kavianpour et al. [9] and Zhang et al. [10] used CNN for predicting earthquakes. Recurrent Neural Networks (RNNs) demonstrate exceptional proficiency in effectively processing sequential data, rendering them ideal for studying time series data within the field of seismology. Temporal dependencies in seismic data have been modeled using them, and they have also been utilized to detect earthquake signals. Vardaan et al. [11], Berhich et al. [12], and Doğan and Demir [13] used RNN for predicting earthquakes. Support Vector Machines (SVMs) have been utilized to classify seismic occurrences and discern between foreshocks, aftershocks, and mainshocks by leveraging waveform properties. Xing et al. [14], Astuti et al. [15], and Asaly et al. [16] used SVM for predicting earthquake events. Ensemble methods, such as random forests and gradient boosting, have been employed to amalgamate the predictions generated by numerous models, enhancing the accuracy and resilience of forecasts. Cui et al. [17], Asim et al. [18], Fernández-Gómez et al. [19], and Marzocchi et al. [20] used ensemble learning techniques for predicting earthquakes.

The field of machine learning has shown promise in earthquake prediction, presenting an opportunity to deepen our understanding of seismic events and improve the precision of forecasts. As ML-based methodologies are further developed and enhanced by researchers, it is imperative to prioritize resolving data quality concerns, enhancing model interpretability, and effectively managing ethical considerations. These factors are crucial in maximizing the efficacy of machine learning for earthquake prediction and early warning systems.

3 Proposed Methodology and Modelling

The dataset utilized in this research comprises earthquake data recorded in Turkey from 1915 through 2023, sourced from the Kandilli Observatory at Boğaziçi University [21]. This study aims to employ machine learning techniques using Python programming to forecast future earthquake intensities in Turkey based on historical earthquake data. The fundamental characteristics used in this study are presented in Table 1. The entire dataset comprises 15 distinct columns and 19,908 rows.

During the feature engineering process, some operations were performed on the dataset. Specifically, the columns labeled "no," "earthquake code," "type," and "location" were identified as redundant for the machine learning application and then eliminated from the dataset.

Table 1 Base features used in earthquake prediction

Feature	Explanation	Data type	Count
No	Sequence no	Integer	19,908
Earthquake code	Earthquake code number	Integer	19,908
Date	Date	Object	19,908
Time	Time	Object	19,908
Lat	Earthquake latitude	Float	19,908
Lon	Earthquake longitude	Float	19,908
Depth	Earthquake depth	Float	19,908
Xm	Maximum magnitude	Float	19,908
MD	Magnitude of duration	Float	19,908
ML	Magnitude of location	Float	19,908
Mw	Magnitude of moment	Float	6927
Ms	Magnitude of surface wave	Float	19,908
Mb	Magnitude of body wave	Float	19,908
Type	Earthquake or possible explosion	Object	19,908
Location	City/Sea/Country name	Object	19,908

Table 2 Statistics of earthquakes at sea in Turkey

Sea	Count
Mediterranean	4610
Aegean	1527
Marmara	225
Black Sea	31
East Mediterranean	2

Table 3 Statistics of earthquakes at land in Turkey

Land	Count
Van	914
Kahramanmaras	756
Mugla	745
Kutahya	653
Manisa	596
Malatya	531
Denizli	493
Izmir	472
Elazıg	407
Bingol	394

During the study of the dataset, it was seen that a total of 13,513 earthquakes transpired on land, while 6395 earthquakes occurred in the water during the time frame spanning from 1915 to 2023. Table 2 presents the frequency of seismic events in the sea surrounding Turkey, whereas Table 3 displays the occurrence of the top 10 earthquakes on land within Turkey.

After completing the data processing and feature engineering steps, the data set was subjected to classical and ensemble learning-based machine learning methods. Table 4 displays the error rates and execution times for each algorithm. The findings presented in Table 4 indicate that the Extra Trees Regressor, Random Forest Regressor, and CatBoost Regressor algorithms exhibit the lowest error rates and highest prediction scores. Furthermore, these algorithms demonstrate comparable performance since their results are closely aligned. In contrast, alternative ensemble learning methods such as Extreme Gradient Boosting, Gradient Boosting Regressor, Decision Tree Regressor, and Light Gradient Boosting Machine exhibit encouraging outcomes.

Upon examining the prediction success rates of several algorithms using their default parameters, the four most successful algorithms underwent hyperparameter modification: Extra Trees Regressor, Random Forest Regressor, CatBoost Regressor, and Extreme Gradient Boosting methods. During this phase, the grid search methodology was employed to identify the optimal parameters for these algorithms. Hence, to explore the impact of various factors on the model's performance,

Table 4 Algorithm's results with their default parameters

Model	MAE	MSE	RMSE	R2	RMSLE	MAPE	TT (Sec)
Extra trees regressor	0.0137	0.0033	0.0516	0.9898	0.0081	0.0030	23.3260
Random forest regressor	0.0103	0.0040	0.0571	0.9878	0.0091	0.0020	53.8840
CatBoost regressor	0.0169	0.0041	0.0587	0.9874	0.0092	0.0036	48.1100
Extreme gradient boosting	0.0166	0.0047	0.0631	0.9853	0.0103	0.0035	15.6430
Gradient boosting regressor	0.0245	0.0055	0.0680	0.9833	0.0110	0.0055	24.9690
Decision tree regressor	0.0113	0.0060	0.0719	0.9817	0.0108	0.0021	1.2570
Light gradient boosting machine	0.0177	0.0063	0.0746	0.9808	0.0113	0.0036	8.7780
Orthogonal matching pursuit	0.2367	0.1120	0.3343	0.6519	0.0624	0.0569	0.4770
Bayesian ridge	0.2387	0.1124	0.3349	0.6506	0.0624	0.0574	0.9940
Ridge regression	0.2411	0.1145	0.3381	0.6440	0.0632	0.0581	0.4810

the learning rates, max_depth, and the number of estimator parameters were integrated and utilized to train independent instances of the model with distinct parameter configurations.

K-Fold Cross-validation is crucial for evaluating the model's capacity to generalize and maintain stability. The K-Fold Cross-Validation method is commonly used in machine learning and statistical analysis to assess a predictive model's performance and generalization ability. The dataset is partitioned into "k" equal segments, enabling the creation and processing of validation data for each segment individually. Hence, every individual data point is utilized as validation data at a minimum of one instance, enhancing the accuracy of the model's overall performance evaluation. The application of tenfold cross-validation was performed on each algorithm using their optimal parameter combinations. The findings are presented in Tables 5, 6, 7, and 8.

The findings of the tenfold cross-validation demonstrate that the Extra Trees Regressor exhibits the lowest Mean Absolute Error and the highest score. Furthermore, the conclusions obtained indicate that the performance score of the Extreme Gradient Boosting algorithm surpasses that of the CatBoost algorithm. Upon concluding the examination of the findings, it was observed that the Extreme Gradient

Table 5 Catboost regressor 10-fold cross-validation results with its best parameters

	MAE	MSE	RMSE	R2	RMSLE	MAPE
0	0.0213	0.0042	0.0647	0.9868	0.0105	0.0044
1	0.0186	0.0022	0.0468	0.9918	0.0079	0.0041
2	0.0241	0.0066	0.0810	0.9811	0.0123	0.0049
3	0.0186	0.0044	0.0662	0.9851	0.0098	0.0040
4	0.0221	0.0075	0.0868	0.9748	0.0124	0.0046
5	0.0205	0.0035	0.0595	0.9883	0.0094	0.0044
6	0.0262	0.0099	0.0995	0.9678	0.0145	0.0053
7	0.0242	0.0056	0.0748	0.9827	0.0120	0.0051
8	0.0182	0.0032	0.0568	0.9895	0.0085	0.0039
9	0.0229	0.0087	0.0931	0.9747	0.0135	0.0046
Mean	0.0217	0.0056	0.0729	0.9822	0.0111	0.0045
Std	0.0026	0.0024	0.0162	0.0073	0.0021	0.0005

Table 6 Extreme gradient boosting regressor 10-fold cross-validation results with its best parameters

	MAE	MSE	RMSE	R2	RMSLE	MAPE
0	0.0281	0.0034	0.0586	0.9892	0.0098	0.0063
1	0.0286	0.0025	0.0499	0.9907	0.0090	0.0067
2	0.0293	0.0055	0.0742	0.9841	0.0112	0.0065
3	0.0285	0.0040	0.0636	0.9862	0.0102	0.0064
4	0.0282	0.0035	0.0594	0.9882	0.0100	0.0064
5	0.0320	0.0045	0.0669	0.9852	0.0111	0.0072
6	0.0306	0.0030	0.0550	0.9902	0.0095	0.0070
7	0.0293	0.0029	0.0538	0.9910	0.0092	0.0067
8	0.0276	0.0023	0.0476	0.9927	0.0084	0.0064
9	0.0315	0.0090	0.0947	0.9739	0.0144	0.0068
Mean	0.0294	0.0041	0.0624	0.9871	0.0103	0.0066
Std	0.0014	0.0019	0.0132	0.0051	0.0016	0.0003

Boosting Machine and Catboost exhibited comparable performance in prediction. However, the Extra Trees Regressor was ultimately selected as the production model.

Table 7 Random forest regressor 10-fold cross-validation results with its best parameters

	MAE	MSE	RMSE	R2	RMSLE	MAPE
0	0.0083	0.0014	0.0375	0.9956	0.0064	0.0016
1	0.0065	0.0006	0.0251	0.9976	0.0040	0.0013
2	0.0101	0.0034	0.0586	0.9901	0.0085	0.0019
3	0.0080	0.0023	0.0484	0.9920	0.0072	0.0015
4	0.0083	0.0022	0.0471	0.9926	0.0068	0.0016
5	0.0095	0.0018	0.0427	0.9940	0.0066	0.0019
6	0.0082	0.0014	0.0380	0.9953	0.0055	0.0016
7	0.0091	0.0024	0.0491	0.9925	0.0080	0.0018
8	0.0086	0.0015	0.0391	0.9950	0.0061	0.0017
9	0.0109	0.0060	0.0772	0.9826	0.0109	0.0020
Mean	0.0087	0.0023	0.0463	0.9927	0.0070	0.0017
Std	0.0012	0.0014	0.0134	0.0039	0.0018	0.0002

Table 8 Extra trees regressor 10-fold cross-validation results with its best parameters

	MAE	MSE	RMSE	R2	RMSLE	MAPE
0	0.0128	0.0015	0.0388	0.9953	0.0066	0.0028
1	0.0125	0.0014	0.0371	0.9949	0.0065	0.0029
2	0.0147	0.0032	0.0566	0.9907	0.0083	0.0031
3	0.0122	0.0014	0.0368	0.9954	0.0059	0.0027
4	0.0123	0.0019	0.0440	0.9935	0.0062	0.0026
5	0.0135	0.0016	0.0402	0.9947	0.0062	0.0030
6	0.0123	0.0014	0.0370	0.9955	0.0055	0.0027
7	0.0109	0.0009	0.0298	0.9973	0.0051	0.0024
8	0.0130	0.0011	0.0325	0.9966	0.0053	0.0029
9	0.0154	0.0058	0.0763	0.9830	0.0110	0.0031
Mean	0.0129	0.0020	0.0429	0.9937	0.0066	0.0028
Std	0.0012	0.0014	0.0131	0.0039	0.0017	0.0002

4 Conclusion and Future Work

This study's objective was to forecast the magnitude of earthquakes in Turkey using data from 1915 through 2023. To achieve this aim, we implemented data preprocessing and feature engineering techniques on our dataset.

Our objective was to enhance the learning capacity of the regression algorithms. Following this procedure's completion, we used traditional and ensemble machine learning algorithms on the preprocessed dataset, utilizing their default parameters. The initial findings indicate that the Extra Trees Regressor, Random Forest Regressor,

CatBoost Regressor, and Extreme Gradient Boosting algorithms exhibit the highest levels of success in the prediction phase.

Given the significance of hyperparameter optimization in machine learning tasks, we optimized hyperparameter using our top four methods: Extra Trees Regressor, Random Forest Regressor, CatBoost Regressor, and Extreme Gradient Boosting. The procedure mentioned above exerts a beneficial influence on Extreme Gradient Boosting algorithms, hence enhancing the efficacy of prediction outcomes.

The findings indicate that the artificial intelligence Earthquake prediction strategy has yielded favorable outcomes. In particular, ensemble learning algorithms such as CatBoost Regressor, Extreme Gradient Boosting, and Extra Trees Regressor exhibit strong predictive capabilities in the final stage. Other ensemble learning algorithms demonstrate promising outcomes, including Extreme Gradient Boosting, Gradient Boosting Regressor, Decision Tree Regressor, and Light Gradient Boosting Machine. The limitation of this study is it depends on the Kandilli Observatory. The prediction system will not generate predictions unless the dataset is updated in real time. The results of these algorithms will be juxtaposed with time-series approaches such as Holt-Winter and ARIMA for future research.

References

1. Ren CX, Hulbert C, Johnson PA, Rouet-Leduc B (2020) Machine learning and fault rupture: a review. Adv Geophys 61:57–107
2. Salam MA, Ibrahim L, Abdelminaam DS (2021) Earthquake prediction using hybrid machine learning techniques. Int J Adv Comput Sci Appl 12(5):654–6652021
3. Jozinović D, Lomax A, Štajduhar I, Michelini A (2020) Rapid earthquake ground shaking intensity prediction using raw waveform data and a convolutional neural network. Geophys J Int 222(2):1379–1389
4. Sipkin SA (1982) Estimation of earthquake source parameters by the inversion of waveform data: synthetic waveforms. Phys Earth Planet Inter 30(2–3):242–259
5. Allen RM, Brown H, Hellweg M, Khainovski O, Lombard P, Neuhauser D (2009) Real-time earthquake detection and hazard assessment by ElarmS across California. Geophys Res Lett 36(5)
6. Funning GJ, Fukahata Y, Yagi Y, Parsons B (2014) A method for the joint inversion of geodetic and seismic waveform data using ABIC: application to the 1997 Manyi, Tibet, earthquake. Geophys J Int 196(3):1564–1579
7. Jena R, Pradhan B, Naik SP, Alamri AM (2021) Earthquake risk assessment in NE India using deep learning and geospatial analysis. Geosci Front 12(3):101110
8. Shafapourtehrany M, Batur M, Shabani F, Pradhan B, Kalantar B, Özener HA (2023) Comprehensive review of geospatial technology applications in earthquake preparedness, emergency management, and damage assessment. Remote Sens 15(7):1939
9. Kavianpour P, Kavianpour M, Jahani E, Ramezani A (2023) A cnn-bilstm model with attention mechanism for earthquake prediction. J Supercomput 1–33
10. Zhang H, Melgar D, Sahakian V, Searcy J, Lin JT (2022) Learning source, path and site effects: CNN-based on-site intensity prediction for earthquake early warning. Geophys J Int 231(3):2186–2204
11. Vardaan K, Bhandarkar T, Satish N, Sridhar S, Sivakumar R, Ghosh S (2019) Earthquake trend prediction using long short-term memory RNN. Int J Electr Comput Eng 9(2):1304–1312

12. Berhich A, Belouadha FZ, Kabbaj MI (2022) A location-dependent earthquake prediction using recurrent neural network algorithms. Soil Dyn Earthq Eng 161:107389
13. Doğan A, Demir E (2022) Structural recurrent neural network models for earthquake prediction. Neural Comput Appl 34(13):11049–11062
14. Xing H, Zhonglin Z, Shaoyu W (2015) The prediction model of earthquake causality based on robust wavelet v-SVM. Nat Hazards 77:717–732
15. Astuti W, Akmeliawati R, Sediono W, Salami MJE (2014) Hybrid technique using singular value decomposition (SVD) and support vector machine (SVM) approach for earthquake prediction. IEEE J Sel Top Appl Earth Obs Remote Sens 7(5):1719–1728
16. Asaly S, Gottlieb LA, Inbar N, Reuveni Y (2022) Using support vector machine (SVM) with GPS ionospheric TEC estimations to potentially predict earthquake events. Remote Sens 14(12):2822
17. Cui S, Yin Y, Wang D, Li Z, Wang Y (2021) A stacking-based ensemble learning method for earthquake casualty prediction. Appl Soft Comput 101:107038
18. Asim KM, Idris A, Martínez-Álvarez F, Iqbal T (2016) Short term earthquake prediction in Hindukush region using tree based ensemble learning. In: 2016 International conference on frontiers of information technology (FIT). IEEE, pp 365–370
19. Fernández-Gómez MJ, Asencio-Cortés G, Troncoso A, Martínez-Álvarez F (2017) Large earthquake magnitude prediction in Chile with imbalanced classifiers and ensemble learning. Appl Sci 7(6):625
20. Marzocchi W, Zechar JD, Jordan TH (2012) Bayesian forecast evaluation and ensemble earthquake forecasting. Bull Seismol Soc Am 102(6):2574–2584
21. BOUN Kandilli observatory webpage. http://www.koeri.boun.edu.tr/sismo/zeqdb/. Accessed 01 July 2023

PESTLE Analysis for Public Transportation: A Fuzzy Approach

Büşra Buran

Abstract The public transportation plays an important role in sustaining mobility in cities. It is seen as a backbone of the developing and developed cities. With the changing conditions such as political changes, technological developments, passenger behaviors, etc., directly affect the system. To understand and handle changing conditions, external analysis is applied to the systems. PESTLE analysis is applied as an external analysis with a fuzzy approach for public transportation in this study. The Intuitionistic Fuzzy Set was used for fuzzy AHP to discover the importance level of PESTLE criteria by transportation experts which are political, economical, social, technological, legal, and environmental. Regarding findings the most important criterion is social factor, the second one is technological issues. This study will guide transportation authorities and operators to adapt with respect to changing conditions. Also, the study will shed on lights which external environmental criteria are more important than for both authorities and operators.

Keywords PESTLE analysis · Fuzzy AHP · Intuitionistic fuzzy set · Public transportation

1 Introduction

Dealing with uncertainty does not lose its importance during time [1, 2]. For the sustainability of the company, it gains more importance. At this point, PESTEL analysis can be used as a strategic tool to understand macro-environment in the business area [3]. It is widely preferred in strategic management [4].

PESTLE analysis which composes of political, economical, social, technological, environmental, and legal factors, is used to discover the external environment [5]. It is applied to a wide range of areas such as startup business, energy, economy, food industry, healthcare sector, construction projects, transportation and tourism

B. Buran (✉)
Management Engineering, Istanbul Technical University, Istanbul, Turkey
e-mail: buran18@itu.edu.tr

F. Çebi et al. (eds.), *Engineering and Technology Management in Challenging Times*, Lecture Notes in Networks and Systems 1137,
https://doi.org/10.1007/978-3-031-70935-7_16

etc. [6–10]. It plays a critical role to handle uncertainty from a business perspective. With fast developing technology, especially technological factor of PESTLE analysis come forward.

Although there are lots of studies about PESTLE analysis about different domains such as energy management, logistics, education in literature, there is a lack of public transportation domain [11–16]. This study will fill the gap by presenting PESTLE analysis for the public transportation systems. The study presents the importance of criteria weights of the analysis with a real case study in Istanbul. Thus, originality of the paper comes from analyzing the most important criterion for public transportation regarding external environmental factors.

The study is constructed as follows. Related works are summarized in Sect. 2. In Sect. 3, the methodology with subsections which are model design, method, and findings are presented respectively. Lastly, the conclusion is represented with future avenue in Sect. 4.

2 Related Works

The PESTLE analysis with transportation, especially public transportation studies are summarized in this section. Analyzes show that PESTLE framework is widely used in transport studies.

Guno et al. presented a study to understand what are barriers and drivers of the transition to sustainable transportation. The authors applied PESTEL analysis to the Philippines as a case. Results showed that economic and technological factors are the main barriers to the adoption of electric public transport. On the other hand, legal and policy are seen as drivers for the transition of transportation [11]. What are the factors that influence the development of hybrid or electric vehicles in the transportation sector in Brazil? To answer this question Sousa and Castañeda-Ayarza presented a study. The study showed that political and legal factors are positive forces. Also, technological and ecological issues are the leading positive influencers [12]. Gheibi et al. presented a study about controlling air pollution in the cities. PESTLE framework was used to understand external factors in the business. According to the results, supporting local governments as a policy factor comes forward in others [13]. Christodoulou and Cullinane proposed the implementation of a port energy management system with SWOT and PESTEL analysis. The authors used real cases which were North-European port authorities. According to the findings, the adoption of the ISO 50001 standard allows a big advantage for integrated energy management for the maritime transport sector [14]. Düerkop and Huth conducted a PESTLE analysis for critical Logistical Infrastructures. They aimed to identify possible threats affecting logistical infrastructures. At the end of the study, the authors suggest that different risks should be categorized regarding PESTEL analysis [15]. For analyzing the competitiveness of unconventional modes of transportation in developing cities, Sobhani et al. proposed a framework using PESTEL analysis. Dhaka, the capital city of Bangladesh was used as a case study. Regarding the study, economic, social, and political factors come forward in others [16]. Radomska and Kartash made a comparative analysis of the environmental and economic efficiency of Kyiv city for transport modernization. Findings showed that there is an urgent need for green and innovative transport vehicles regarding environmental and technological factors taking into cost [17].

3 Methodology

3.1 Model Design

To understand the external environment, PESTLE analysis is conducted for public transportation in the study. While regulation, tax policy, labor law, and government structure take part under policy issue [18], petrol price, consumer disposable income, private car ownership, gross domestic product, average wage, exchange rate, and inflation rate are located under economic factors [19]. Growing population, aging population, mobility, transportation culture, and lifestyle are seen as social factors [20]. For technological factors, rapid development in technological areas and faster innovation cycles are taken into account [18]. For legal factors, government regulations, health and safety law, and data protection are taken into account [19]. Lastly, for environmental factors, climate change and environmental protection issues are considered in the transportation domain [18].

The model is constructed with one level which is presented in Fig. 1. PESTLE analysis which are political, economical, social, technological, legal, and environmental are taken into account in the proposed model.

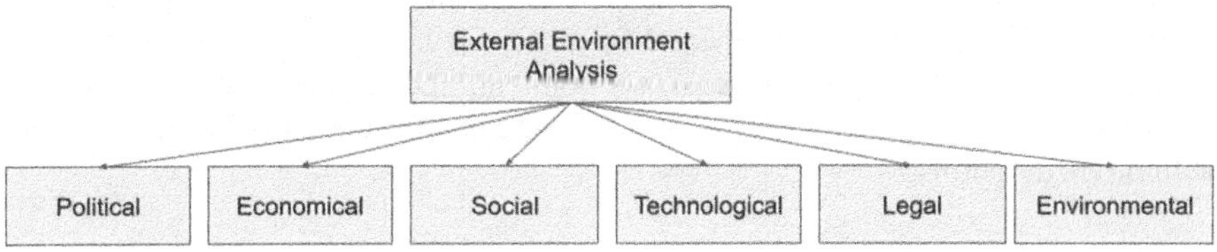

Fig. 1 Proposed model structure

3.2 Method

Intuitionistic Fuzzy Set (IFS) was proposed by Atasanow is used widely in different fields in the literature such as education, banking, healthcare, and information technology [21–25]. Basic operations of IFS such as addition, multiplication, and multiplication by a scalar, etc. were defined in the study of Atanassov's study in 2017 [26].

In this study, IFS is applied to the proposed model which is presented in Fig. 1. To discover which criterion is the most important regarding PESTLE analysis. The method framework is shown in Fig. 2. According to Figure 2, the proposed model is assessed by transportation experts with different expertise that are strategy, information technology, and quality sides. Also, detailed information about experts is presented in Table 1. Firstly, with the aim of the study, construct hierarchical structure. After constitute pairwise matrices regarding criteria. Using IFS, the matrices are evaluated by experts. After evaluation process, check consistency ratio (CR). If CR is lower than 0.1, the calculate weights of criteria, otherwise back to step 3. IFWA operator is used to find the score function. Finally, the global weights are obtained and rank criteria regarding global weights.

3.3 Findings

The results of the model are summarized in Table 2. Regarding results, social factors with 0.218 weight are seen as the most important factor for the public transportation taking into account external environment. The second important criterion is the technology with 0.172 weight. The third one is political, the fourth one is legal factors and the fifth one is economic with close values that are 0.156, 0.153, and 0.152 respectively. Finally, the environmental factor comes last with 0.150 weight.

Table 2 shows that political, legal, economical, and environmental factors take similar weights which are around 0.150. But there is different view for the social and technological factors.

4 Conclusion

External analysis plays a critical role in the understanding of a company in the business area. Also, it contributes sustainability of business. PESTLE analysis which is a tool for the external environment, is modeled with a hierarchy structure for public transportation in the study. Fuzzy logic is applied as a method using an Intuitionistic Fuzzy Set to handle the vagueness of the problem. Transportation experts evaluate the model regarding political, economical, social, technological, legal, and environmental criteria. Results showed that social and technological factors come first and

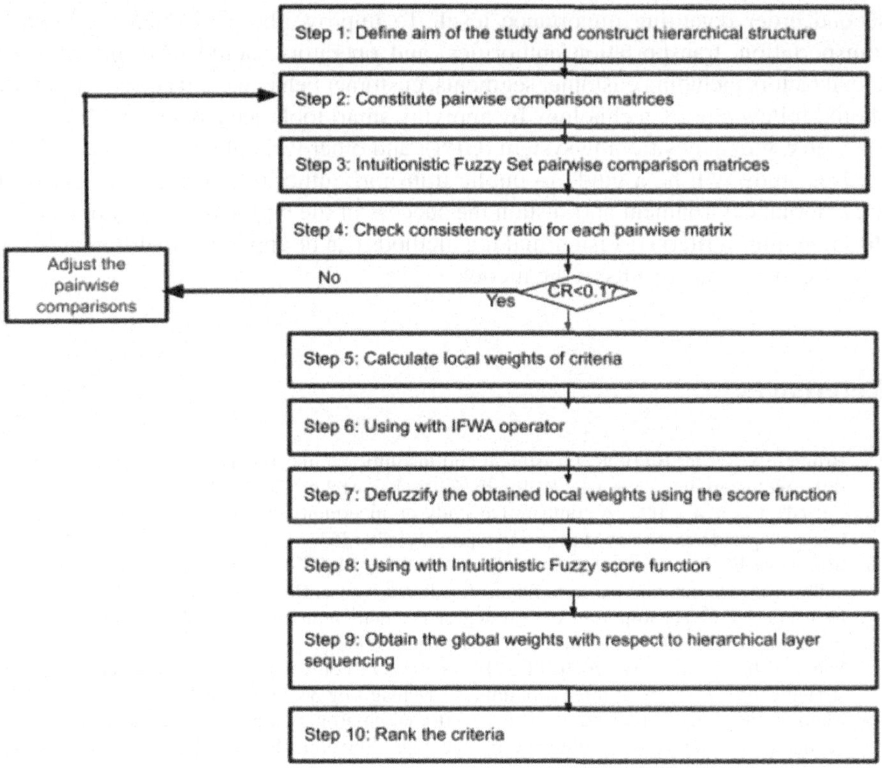

Fig. 2 Proposed method

Table 1 Experts' information

Expertise domain	Experience in YEARS
Strategy	13
Technology and information	12
Quality standards	20

Table 2 Global criteria weights for the proposed model

Criteria	Global weights
Social	0.218
Technological	0.172
Political	0.156
Legal	0.153
Economical	0.152
Environmental	0.150

second order regarding importance level. To improve the performance of public transportation, transportation authorities, and operators should take into account social factors including customer segments, customer behavior, and needs. Also, they should follow closely technology by applying smart tools such as sensors, cameras, telemetry, global positioning system (GPS), and other technologies.

This study will be a guide to public transport authorities/operators to discover the external environment and sustain the success of the business. For future studies, different multi-criteria decision-making methods can be applied with different fuzzy sets to compare the results of the model.

References

1. Brunetti A, Weder B (1998) Investment and institutional uncertainty: a comparative study of different uncertainty measures. Weltwirtschaftliches Archiv 134(3):513–533
2. Lee SH, Chen W (2009) A comparative study of uncertainty propagation methods for black-box-type problems. Struct Multidiscip Optim 37:239–253
3. Shatskaya E, Samarina M, Nekhorosheva K (2016) PESTEL analysis as a tool of strategic analysis in international markets. Scope Academic House. B&M Publishing, p 47
4. Menet G (2016) The importance of strategic management in international business: expansion of the PESTEL method. Int Bus Glob Econ 35(2):261–270
5. Pan W, Chen L, Zhan W (2019) PESTEL analysis of construction productivity enhancement strategies: a case study of three economies. J Manag Eng 35(1):05018013
6. Matović IM (2020) PESTEL analysis of external environment as a success factor of startup business. In: ConScienS conference proceedings. Scientia Moralitas Research Institute, pp 96–102
7. Igliński B, Iglińska A, Cichosz M, Kujawski W, Buczkowski R (2016) Renewable energy production in the Łódzkie Voivodeship. The PEST analysis of the RES in the voivodeship and in Poland. Renew Sustain Energy Rev 58:737–750
8. Alava RP, Murillo JM, Zambrano RB, Vélez MIZ, Vazquez ML (2018) PEST analysis based on neutrosophic cognitive maps: a case study for food industry. Neutrosophic Sets Syst 21:84–92
9. Soundari BP (2016) PEST analysis of the health care sector in Coimbatore district. Scholars J Econ Bus Manage 3(5):271–274
10. Barkauskas V, Barkauskienė K, Jasinskas E (2015) Analysis of macro environmental factors influencing the development of rural tourism: Lithuanian case. Procedia Soc Behav Sci 213:167–172
11. Guno CS, Collera AA, Agaton CB (2021) Barriers and drivers of transition to sustainable public transport in the Philippines. World Electr Veh J 12(1):46
12. de Sousa GC, Castañeda-Ayarza JA (2022) PESTEL analysis and the macro-environmental factors that influence the development of the electric and hybrid vehicles industry in Brazil. Case Stud Transp Policy 10(1):686–699
13. Gheibi M, Karrabi M, Mohammadi A, Dadvar A (2018) Controlling air pollution in a city: a perspective from SOAR-PESTLE analysis. Integr Environ Assess Manag 14(4):480–488
14. Christodoulou A, Cullinane K (2019) Identifying the main opportunities and challenges from the implementation of a port energy management system: a SWOT/PESTLE analysis. Sustainability 11(21):6046
15. Düerkop S, Huth M (2017) Transportation under threat—a PESTLE analysis for critical logistical infrastructures. ICTA2017, p 14
16. Sobhani MG, Imtiyaz MN, Azam MS, Hossain M (2020) A framework for analyzing the competitiveness of unconventional modes of transportation in developing cities. Transp Res Part A Policy Pract 137:504–518

17. Radomska M, Kartash Y (2019) The comparative analysis of the environmental and economic efficiency of the Kyiv city public transport modernization. Adv Inf Syst 3(2):122–126
18. Kashi IDK (2014) MCDM methods in practice: determining importance of PESTEL analysis criteria. In: The 8th international days of statistics and economics, Prague, pp 334–344
19. Buran B, Erçek M (2021) Convergence or divergence among business models of public bus transport authorities across the globe: a fuzzy approach. Sustainability 13(19):10861
20. Strzelczyk M, Chłąd M (2017) Use of PESTEL analysis for assessing the situation of polish transport enterprises (Part I). Zeszyty Naukowe Politechniki Częstochowskiej Zarządzanie 27(2):161–168
21. Çitil M (2019) Application of the intuitionistic fuzzy logic in education. Commun Math Appl 10(1):131–143
22. Puri J, Yadav SP (2015) Intuitionistic fuzzy data envelopment analysis: an application to the banking sector in India. Expert Syst Appl 42(11):4982–4998
23. Bharati SK (2021) Transportation problem with interval-valued intuitionistic fuzzy sets: impact of a new ranking. Progr Artif Intell 10:129–145
24. Otay İ, Oztaysi B, Onar SC, Kahraman C (2017) Multi-expert performance evaluation of healthcare institutions using an integrated intuitionistic fuzzy AHP&DEA methodology. Knowl-Based Syst 133:90–106
25. Biswas PRANAB, Pramanik SURAPATI, Giri BC (2014) A study on information technology professionals' health problem based on intuitionistic fuzzy cosine similarity measure. Swiss J Stat Appl Math 2(1):44–50
26. Atanassov KT (2017) Type-1 fuzzy sets and intuitionistic fuzzy sets. Algorithms 10(3):106

Review of AI in Project Management and Conceptual Analysis on Risk Management

Ercan Öztemel and Yasemin Tuncer

Abstract Project management is becoming a critical activity for the enterprises as the resources are getting limited and scarce. It seems Artificial Intelligence (AI) will be providing effective utilization of resources and enabling easy operations with human like operations, machine learning, intelligent agents, fuzzy logic and other AI methodologies provide effective data analysis and automation to enhance of project planning, execution, and monitoring. The paper reviews some of the AI application in these areas. AI cam also play a critical role in risk management and contingency planning, in order to identify and mitigate potential obstacles. A conceptual analysis along this line is also provided in the paper. Since AI is well known for automating routine tasks, streamlining workflows, and improving resource utilization, easing resource allocation and improving both efficiency and effectiveness for other domains, it could very well be implemented in project management domains. As explained in the paper respective literature and previous research encourage this kind of studies. While AI holds immense potential in project management, some technical improvement is necessary. There is a need for defining conceptual analysis and implementable models for basic project management activities. It is also very imminent that the ethical considerations, data privacy, and the need for human-AI collaboration should not be overlooked. In conclusion, AI-based PM represents a paradigm shift in the field. This has to be formulated to easy the respective development activities. Considering the complexity in risks and uncertainties, resource controls, schedule, stakeholder and scope management, the research shows that using artificial intelligence in project management will provide many benefits to project managers. This article is an attempt to explore the multifaceted role of AI in PM, investigating its core components, applications, benefits, and challenges. It specifically takes the attention to risk management with respective agent architecture and responsibilities.

E. Öztemel · Y. Tuncer (✉)
Department of Industrial Engineering, Faculty of Engineering, Marmara University, Istanbul, Turkey
e-mail: yasemintuncer@marun.edu.tr

E. Öztemel
e-mail: eoztemel@marmara.edu.tr

F. Çebi et al. (eds.), *Engineering and Technology Management in Challenging Times*, Lecture Notes in Networks and Systems 1137,
https://doi.org/10.1007/978-3-031-70935-7_17

Keywords Artificial intelligence · Project management · Intelligent agents

1 Introduction

Artificial intelligence (AI) has been a transformative force in modern project management, offering innovative solutions to the challenges posed by increasingly complex and large-scale projects. The use of AI-based project management tools has the potential to streamline processes, enhance decision-making, and improve project outcomes across various industries and domains. By leveraging the capabilities of AI, project managers can navigate the intricacies of their projects with greater agility, precision, and foresight.

Integration of AI in project management encompasses a broad range of applications throughout the project lifecycle. At the planning stage, AI-driven tools can assist in defining project objectives, estimating timelines, and developing comprehensive project plans. AI's predictive analytics capabilities enable project managers to anticipate potential risks and devise proactive risk mitigation strategies. Moreover, AI-based scheduling tools can optimize resource allocation and balance workloads to improve team efficiency and productivity.

In the execution phase, AI facilitates real-time monitoring and control of project activities. By analyzing large volumes of data from various sources, AI can provide project managers with valuable insights into project performance and progress. This data-driven approach allows for immediate adjustments and course corrections, ensuring that projects stay on track and meet their objectives. Additionally, AI's ability to identify patterns and anomalies can help project managers spot issues early and take corrective action. The impact of AI extends beyond the technical aspects of project management and into the realm of team dynamics and collaboration. AI-powered collaboration platforms can enhance communication and coordination among team members, stakeholders, and clients. These platforms often incorporate natural language processing (NLP) features, enabling more efficient and effective interactions. Furthermore, AI-driven decision support systems offer project managers data-backed recommendations, helping them make informed choices in complex situations.

Research in AI-based project management continues to expand as organizations seek to harness the full potential of AI to achieve their project goals. Scholars are investigating AI's role in various facets of project management, such as cost estimation, quality control, and change management. AI's ability to process unstructured data, including text and images, opens up new avenues for research in areas like automated documentation and reporting. Another area of interest in AI-based project management research is the ethical and societal implications of AI integration. As AI becomes more prevalent in project management, concerns about data privacy, bias, and accountability arise. Researchers are exploring best practices for responsible AI usage and the development of ethical frameworks to guide AI implementation in project management. Moreover, the rise of AI-driven tools presents opportunities for the continuous improvement of project management methodologies. Agile and Scrum, for instance, can benefit from AI's iterative feedback and adaptive learning capabilities. AI can also contribute to the evolution of project management standards and certifications, ensuring that professionals stay abreast of emerging technologies and best practices.

In today's fast-paced and dynamic business environment, efficient project management is essential for organizations to stay competitive and deliver successful outcomes. Traditional project management methods have served us well for decades, but they are increasingly facing challenges in keeping up with the complexity and speed of modern projects.

Artificial Intelligence (AI) is revolutionizing the way we approach project management. AI-Based PM, often abbreviated as AI-PM, harnesses the power of machine learning, data analysis, and automation to enhance project planning, execution, and monitoring. This innovative approach promises to streamline processes, improve decision-making, and ultimately deliver projects more effectively and efficiently [31].

This article investigates the exciting world of AI-Based PM. The key concepts, benefits, and applications of AI in PM are explored, as well as the challenges and considerations that come with implementing AI-driven solutions. A conceptual framework is proposed for possible system development. It is assumed that for the newcomers interested in the potential of AI in PM, this article aims to provide valuable insights and guidance.

In the following sections, the foundations of AI-Based Project Management, its core components, and how it can be integrated into your organization's project management practices are elaborated. The conceptual model proposed will also be discussed in addition to real-world examples of AI-driven project management success stories and some ethical considerations.

Note that, AI-based project management represents a dynamic and rapidly evolving field with the potential to revolutionize traditional project management practices. As research delves into the intersection of AI and project management, the insights gained will pave the way for more efficient, effective, and innovative project management strategies. By embracing AI's transformative power, organizations can drive project success, achieve their strategic objectives, and maintain a competitive edge in an ever-changing business landscape.

After the introduction the paper continues with a comprehensive literature review indicting possible implementation of AI in project management in general. A brief description of Artificial Intelligence and some basic information regarding project management are then followed. The attention of the reader is than taken on Research in the Applications of Artificial İntelligence For Project Management with respect to project planning and prediction, risk management and prevention, Resource management and allocation, as well as real time monitoring of project as well as some decision support system capabilities. Some information regarding future research areas are also provided after this. The paper then, introduces the conceptual model for agent based intelligent project management describing a multi agent risk management system as well as the basic functionalities and structures of the respective agents. Paper end with a conclusion addressing possible future work.

2 Literature Review

When examining studies on the use of artificial intelligence techniques in project management software, many works across different fields can be found. A web-based system has been developed using artificial neural networks to minimize potential risk factors that may be encountered in software projects. The system, developed using project management, knowledge management, and artificial intelligence techniques, offers intelligent support to users in managing large projects and dealing with complex decision-making situations [29]. This support is provided through the integration of daily site report modules, program control modules, document management modules, and the identification of critical topics using artificial intelligence techniques.

The belief that increasingly complex technology applications will help modernize existing industry standard procedures in project management has led to a focus on different applications that aim to simplify project processes and increase efficiency through the use of artificial intelligence. An AI-supported communication model has been established to assist managers in energy projects by providing virtual communication support and intermediary units that facilitate the integration of subscribers and stakeholders into the system [6]. Various optimization processes using machine learning are proposed for critical areas in project management, such as the estimation of defined tasks, the rate at which teams complete their assigned tasks, and the estimation of completion times. Manik Reddy and Iyer [24] have also made approaches to improve workflows in global systems and to engage with smart process systems.

Recent studies in project management have demonstrated the significant role of individuals' entrepreneurial orientation in project performance. While the identification of the role of entrepreneurial orientation as a critical success factor in project performance is seen as an important topic, Sabahi and Parast [32] also argue that it is essential to develop a measurement system to predict performance based on the degree of an individual's entrepreneurial orientation. Their work uses predictive analytics and machine learning methods to anticipate the performance of individuals in projects based on their different dimensions of orientation and attitudes towards entrepreneurship, as assessed through various measurements.

In their exploration of how machine learning algorithms should be used in project management, the authors emphasize that outcomes can vary depending on factors such as the diversity of the sample set. Given the current dynamic software development environment, which values agility, prototyping, and rapid delivery, the study concludes that software incorporating machine learning methods is more effective in predicting project processes and outcomes [30].

Fridgeirsson's goal is to find out how intelligent computer programs improve the discipline of PM in each of the ten categories of project management knowledge areas identified in his PMBOK article [12]. He argued that AI is an integral step in future PM practice and will have an impact on PMBOK knowledge areas in the near future. He noted that AI systems can help project managers estimate costs, manage schedules, track progress, provide reminders and follow-ups, manage tasks and allocate resources, with significant positive impacts on cost, schedule and risk management. On the other hand, AI may have less impact in knowledge areas and methods that require human guidance, such as team development, management and stakeholder management.

AI in PM aims to facilitate and increase the efficiency of project operations, as demonstrated by several applications. Its uses and approaches, which are generally useful, ranged from complexity to risk management, assessments, strategic planning, key performance indicator (KPI), forecasting and testing. Ong and Uddin discussed and revealed the current state of data science and artificial intelligence in the PM area [29]. Potential future implementations and functions are identified, multiple patterns are identified, and the gathered information is combined to provide an accurate picture of the potential landscape of project management. It has been hypothesized that advanced implementations of AI will help simplify current industry standard practices. Several data science positions are bound to emerge during the technology development process. As the most mundane tasks are performed by technology, they are inevitably overtaken by more advanced positions. Another study investigating the impact of artificial intelligence on PM was conducted by Garcia et al., who examine the progress and challenges in project management over the past decades using computer intelligence methods, schools and technological tools for PM, and the use of open source. Software Computer Intelligence (CI) approaches [13]. Garcia argued that project management is a difficult endeavor with a lot of conceptual and informational ambiguity, which is why CI methods prove useful. He suggests that the use of machine learning techniques in CI approaches to project assessment improves organizational agility when management models change due to maturity gained through continuous development. The information presented in the article can help project-oriented companies make better decisions. In another article that developed and applied a situational model of human and robotic resource management, Tunc shared his thoughts and explanations about the interaction between artificial intelligence and humans in the work environment and its effects [26]. In his article, Jüngen and Kowalczyk [19] provide a broad basis for building complex systems for activities such as PM, dynamic planning, and scheduling.). The concept is based on the idea of dealing with meta-boundary satisfaction problems dynamically and incorporates various AI concepts that combine an advanced feedback mechanism into a fully automated system that generates solutions without user intervention. Since the approach has a clear logical basis, it is possible to combine systems based on it with data-based systems.

Although AI has a lot to offer the domain, there are certain requirements for project managers to take full advantage of it. Niederman analyzes the area to learn how artificial intelligence affects and inputs into PM techniques [28]. He suggests that normal project managers probably don't produce AI/analytics themselves, so they don't need to master the technicalities of machine learning, so software vendors can offer PMs as a product. In the second study, Hazir examines project management system challenges, methods, and analytical models, as well as future research extensions explored in the literature on earned value analysis (EVA) optimization techniques and decision support system (DSS) architecture. They can help project managers plan and manage in unpredictable project contexts [16]. He emphasized the need to create analytical models to assess the performance of the project more accurately, especially research on performance forecasting using new full EVA metrics and statistical control tools. Lauras et al. [20] reflects on the difficulties of managing a complex project due to many performance indicators. He sought to give project managers more control over the execution of their projects. According to Lauras, published literature shows that there is no defined instrument that would support global research on methods and tools to analyze the performance of the project according to different dimensions (cost, time, quality, risk, etc.). This paper presented a multi-criteria method based on the performance cube, which has three dimensions such as project activities or tasks, categories of performance indicators, and a summary of the performance.

Project managers have several tools at their disposal to help them better handle projects in the literature. Marques et al. [25] focus on decision making in service and product development projects and present a new multidimensional project performance measurement system PPMS helping project managers handle large volumes of data. PPMS incorporates the unique characteristics of each project (tasks, priorities, personalities and expertise of decision makers) and several best practices for global aspects of project management and performance evaluation. On the other hand, non-technical issues are equally important for the project. Forsythe and Buchanan [11] suggest that non-technical problems can go unnoticed by technical people and that these errors and mistakes can ruin a project just as easily as technical ones. It is argued that managers and researchers, especially information engineers, focus more on technical problems than technical answers. On the other hand, non-technical issues are just as important and can be just as annoying as technical ones. In conclusion, conventional thinking in developing expert systems oversimplifies the process, neglecting concerns that go beyond the machine-readable representation of expert knowledge. Information technology research suggests that expert system project managers should focus beyond what conventional wisdom says. Carroll [8] examines a project office that designed and implemented a structured and comprehensive project permit application tool. He suggests that the availability of a Project Authorization Request tool can significantly contribute to the training and development of project and organizational personnel and act as a central source of information exchange. From delay to cash flow forecasting to project duration, AI has added to the project management literature, providing project managers with superior computing power and reducing reliance on human input. Cheng and Roy [9] suggest that artificial intelligence can reduce the reliance on human input in cash flow forecasting systems. They propose an evolutionary fuzzy support vector machine inference model for time series data, an artificial intelligence hybrid system for predictive project performance management that focuses on cash flow forecasting. Trappey et al. [34] plan to create an intelligent workflow management system capable of managing complex project processes, controlling project processes, and dynamically allocating available resources for project execution across business boundaries. In research, processes and resource allocations are intelligently justified using a virtual web-based environment and controlled by the reasoning engine of an expert system. The proposed system makes it possible to accurately manage project outcomes, to satisfy the project's objectives and deliverables.

Wauters and Vanhoucke [36] discuss five AI approaches for predicting project duration using Monte Carlo simulation, Principal Component Analysis, and cross-validation as an approach. They proved that, when the training and test sets are at least similar, the AI approaches outperform the Earned Value Management (EVM)/ Earned Schedule (ES) methods. Furthermore, the AI approaches produce outstanding predicting outcomes at the early and mid-stages. A robustness experiment steadily widens the gap between the training and test sets, revealing the limits of the recently suggested AI approaches.

2.1 *Artificial Intelligence*

The term "Artificial Intelligence," commonly referred to as AI, encapsulates the emulation of human-like cognitive processes through computer systems and algorithms. It involves the creation of software and hardware systems that enable machines to imitate human behavior, decision-making, learning, and problem-solving. AI technologies aim to provide computers with the ability to process and interpret data, make informed judgments, adapt to changing situations, and improve their performance over time.

AI is a multidisciplinary field that draws inspiration from various domains such as computer science, mathematics, psychology, neuroscience, and linguistics. The ultimate goal of AI is to develop intelligent agents capable of interacting with their environment in a way that exhibits human-like intelligence. This encompasses tasks such as understanding natural language, recognizing patterns in data, playing complex games, driving vehicles autonomously, and making informed recommendations based on historical information [18].

The roots of AI can be traced back to the mid-twentieth century when researchers began exploring the possibility of creating machines that could simulate human reasoning. The term "artificial intelligence" was coined in 1956, marking the official beginning of the field. Since then, AI has evolved significantly, experiencing periods of both progress and stagnation. Breakthroughs in machine learning, neural networks, and deep learning have led to remarkable achievements in recent years, propelling AI into the forefront of technological advancement [7].

AI applications are ubiquitous in today's world, impacting a wide range of industries such as healthcare, finance, manufacturing, entertainment, and more. From medical diagnostics to fraud detection, recommendation systems to autonomous drones, AI technologies are transforming the way we live and work. However, AI also raises ethical and societal concerns, particularly related to data privacy, bias in algorithms, and the potential displacement of certain jobs [18].

In conclusion, artificial intelligence represents a visionary pursuit to create machines that can replicate, and even surpass, human cognitive abilities. The journey of AI development continues to shape the technological landscape, promising unprecedented opportunities and challenges as we endeavor to bridge the gap between human and machine intelligence.

2.2 Project and Project Management

In the realm of business and organizational management, a project is a temporary endeavor undertaken to achieve a unique goal, product, service, or result. Projects are defined by a clear beginning and end, scope, objectives, and allocated resources. Project management, in turn, refers to the application of knowledge, skills, tools, and techniques to project activities to meet the project requirements effectively. The discipline of project management encompasses a range of methodologies, best practices, and frameworks designed to guide project teams in delivering successful outcomes within the constraints of time, budget, and scope [15].

Project management is built on several core concepts that underpin the successful planning, execution, and closure of projects. These concepts include scope, which defines the work that needs to be done; time, which establishes a timeline for project activities and milestones; cost, which manages the budget and expenses; quality, which ensures deliverables meet standards; risk, which identifies potential challenges; communication, which facilitates information flow; and resources, which involves managing human, material, and financial assets [17].

Projects typically follow a structured life cycle consisting of several phases. The initiation phase involves conceptualizing the project and assessing its feasibility. The planning phase entails developing detailed project plans, including schedules and budgets. In the execution phase, project activities are carried out, and progress is monitored. Monitoring and controlling involve tracking performance and making necessary adjustments to keep the project on track. Finally, the closure phase finalizes the project, hands over deliverables, and conducts a post-project review to assess performance and identify lessons learned. Project management offers a variety of methodologies and frameworks to suit different project types and environments. These include the Waterfall approach, which follows a linear sequence of phases; Agile, an iterative approach that emphasizes flexibility and customer feedback; Scrum, a framework within Agile that focuses on collaboration and incremental delivery; PRINCE2, a structured methodology emphasizing clear project governance; and the Critical Path Method (CPM), which schedules and identifies the longest sequence of tasks in a project [21].

The field of project management is continuously evolving, with emerging trends and technologies reshaping traditional practices. Digital transformation integrates cloud computing, artificial intelligence, and data analytics into project management processes. Remote and distributed teams present new challenges and opportunities for managing geographically dispersed project teams. Sustainable project management focuses on environmental and social sustainability in project planning and execution. Change management effectively handles organizational change to ensure successful project outcomes [22].

In conclusion, project management is a multifaceted discipline that plays a crucial role in the successful delivery of projects across industries. By leveraging best practices, methodologies, and emerging technologies, project managers can navigate the complexities of modern projects and deliver value to stakeholders. As the landscape of business and technology continues to evolve, project management will remain an essential function for achieving organizational goals and driving innovation.

In recent years, the integration of Artificial Intelligence (AI) into PM practices has ushered in a new era of efficiency and innovation. AI's capacity to process vast amounts of data, make predictions, automate tasks, and enhance decision-making processes has positioned it as a transformative force in the field of project management. This article explores the multifaceted role of AI in PM, shedding light on how it is revolutionizing the way projects are planned, executed, and monitored.

One of the most significant contributions of AI to PM is its ability to provide data-driven insights. AI algorithms can analyze historical project data, identify patterns, and offer valuable recommendations for project planning and resource allocation. This data-driven approach empowers project managers to make informed decisions that increase the likelihood of project success [23].

Figure 1 illustrates the integration of artificial intelligence (AI) in project management and how it enhances the effectiveness of managing projects.

Fig. 1 Use of artificial intelligence in project management [27]

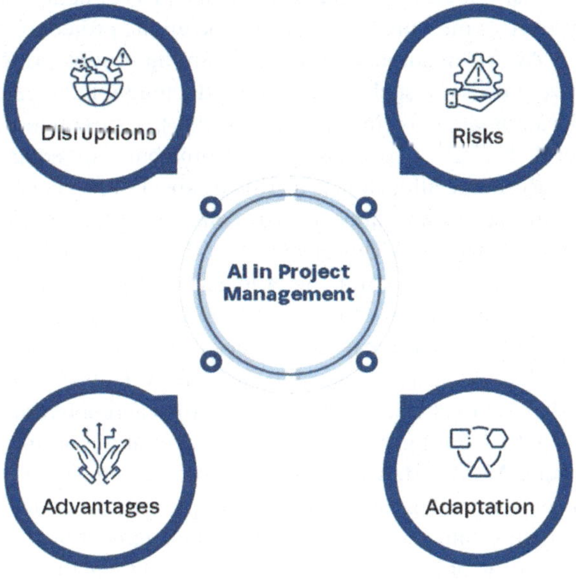

In project management, AI plays a crucial role across various dimensions. It assists in identifying potential disruptions by analyzing vast datasets and patterns, thereby enabling proactive mitigation strategies. Additionally, AI enhances risk management through predictive analytics, empowering project managers to anticipate and address risks before they escalate. The integration of AI within project management also yields numerous advantages, including improved decision-making, resource optimization, and real-time insights, consequently fostering efficiency and productivity. Furthermore, AI facilitates adaptation by dynamically adjusting project plans based on changing circumstances and evolving requirements, ensuring agility and resilience in project execution.

AI aids in defining project objectives and optimizing planning processes during the project planning phase. This results in more accurate and efficient project plans. AI supports effective allocation of human and material resources in projects, helping project managers make better-informed decisions. AI can be used to identify and analyze potential risks in advance, enabling proactive risk mitigation and management.

Monitoring project progress in real time and evaluating performance against goals is made easier with AI. This allows project managers to intervene in a timely manner when necessary. AI analyzes large amounts of data to provide project managers with data-driven recommendations, making decision-making processes more effective and efficient. AI facilitates comprehensive and accurate analysis of project outcomes and assessments of successes. This contributes to continuous improvement and increased success.

AI's predictive capabilities are invaluable for project managers seeking to mitigate risks and optimize resource utilization. By analyzing project data and external factors, AI can forecast potential roadblocks, project delays, and budget overruns, allowing proactive measures to be taken to keep the project on track [31].

AI-driven automation is transforming routine project management tasks. Repetitive processes, such as data entry, document management, and status reporting, can be automated, freeing up project teams to focus on more strategic and creative aspects of their work. This not only saves time but also reduces the risk of human errors [1].

Resource allocation is a critical aspect of project management. AI algorithms can optimize resource allocation by considering factors such as skillsets, availability, and project priorities. This ensures that the right people are assigned to the right tasks at the right time, maximizing efficiency and productivity [31].

AI-based tools enable real-time project monitoring and performance tracking. Project managers can access up-to-the-minute data on project progress, identify bottlenecks, and make immediate adjustments as needed. This agility is crucial for adapting to changing project conditions and delivering results promptly [23].

AI's learning capabilities support continuous improvement in project management. As AI systems accumulate more data and learn from past projects, they become even more adept at predicting and preventing issues. This iterative process fosters ongoing enhancements in project management practices.

Artificial Intelligence is also instrumental in the development of Decision Support Systems (DSS) for project management. These systems use AI algorithms to provide project managers with real-time insights and recommendations. DSS can analyze project data, assess the impact of various decisions, and suggest the most optimal course of action. This capability empowers project managers to make more informed and strategic decisions throughout the project lifecycle [2].

NLP is a subset of AI that is making communication and collaboration in project management more efficient. AI-powered chatbots and virtual assistants can understand and respond to natural language queries, facilitating smoother communication among team members and stakeholders. NLP also enables automated document summarization and sentiment analysis, helping project managers gauge the overall sentiment and feedback related to a Project [4].

AI plays a crucial role in risk management and contingency planning for projects. By analyzing historical data and external factors, AI algorithms can identify potential risks and recommend strategies for mitigating them. This proactive approach ensures that project managers are better prepared to handle unexpected challenges, reducing the likelihood of project disruptions and ensuring project success [31].

In conclusion, AI is not merely a buzzword in project management; it is a powerful ally that empowers organizations to deliver projects more efficiently, on time, and within budget. The integration of AI tools and techniques into PM practices represents a significant step toward achieving greater project success rates and ensuring a competitive edge in today's rapidly evolving business landscape. As AI continues to advance, its role in PM will undoubtedly become even more prominent, reshaping the way projects are conceived, executed, and completed.

2.3 Research in the Applications of Artificial İntelligence for Project Management

Possible applications of AI in project management are categorized and reviewed as following.

Project Planning and Predictions Artificial Intelligence plays a significant role in assisting project managers with complex project planning and predictions. It can analyze past project data and provide recommendations to optimize processes for future projects. Particularly in large-scale projects, AI helps improve resource allocation, timelines, and cost estimates [33].

Risk Management and Prevention AI is critical in identifying and managing project risks. AI algorithms can identify potential risks based on historical data and external factors, offering strategies to reduce or prevent these risks. This helps project managers be better prepared for unexpected challenges and ensures project success [14].

Resource Management and Allocation Artificial Intelligence provides a substantial advantage in resource management for project managers. By considering factors such as skills, availability, and project priorities, AI can optimize resource allocation. This ensures that the right people are assigned to the right tasks at the right time, leading to increased efficiency and productivity [33].

Real-time Monitoring of Project Progress AI-based tools allow real-time monitoring of project progress and performance. Project managers can access real-time progress data, identify issues, and make quick adjustments as necessary. This agility is crucial for adapting to changing project conditions and delivering results on time [28].

Decision Support Systems (DSS) AI also plays a significant role in the development of DSS for project management. These systems utilize AI algorithms to provide project managers with real-time insights and recommendations. DSS can analyze project data, assess the impact of various decisions, and suggest the most optimal course of action. This capability empowers project managers to make more informed and strategic decisions throughout the project lifecycle [3].

In conclusion, the application of AI in PM extends well beyond data analysis and automation. AI is reshaping decision-making processes, enhancing communication, and strengthening risk management strategies. As organizations continue to embrace AI technologies, PM practices will evolve to become more adaptable, efficient, and responsive to the demands of today's complex projects. The collaboration between human expertise and AI capabilities is poised to redefine the future of project management.

2.4 Research Topics for Future Studies

Similar to the previous section, the following areas require further research for the sake of implementing more efficient and effective project management.

Risk Prediction and Analysis AI serves as a powerful tool in risk prediction and analysis within the domain of PM. Project managers can utilize historical project data to identify and analyze potential future risks. AI algorithms can examine this data, allowing for the prediction of specific risk factors and potential outcomes. This aids in better project preparation, risk mitigation, and resource management [2].

Project Resource Management AI plays a significant role in the effective management of project resources. AI can optimize resource allocation by considering the skills, availability, and job priorities of project personnel. This ensures that the right individuals are assigned to the right tasks, ultimately enhancing project success [10].

Time Schedule Management Effectively managing time is a crucial aspect of PM success. AI can assist in managing dates and timelines efficiently. AI can track project progress and predict delays or issues as deadlines approach, enabling proactive measures to ensure timely project completion [35].

Automated Reporting and Documentation AI can automate routine reporting and documentation tasks in PM. Project managers and teams can streamline bureaucratic processes and eliminate manual data entry. This not only saves time but also reduces the likelihood of errors and inconsistencies [23].

Collaboration and Communication Enhancements AI can enhance collaboration and communication among project teams. Chatbots and virtual assistants can answer team members' and stakeholders' questions and provide rapid access to information. Additionally, through natural language processing (NLP), AI can analyze the tone and sentiment in communication, fostering a positive and effective collaborative environment [31].

Data Analysis and Insights AI can be employed to analyze vast amounts of project data. Data mining and machine learning techniques can yield crucial insights related to projects. This data-driven approach can be used for continuous project improvement and as a basis for future decision-making [5].

3 Conceptual Model for the Agent Based Intelligent Project Risk Management

Developing an intelligent risk management system for executing the projects is conceptually defined as the following possible components of this system is given in Fig. 2. As seen in the Figure, the system is based on intelligent agents. Several agents are considered to be enriched with intelligent capabilities and integrated to work together. The system is capable of identifying, planning, eliminating, and monitoring project risks.

Proposed framework for running risk management process starts with the identification of a list of potential risks by agent 1. In order to define a possible risk,

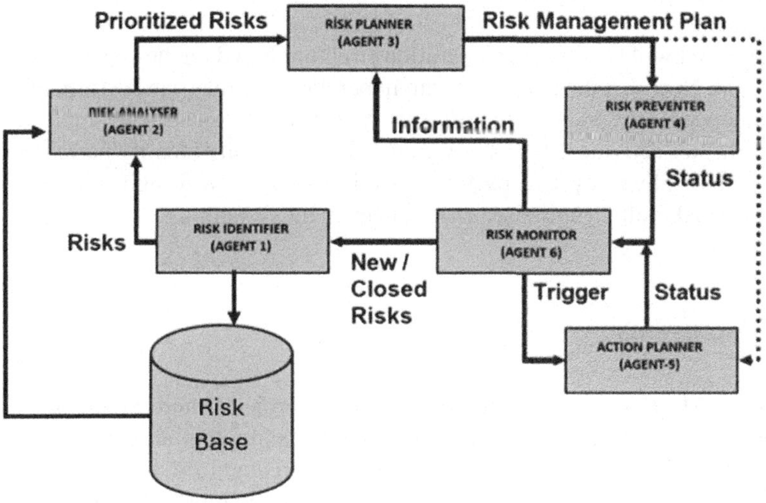

Fig. 2 AI based risk analysis in project management

environment in which project is running should be assessed by another agent (situation analyzer). A risk base is generated and risk with respective information is stored.

Each of these risks is then analyzed and prioritized by agent 2. Once risks are prioritized, Agent 3 then creates a risk management plan that identifies containment actions that will reduce the probability of the risk occurring and/or reduce the impact if the risk turns into a problem. The plan also includes contingency actions that will be taken if the risk turns into a problem and the associated triggers (indicators that the risk is turning into a problem). The containment part of the plan is then implemented by agent 4 and actions are taken by agent 5. On the other hand, the Agent 6 runs monitoring and tracking process. The tracking step involves monitoring the status of known risks as well as the results of risk reduction actions. If a trigger indicates the onset of a problem, the corresponding action plans is generated and are implemented by agent 5. As new status and information are obtained, the risk management plans are updated accordingly. Tracking may also result in the addition of newly identified risks or in the closure of known risks.

Each RM agent could have the following structure as resembled in Fig. 3. As indicated each agent is running in an environment where agent related data is produced. Data is collected and filtered by the situation assessment module with respect to the goal of the agent. Data are stored in the respective databases. Each agent has its own goal which enables the agent to perform respective operations. The goal analyzer defines the focus of attention of the agents Based on its operational requirements agents may perform learning, event prioritization as well as other intelligence capabilities. Then the agent perform respective reasoning to define the action or decision which is to be generated with respect to the data perceived by the situation analyzer. The decision and action are carried out which will than have an effect over the environment.

Each agent will have its own operation structure based on the specific tasks to be carried out. Note that the agents can run in parallel, and they can perform their operation independently. Most of the time every agent is designed with some sub agent to perform sub-activities. This study continues to formally identify basic operations and systematically analyze respective capabilities in order to design fully intelligence and integrated multi agent-based risk management system.

4 Conclusions

AI-based PM offers the opportunity to manage projects more efficiently, rapidly, and predictably compared to traditional methods. In this article, the roles and benefits of AI in PM is explored. It has been recognized that AI is generally used in areas such as data analysis, prediction, automation, and decision support systems. Additionally, AI contributions are highlighted in critical areas such as risk management, resource management, time schedule management, and communication enhancements. However, the successful implementation of AI-based PM comes

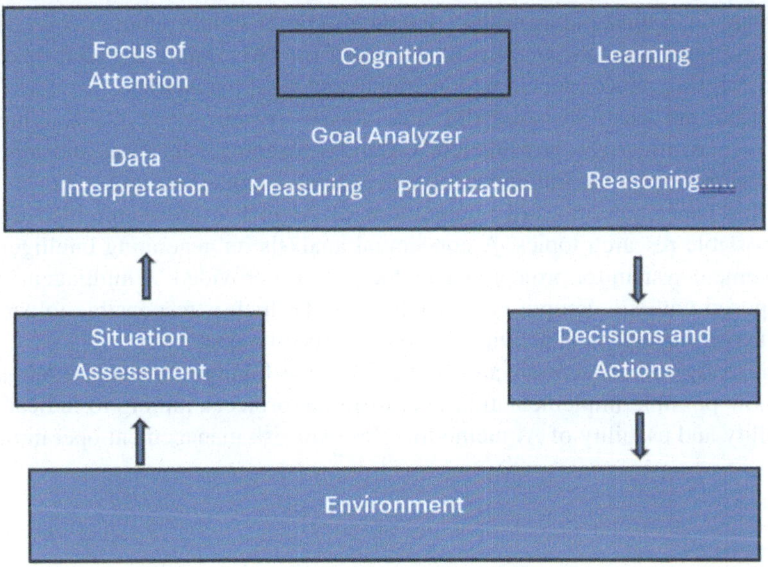

Fig. 3 A general components of a risk management agent

with some challenges and important considerations. Factors such as data privacy, ethical concerns, and human-AI collaboration should carefully be addressed in this context. The following issues seem to be attracting the attention of the research community more and more every day.

Education and Skill Development: Project managers and teams should receive training to effectively utilize AI tools. Ongoing education and skill development programs are essential to enhance knowledge and skills in AI for the sake of better project management. There is a need to define systematic training program for this purpose.

Data Quality and Security: Data quality and security are paramount for AI projects. Organizations should manage their data effectively, protect privacy, and implement data security measures. Standard meta data formats need to be developed and well-used by the research community in order to enrich possible deviations of research based on structured data.

Human-AI Collaboration: AI should be designed to collaborate with humans rather than replace them entirely. Tools and processes that enhance interaction between humans and AI for PM still needs to be searched and easy to use models should be developed.

Research and Development: Continuous research and development efforts should be undertaken to keep pace with new technologies and methods in AI for project management. This would help organizations maintain a competitive advantage and better utilized available resources.

Ethical Compliance: It is certain that the use of AI should adhere to ethical standards and legal regulations. This is also valid for PM. Data usage and operations should carefully be conducted within ethical and legal boundaries.

Last but not least, AI-based PM should assist organizations in managing their projects more effectively and efficiently. Careful planning, education, and adherence to ethical principles are imminent for the successful implementations.

In this paper, possible implementations of AI in PM are highlighted together with possible research topics. A conceptual analysis on generating intelligent risk management system for project management is also provided. A multiagent system is proposed with are composed of agent each of which is responsible to carry out specific risk management activity serving to a specific goal.

This study will concentrate and continue on developing risk management agents and show possible implementation over different projects, aiming to indicate both feasibility and usability of AI methodologies over risk management operations.

References

1. Akinwale P (2023) Artificial intelligence for project managers: a guide for applying artificial intelligence to traditional, hybrid and agile projects through Chatgpt & Google Bard, 2nd edn. Praizion Media
2. Alzoubi HM (2022) Bım as a tool to optimize and manage project risk management. Int J Mech Eng 7(1)
3. Auth G, JokischPavel O, Dürk C (2019) Revisiting automated project management in the digital age–a survey of AI approaches. Online J Appl Knowl Manage (OJAKM) 7(1):27–39
4. Baryannis G, Validi S, Dani S, Antoniou G (2019) Supply chain risk management and artificial intelligence: state of the art and future research directions. Int J Prod Res 57(7):2179–2202
5. Boudreau P (2019) Applying artificial intelligence to project management. KDP Print Us.
6. Buah E, Linnanen L, Wu H, Kesse MA (2020) Can artificial intelligence assist project developers in long-term management of energy projects? The case of CO_2 capture and storage. Energies 13(23):6259. https://doi.org/10.3390/en13236259
7. Buiten MC (2019) Towards intelligent regulation of artificial intelligence. Eur J Risk Regul 10(1):41–59
8. Carroll CT (2004) Structured project requests for control, analysis, and training. Inf Syst Manag 21(2):9–15. https://doi.org/10.1201/1078/44118.21.2.20040301/80416.2
9. Cheng MY, Roy AFV (2011) Evolutionary fuzzy decision model for cash flow prediction using time-dependent support vector machines. Int J Project Manage 29(1):56–65. https://doi.org/10.1016/j.ijproman.2010.01.004
10. Dam HK, Tran T, Grundy J, Ghose A, Kamei Y (2019) Towards effective AI-powered agile project management. In: 2019 IEEE/ACM 41st international conference on software engineering: new ideas and emerging results (ICSE-NIER). IEEE, pp 41–44
11. Forsythe DE, Buchanan BG (1992) Nontechnical problems in knowledge engineering: implications for project management. Expert Syst Appl 5(3–4):203–212. https://doi.org/10.1016/0957-4174(92)90005-D
12. Fridgeirsson TV, Ingason HT, Jonasson HI, Jonsdottir H (2021) An authoritative study on the near future effect of artificial intelligence on project management knowledge areas. Sustainability (Switzerland) 13(4):1–20. https://doi.org/10.3390/su13042345
13. García JAL, Peña AB, Pérez PYP, Pérez RB (2017) Project control and computational intelligence: trends and challenges. Int J Comput Intell Syst 10(1):320–335. https://doi.org/10.2991/ijcis.2017.10.1.22

14. Gil J, Martinez Torres J, González-Crespo R (2021) The application of artificial intelligence in project management research: a review
15. Ginger L (2018) Project integration management. In: Pmp® exam preparation. Auerbach Publications
16. Hazir Ö (2015) A review of analytical models, approaches and decision support tools in project monitoring and control. Int J Project Manage 33(4):808–815. https://doi.org/10.1016/j.ijproman.2014.09.005
17. Hornstein HA (2015) The integration of project management and organizational change management is now a necessity. Int J Project Manage 33(2):291–298
18. Jackson PC (2019) Introduction to artificial intelligence. Courier Dover Publications
19. Jüngen FJ, Kowalczyk W (1995) An intelligent interactive project management support system. Eur J Oper Res 84(1):60–81. https://doi.org/10.1016/0377-2217(94)00318-7
20. Lauras M, Marques G, Gourc D (2010) Towards a multi-dimensional project performance measurement system. Decis Support Syst 48(2):342–353. https://doi.org/10.1016/j.dss.2009.09.002
21. Layton MC, Ostermiller SJ, Kynaston DJ (2020) Agile project management for dummies. Wiley & Sons
22. Locatelli G, Ika L, Drouin N, Müller R, Huemann M, Söderlund J, Clegg S (2023) A manifesto for project management research. Eur Manag Rev 20(1):3–17
23. Lock D (2020) Project management. Routledge, Hampshire England
24. Manik Reddy P, Iyer J (2018) Effective collaboration across the globe through digital dashboards and machine learning. In: 2018 IEEE/ACM 13th international conference on global software engineering (ICGSE), pp 30–34
25. Marques G, Gourc D, Lauras M (2011) Multi-criteria performance analysis for decision making in project management. Int J Project Manage 29(8):1057–1069. https://doi.org/10.1016/j.ijproman.2010.10.002
26. Medeni TD (2018) Selected reflections and elaborations on situational leadership and management: proposing situation-action model for project contextualization and human and robot resource management, Postgraduate, Halu Oleo University, Kendari, Indonesia. World J Adv Res Rev 2023, 17(03):632–643
27. Millawati N (2023) Artificial intelligence in project management: disruptions, risks, advantages, and adaptation. Institute of Project Management. https://instituteprojectmanagement.com/blog/artificial-intelligence-in-project-management-advantages-disruptions-and-adaptation/. Accessed 20 Apr 2024
28. Niederman F (2021) Project management: openings for disruption from AI and advanced analytics. Inf Technol People 34(6):1570–1599. https://doi.org/10.1108/ITP-09-2020-0639
29. Ong S, Uddin S (2020) Data science and artificial intelligence in project management: the past, present and future. J Mod Project Manage 7(4):26–33. https://doi.org/10.19255/JMPM02202
30. Pospieszny P, Czarnacka-Chrobot B, Kobylinski A (2018) An effective approach for software project effort and duration estimation with machine learning algorithms. J Syst Softw 137:184–196. https://doi.org/10.1016/j.jss.2017.11.066
31. PMI (2017) A guide to the project management body of knowledge (Pmbok® Guide), 6th edn. Project Management Institute
32. Sabahi S, Parast MM (2020) The impact of entrepreneurship orientation on project performance: a machine learning approach. Int J Prod Econ 226:107621. https://doi.org/10.1016/j.ijpe.2020.107621
33. Taylor P (2021) AI and the project manager. Taylor & Francis, New York, USA
34. Trappey AJC, Chiang TA, Ke S (2006) Developing an intelligent workflow management system to manage project processes with dynamic resource control. J Chin Inst Ind Eng 23(6):484–493. https://doi.org/10.1080/10170660609509345
35. Wang Q (2019) How to apply AI technology in project management1, 2. PM World J VIII(III)
36. Wauters M, Vanhoucke M (2016) A comparative study of artificial intelligence methods for project duration forecasting. Expert Syst Appl 46:249–261. https://doi.org/10.1016/j.eswa.2015.10.008

Structural Modelling of Engineering Performance Under Crisis: The Case of Covid-19 Pandemic

Omar Al-Kaff and Zeynep Gergin

Abstract Employees' performance has been widely assessed especially during challenging and uncertain times like the COVID-19 pandemic, as organizational success and sustainability crucially rely on the performance of its employees. However, greater responsibilities could lead to critical consequences, therefore, this paper explores the COVID-19 pandemic effect on engineers' performance, since the literature lacks focused metrics for engineers' performance. Partial Least Squares Structural Equation Modelling (PLS-SEM) is applied for the analysis of 100 engineers' responses during the COVID-19 crisis. The validity and reliability of the measurement model are established according to the newly suggested assessment method. The results represented a total positive effect of the COVID-19 pandemic on engineers' performance through Organizational Change, Work-Life Balance, Job Satisfaction and Job Stress as mediating factors. The perceived work stress result of the tested engineers hints at boosting the performance rather than diminishing, and engineers with extended experience along with being unmarried showed relatively better performance compared to others.

Keywords Engineer's performance · COVID-19 pandemic · Structural equation modelling

1 Introduction

The employees' performance is a critical factor of success in any organization, especially when they are responsible for taking crucial decisions and performing imperative executions, which probably is also the case for most engineers. Assessment of the workforce performance is a vital management task that helps the organizations

O. Al-Kaff (✉) · Z. Gergin
Industrial Engineering Department, Istanbul Kültür University, Istanbul, Turkey
e-mail: oyk@outlook.sa

Z. Gergin
e-mail: z.gergin@iku.edu.tr

F. Çebi et al. (eds.), *Engineering and Technology Management in Challenging Times*,
Lecture Notes in Networks and Systems 1137,
https://doi.org/10.1007/978-3-031-70935-7_18

in evaluating the employees' contribution to reach the organization goals [1]. During the recent Corona Virus of 2019 (COVID-19) pandemic, many questions were raised about its impact on organizations and businesses due to the obvious effects on the economy and the social life. One of these questions focuses on the impact of the pandemic on the employees' performance after the changes in the working environments due to the preventative actions taken against the virus spread. Probably many companies applied the online-working method, and mostly all the organizations have taken the social-distance into consideration, besides other preventions. Related to the previous concern, Bustamam et al. [2] believed that the employees as human beings are influenced by their emotional and psychological aspects, and the balance between their life activities and work plays a major role in their performance. Thus, these sudden changes resulted by the pandemic, economical and health risks, and the sudden changes in social life activities in general, may have affected the engineers psychologically, and their performance consequently.

With this motivation, the main objective of this research is to investigate the impact of COVID-19 pandemic on the Engineers' Performance (EP) that implicates sub-objectives to achieve which are:

1. Proposing factors which relate the pandemic with the engineers' performance.
2. Constructing a valid and reliable model that relates the COVID-19 pandemic with the engineer's performance.
3. Testing ten hypotheses (H1–H10) with the data collected with the developed questionnaire.

The significance of this research may present in connecting the extraordinary new case of COVID-19 and its effect on the engineers' performance by providing a reliable and valid model that relate probable effective factors in such pandemics, which may enrich the literature by providing valuable experience for similar future situations.

The rest of the paper is organized as follows; the theoretical context underlying the research hypotheses is discussed in the following Literature Review. The structural model showing the hypotheses and the implementation results are provided under the third section as Methodology and Implementations. The paper is finalized with the findings, limitations and implications in the last Conclusion section.

2 Literature Review

Throughout literature, there is no significant focus on the impact of COVID-19 pandemic relevant to engineers' performance, though almost all the literature studies consider the employees performance topic, each in its specific industry as a workforce assessment tool [3–5]. Generally, many factors proposed in the literature have a direct or indirect effect on the employees' performance; most probably their importance varies from one case to another, and between different organizations and situations. Alefari et al. [6] studied for performance through the literature in the last decade and reported that the most frequently mentioned factors are observed as leadership style, organizational culture, employee wellbeing, and training; while the least frequent are flexible working arrangement, work-life balance, career development, and external environment.

Almost globally, transformational organizational change has been a mandatory action taken by businesses to survive and cope during COVID-19 pandemic [7]. Irawanto et al. [8] supports the negative impact of stress on job satisfaction, and Saifer et al. [9] believes that increasing the extent of organizational changes and workers' stress decreases employees' job satisfaction. Halkos and Dimitrios [10] discusses that organizational changes may escalate employees' stress, but when employees develop a clear understanding of the reasons and goals behind the implemented changes their performance may increase. Work arrangements and technological changes are suggested to impact work-life balance [11], both variables may be included under organizational changes during COVID-19 pandemic, and the pandemic related organizational changes have been assumed to decrease work-life balance. Bataineh [12] explores the effect of work-life balance and happiness on the pharmaceutical employees' performance in Jordan and argues that enhancing work-life balance probably improves the workers' performance. Additionally, during COVID-19 the case studied by Wolor et al. [5] indicates the same positive effect as [12]. Halkos and Dimitrios [13] argued that stress among employees commonly developed due to the imbalance between work and social life, which affects the performance negatively, while suggesting a weaker effect of stress on the produced work quality. Similarly, according to [14], work-life balance has a significant positive effect on job satisfaction directly and indirectly through motivation. Pandey [15] also discussed how stress is a common psychological factor and supported the idea of inverse relationship between stress and performance of workers, which means higher stress decreases the performance. Many studies verified with research outputs that higher job satisfaction leads to higher improvement in worker performance [16–18]. On the other hand, most studies focused on the stress and other mentioned factors as independent variables and examined the impact on performance, and none focused specifically on engineers' performance. Also, many discussed the organizational changes due to strategic decisions. Hence, this study is initiated to examine the impact of organizational changes based on unforeseen crisis—i.e., the pandemic—on a specific career area considering the mediating variables—i.e., work-life balance, job stress and job satisfaction.

3 Methodology and Implementation

3.1 Research Model and Hypotheses

The proposed factors which are suggested in the literature to impact the employees' performance in general and possibly may have a connection with the *COVID-19 (CP)* widespread are identified for this study as: *Organizational Changes (OC), Work—Life Balance (WLB), Job Satisfaction (JS), and Job Stress (S)*. After choosing the factors, 10 hypotheses have been developed as following:

H1: *COVID-19 Pandemic (CP)* increases *Organizational Changes (OC).*

H2: The *Organizational Changes* impact the *Job Satisfaction (JS)* negatively.

H8: *Job Stress* decreases *Job Satisfaction.*

H3: Organizational *Changes* increases engineers' *Job Stress (S).*

H4: Organizational *Changes* decreases *Work-life Balance (WLB).*

H5: *Work—Life Balance* has a direct positive impact on *Engineers' Performance (EP).*

H6: *Work—Life Balance* decreases *Job Stress.*

H7: *Work—Life balance* increases *Job Satisfaction.*

H9: *Job Stress* has a negative impact on *Engineers' Performance.*

H10: *Job Satisfaction* increases *Engineers' Performance.*

Figure 1 represents the constructed theoretical model using the developed hypotheses. The model may contain different indirect path alternatives and mediation effects, so these paths have not been hypothesized due to prediction complexity of plentiful indirect path possibilities in the model.

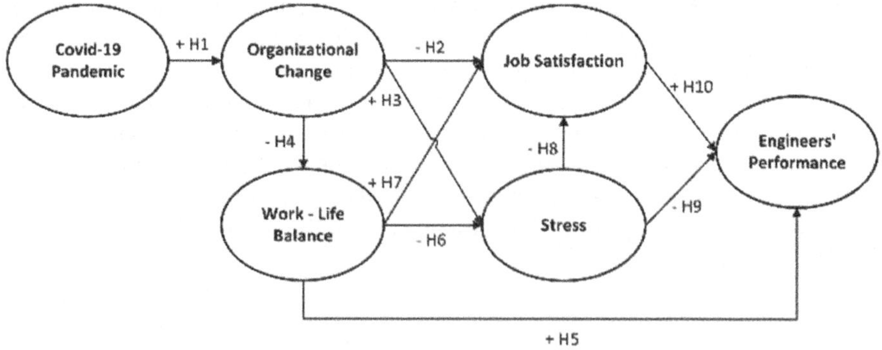

Fig. 1 Theoretical structural model before modification

3.2 Analytical and Measurement Tools

A questionnaire is designed, and survey method has been used to collect the data, Structural Equation Modeling (SEM) has been performed for the analysis. As a brief comparison of the main types of SEM, Partial Least Square SEM (PLS-SEM) is more suggested for exploratory studies, non-normal distributed data, and small sample size cases, while Covariance Based SEM (CB-SEM) is suggested for confirmatory studies, normal distributed data, and larger sample size cases [19, 20]. PLS-SEM can also be applied to confirmative research because of the difficulty of collecting relatively large sample sizes [21]. Thus, PLS-SEM has been applied in this research via SmartPLS software, taking the advantage of avoiding the normality limitations. Furthermore, conducting measurement model analysis is mandatory to ensure validity and reliability of the measurement model, where this study has adapted from [21], the newly proposed method Hair's Confirmatory Composite Analysis (HCCA) which performed within PLS-SEM.

Items for each factor for the survey have been either generated or adopted from the literature [22–35]. In addition, items for each construct have been assessed by a panel of experts where face validity has been established. Since the chosen factors are considered latent, the items have been chosen according to some representative variables rather than selecting items randomly during survey design process. Because of the increased number of measures in this research, only three dimensions of employee performance (EP) have been used, namely Task Performance, Adaptability, and Counter Productivity [36], and Task Performance is defined by Quality and Delivery [37]. Due to the fuzzy nature of organizational change, a simple construct has been developed to serve the purpose of this research in the extraordinary crisis of COVID-19. All factors are measured using 7-point Likert scale and assumed ordinal, as [38] believes that Likert scale generates ordinal data or can be assumed interval in some cases. The developed survey contains of two main parts; first 7 questions are designed to collect data on demographic information, such as *Gender, Work Experience, Marital Status, Responsibility of Child/Children, Engineering Discipline, Industry, and Country*; beside the other 30 items for analyzing the model's variables.

3.3 Implementation

The survey has been implemented via Google Forms to reach the respondents online during the COVID-19 crisis, and to ease the data collection. Snowball sampling method have been used to reach respondents (engineers) through engineers' groups or contacts using Facebook and LinkedIn websites and 100 contributions have been received in two months period (from 25/2/2022 to 26/4/2022). After data collection the data cleaning has been performed in terms of individual's non variate responses

Table 1 Sample demographic information

Control variable	Answer	Frequency
Gender	Male	48
	Female	52
Experience	5 years or less	69
	6–10 years	13
	11 years or more	18
Marital status	Single	73
	Married	27
Responsibility of children	No	78
	Yes	22

and univariate outliers, the Standard Deviation (STD) for all items of each respondent have been calculated to eliminate non variate responses which probably are invalid. Box plots of each construct have been generated to investigate outliers. Fortunately, neither non variate observations nor outliers have been found and the number of responses kept as one hundred. Some of sample demographic characteristics which are used in the SEM analysis, have been presented in Table 1 as frequencies. The number of both genders were approximately divided equally, while most engineers were single, less than 5 years experienced, and had no children. Additionally, demographic information of respondents in terms of region display that 59% of the respondents are from Middle East countries while 20% are from Europe, 6% are from America and 7% are from other countries. Furthermore, the variety of the engineering disciplines provided balanced figures with 16% Architectural, 16% Civil, 15% Chemical, 14% Industrial, 13% Mechanical, 9% Electrical, 7% Computer and Software, and 10% other disciplines.

4 Results and Discussion

Initially, the structural and measurement models have been set in SmartPLS software completely as hypothesized in the theory before any applied modifications as displayed in Fig. 2.

The first step on HCCA requires items loadings assessment, each loading should be $\geq$0.708 [21]. After the first run, items below the threshold have been omitted one by one and loading reassessment performed after each deletion until the passing criteria is achieved. 11 iterations are applied and deleted items are: CP1, CP2, E5, E6, JS1, JS5, OC3, OC5, WLB3, WLB4, and WLB5; item WLB2 have been omitted lastly to achieve the discriminant validity. Discriminant validity assessment is applied using Heterotrait—Monotrait (HTMT) Ratio of Correlations, and HTMT values should be $<$0.90 to ensure discriminant validity [21]. HTMT value of work-life balance with job stress is found above 0.90 which implies HTMT modification necessity. WLB2 item

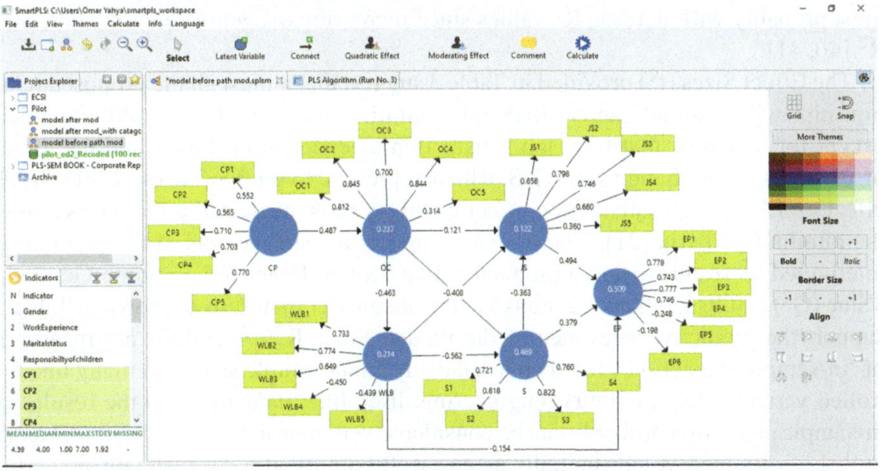

Fig. 2 Structural and measurement model before modification using SmartPLS software

showed the highest cross loading value on job stress, so it has been eliminated and HTMT values have been improved from 0.942 to 0.724, and all the constructs met the acceptance HTMT criteria of discriminant validity as shown in the HTMT matrix Table 2. Additionally in Table 2, composite reliability values are above 0.70 which supports the constructs' reliability, and all values of Average Variance Extracted (AVE) exceed 0.50, so the convergent validity have been achieved [21].

Coefficient of Determination (R^2) results have been generated after testing measurement model reliability and validity, as represented in Table 3, and all the values are significant at $P < 0.05$ except for job satisfaction which have been considered significant at $P < 0.10$. Since R^2 values of engineers' performance and job stress are around 0.50, probably these two endogenous factors have been described moderately by the other factors in the model. In other words, 44.3% of engineers' performance and 50.4% of job stress have been described by the modified model, while other factors are insufficiently described using the study sample where R^2 values are below 0.25 [39]. Thus, adding more factors to predict JS, OC, and WLB

Table 2 Convergent validity, construct reliability, and discriminant validity for the modified model

Construct	Cronbach's Alpha	Composite reliability	AVE	HTMT					
CP	0.707	0.833	0.626		CP	EP	JS	OC	S
EP	0.795	0.867	0.619	EP	0.412				
JS	0.695	0.826	0.614	JS	0.142	0.68			
OC	0.823	0.892	0.734	OC	0.509	0.546	0.164		
S	0.786	0.862	0.611	S	0.685	0.451	0.237	0.611	
WLB	1	1	1	WLB	0.479	0.229	0.214	0.346	0.724

most probably will increase R^2 values since increasing exogenous factors increases R^2 [40, 41].

The Effect Sizes (f^2) provided in Table 3 suggest that almost all the proposed relationships will be significantly affected by omitting any path in the modified model, except for the paths that treat job satisfaction as endogenous factor and the effect of organizational change on WLB which represent low effect sizes, while WLB effect on EP and job satisfaction effect on stress show neglected effect sizes below the minimum criteria [21]. These results support the suggestions of the previously interpreted R^2 results to add more factors that explain JS and WLB. Stone-Geisser's Value (Q^2) results of all endogenous factors are greater than zero as shown in Table 3 support the predictive relevance of the model. Since R^2, f^2, and Q^2 are measures of 'in-sample prediction', for such a relatively small sample size with many uncontrolled variables (e.g., country, engineering discipline, industry, etc.) the results of in-sample prediction probably can be considered very reasonable, and the model most probably will improve substantially using a well-controlled and statistically powerful sample. In this stage after conducting all HCCA steps, the measurement model results show acceptable reliability and validity despite the probable weaknesses of the used sample.

As seen in Fig. 3, in addition to the modified model, some '*control variables*' have been added for further examination, such as *Gender, Work Experience, Responsibility of children, and Marital status. F*ollowing the implementation of [42] which suggest regressing the control variables on the main dependent factor of the study and perhaps on other factors that probably suspected to be influenced by control variables; all control variables have been regressed on EP factor, and both *responsibility of children* with *marital status* have been regressed on WLB. Before testing the paths and hypotheses, all Variance Inflated Factor (VIF) results are less than 3 which refute the existence of multicollinearity problem between the constructs.

Table 3 Coefficient of determination, Stone-Geisser's values, and Effect Sizes for the modified model

Factor	R^2	P value	Q^2	Path	f^2
EP	0.443	0	0.284	CP -> OC	0.221
JS	0.073	0.090	0.012	JS -> EP	0.552
OC	0.181	0.005	0.113	OC -> JS	0.039
S	0.504	0	0.290	OC -> S	0.194
WLB	0.107	0.026	0.062	OC -> WLB	0.120
				S -> EP	0.131
				S -> JS	0.000
				WLB -> EP	0.006
				WLB -> JS	0.035
				WLB -> S	0.508

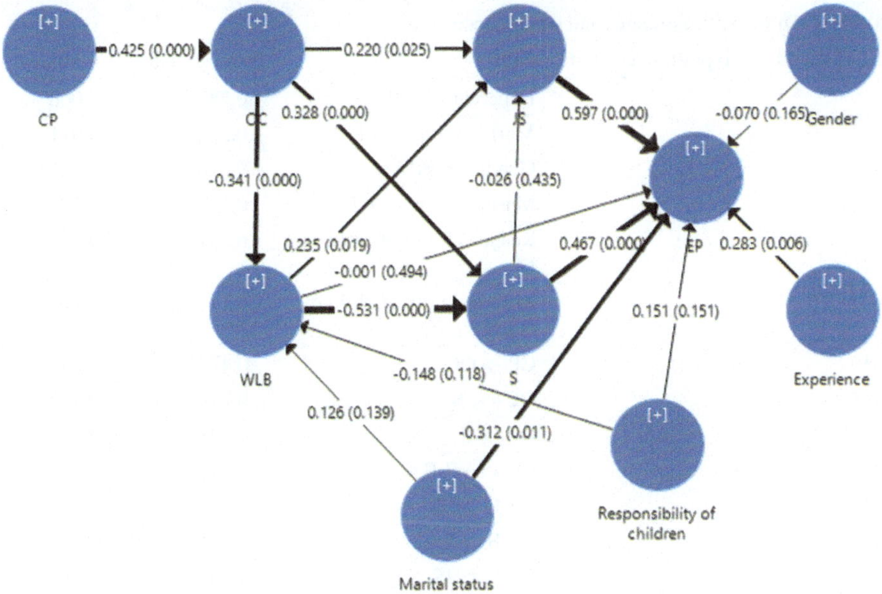

Fig. 3 Modified structural model with categorical variables using SmartPLS software

Table 4 represents each hypothesis with its corresponding path estimates for testing, gender on engineers' performance, marital status on work-life balance, and responsibility of children on both engineers' performance and work-life balance are found to be insignificant. As marital status—as a dump variable—shows negative impact on performance, and being single is coded zero as the marital status, the results express that single engineers perform higher than married. Work experience variable has positive path effect which suggest relatively increased performance for more experienced engineers. Regarding hypotheses testing, H5 and H8 are insignificant paths, so the direct effect of WLB on EP and the direct effect of job stress on JS cannot be justified by collected sample results.

On the other hand, the paths of H2 and H9 show opposite effect of what have been expected, H2 assumed the negative effect of organizational changes on job satisfaction but the result suggests a relatively weak positive impact. However, as [43] results also suggest the positive impact of OC on the employees' job satisfaction, the supportive culture of organizations may overcome the negative consequences of organizational change and result a positive impact on job satisfaction [44] and probably there are other factors mediated and caused the positive impact of OC and JS path.

In addition, job stress was expected to affect EP negatively via H9 but actually it enhanced EP. This result reaches an agreement with the idea that the proper amount of job stress could motivate and push employees to accomplish their work adequately, meaning an increase in their performance [45, 46]. Most probably majority of the examined sample have experienced well managed conditions in their organizations

Table 4 Direct path estimates and significance

Hypothesis	Hypothesized sign	Path	Path estimates	P Values
		Experience -> EP	0.283	0.006
		Marital status -> EP	−0.312	0.011
		Gender -> EP	−0.070	0.165
		Marital status -> WLB	0.126	0.139
		Responsibility of child -> EP	0.151	0.151
		Responsibility of child -> WLB	−0.148	0.118
H1	+	CP -> OC	0.425	0
H2	−	OC -> JS	0.220	0.025
H3	+	OC -> S	0.328	0
H4	−	OC -> WLB	−0.341	0
H5	+	WLB -> EP	−0.001	0.494
H6	−	WLB -> S	−0.531	0
H7	+	WLB -> JS	0.235	0.019
H8	−	S -> JS	−0.026	0.435
H9	−	S -> EP	0.467	0
H10	+	JS -> EP	0.597	0

during the pandemic which enabled raising just the proper amount of motivational job stress to improve the performance.

H1, H3, H4, H6, H7, and H10 are supported by path results; H1 path suggests that COVID-19 pandemic increased the organizational changes moderately by factor of 0.425, which seems very reasonable since the pandemic and the major difficult situations presumably require adaptation and survival actions taken by organizations. Furthermore, organizational changes increased job stress slightly as H3 result conducts, but probably that job stress was well managed by the organizations along with their engineers as it improved the performance rather than harming it. Although H4 path suggests that OC decreased work-life balance slightly, it seems it did not reach to any harmful level since engineers were still satisfied slightly by their WLB as hinted by H7. Additionally, despite the slightly decreased level of WLB it was enough not only to satisfy engineers but also to decrease their job stress moderately as provided in H6. The result of H10 path suggests that job satisfaction increased the EP considerably, so most of examined engineers were managed to be satisfied after all.

To overview the key finding of the path results, the most important interpretation that can be concluded is that the total effect of COVID-19 pandemic on engineers' performance is a weak positive effect (+0.134) based on the examined sample. That result could be interpreted due to the fact that COVID-19 increased organizational change implementations moderately (+0.425), and probably the implementation of these organizational changes was successful as the total effect result of organizational change on engineers' performance suggested a net positive impact (+0.314).

5 Conclusion

The sudden development of COVID-19 pandemic evoked the need of employees' performance assessment to investigate the effect of this crisis. This drove the research to explore the impact of the sudden crisis of COVID-19 pandemic on the engineers' performance specifically by developing a valid and reliable model that also indicates the moderating variables. The newly proposed method of Hair et al. [21] is used to confirm the constructed measurement model quality in terms of reliability and validity within the PLS-SEM method, which after modification showed reliable and valid model with predictive features. Hypothesized direct effects of "work life balance on employee performance" and "job stress on job satisfaction" (H5 and H8 respectively) are insignificant and cannot be justified by the research sample, while "organizational change effect on job satisfaction" and "job stress effect on engineers' performance" (H2 and H9 respectively) have shown opposite sign of what is expected. The rest of the hypotheses are supported by the path results, and the net effect of total paths suggested a weak positive impact of COVID-19 epidemic on engineers' performance (0.134). It can be implied from these results that engineers may show resilience or adaptability at tough times since performance showed overall sustainable results. Thus confirming these attributes of engineers on broader population and understanding the drivers behind these characteristics could lead to fostering benefits beside engineers. Another insight which can be explored is how stress can be managed and controlled within harmless motivating limits, and this may be done by further research on the mediators between stress and performance in the proposed model, by studying the most effective aspects of management in such situations. Moreover, especially with the scarcity of studies on organizational change-inclusive models during crisis periods, the created structural model may serve as a strong basis for future outbreaks, and engineering-centered research.

The main limitation of the study is the sample size and the generalizability of the results, since collecting large and well representative data of a global population of engineers was quite difficult to achieve during the period. Thus, conducting further research using this constructed model on more representative sample can be suggested. Furthermore, since the variance of job satisfaction, WLB, and OC have not been explained sufficiently by the represented factors in the model according to the small R^2 and f^2 values, it is recommended to add exogenous factors that probably impact job satisfaction and WLB directly to improve the model as organizational culture and management related factors.

References

1. Fekete M, Rozenberg I (2014) The practical model of employee performance. In: Human capital without borders: knowledge and learning for quality of life; Proceedings of the management, knowledge and learning international conference, Slovenia
2. Bustamam NM, CheChoh NF, Shaari J, Muda R (2021) Factors affecting the performance of employee in workplace. Int J Mod Trends Soc Sci 3(11):69–79
3. Ahmed S, Al Haderi SMS, Ahmad FB, Jaaffar AR, Walter J, Al Douis GAA (2017) Employee job security and performance relationship in developing economy through employee engagement: critical analysis with PLS-SEM. Int J Econ Res 14(19):133–147
4. Haque A (2021) The COVID-19 pandemic and the role of responsible leadership in health care: thinking beyond employee well-being and organizational sustainability. Leadersh Health Serv 34(1):52–68
5. Wolor CW, Solikhah S, Fidhyallah NF, Lestari DP (2020) Effectiveness of e-training, e-leadership, and work life balance on employee performance during COVID-19. J Asian Fin Econ Bus 7(10):443–450
6. Alefari M, Almanei M, Salonitis K (2020) A system dynamics model of employees' performance. Sustainability 12(16):6511
7. Amis JM, Greenwood R (2020) Organizational change in a (Post-) Pandemic world: rediscovering interests and values. J Manage Stud 58(2):582–586
8. Irawanto D, Novianti K, Roz K (2021) Work from home: measuring satisfaction between work-life balance and work stress during the COVID-19 pandemic in Indonesia. Economies 9(3):96
9. Saifer AG, VanderWielen JJ, Nebeker DM (1995) Organizational change, stress and job satisfaction: three empirically derived models. Proc Hum Factors Ergon Soc Annu Meet 39(13):859–863
10. Halkos GE, Dimitrios B (2012) Importance and influence of organizational changes on companies and their employees. J Adv Res Manage 3(2):90–103
11. Shobitha P, Sudarsan N (2014) Work life balance: a conceptual review. Int J Adv Manage Econ 3(2)
12. Bataineh KA (2019) Impact of work-life balance, happiness at work, on employee performance. Int Bus Res 12(2):99–112
13. Halkos GE, Dimitrios B (2010) The effect of stress and satisfaction on productivity. Int J Product Perform Manag 59:415–431
14. Alfatihah I, Nugroho AS, Haessel E, Maharani A (2021) The influence of work-life balance with work motivation as mediating factor on job satisfaction a prediction toward transition to new normal situation. Manage J Binaniaga 6(1):79
15. Pandey D (2020) Work stress and employee performance: an assessment of impact of work stress. Int Res J Hum Resour Soc Sci 7:124–135
16. Badrianto Y, Ekhsan M (2020) Effect of work environment and job satisfaction on employee performance in pt. Nesinak industries. J Bus Manage Account 2:85–91
17. Mahajan S, Kumar ES (2018) A study on impact of job satisfaction on employee performance in organisation. Asian J Manage 9(3):1046–1054
18. Ratna R, Yezer K, Rasaily BB, Tenzin S, Ghalley L, Lhamo T (2021) Job satisfaction and employee performance at Bhutan board product limited. Effulgence Manage J 19(2):39–52
19. Hair JF, Matthews LM, Matthews RL, Sarstedt M (2017) PLS-SEM or CB-SEM: updated guidelines on which method to use. Int J Multivar Data Anal 1(2):107–123
20. Fan Y, Chen J, Shirkey G, John R, Wu SR, Park H, Shao C (2016) Applications of structural equation modeling (SEM) in ecological studies: an updated review. Ecol Process 5(1), Article 19
21. Hair JF, Howard MC, Nitzl C (2020) Assessing measurement model quality in PLS-SEM using confirmatory composite analysis. J Bus Res 109:101–110

22. Adenubi O, Adebowale O, Adesokan H, Oloye A, Bankole N, Fadipe O, Ayo-Ajayi P, Akinloye A (2021) Knowledge, attitude and perception towards COVID-19 pandemic among veterinary professionals and impacts: a cross-sectional nationwide-based survey. COVID J 1(3):645–665
23. Anyaoku EN, Osuigwe NE, Oguaka CN (2015) Technology and job satisfaction in academic libraries: role of leadership style and librarians' attitude. Int J Lib Sci 4(4):73–80
24. Bernas KH, Major DA (2000) Contributors to stress resistance: testing a model of women's work-family conflict. Psychol Women Q 24(2):170–178
25. Brough P, Timms C, O'Driscoll MP, Kalliath T, Siu OL, Sit C, Lo D (2014) Work–life balance: a longitudinal evaluation of a new measure across Australia and New Zealand workers. Int J Hum Resour Manage 25(19):2724–2744
26. Conway LG, Woodard SR, Zubrod A (2020) Social psychological measurements of COVID-19: coronavirus perceived threat, government response, impacts, and experiences questionnaires
27. Gupta KP, Saurabh S, Bhaskar P (2018) Measuring job satisfaction of customer-facing employees: scale development and validation in the context of organised retail sector. Int J Environ Workplace Employ 4(4):314–346
28. Koopmans L, Bernaards C, Hildebrandt V, Buuren S, Beek A, Vet H (2013) Improving the individual work performance questionnaire using Rasch analysis. Occup Environ Med 70(1)
29. Koopmans L, Bernaards C, Hildebrandt V, Buuren S, Beek AJ, Vet H (2012) Development of an individual work performance questionnaire. Int J Product Perform Manag 62(1):6–28
30. Parlalis S (2011) Organizational changes and job satisfaction among support staff. J Soc Serv Res 37(2):197–216
31. Pradhan RK, Jena LK (2016) Employee performance at workplace: conceptual model and empirical validation. Bus Perspect Res 5(1):69–85
32. Ramos-Villagrasa PJ, Barrada JR, FernándezdelRíoand E, Koopmans L (2019) Assessing job performance using brief self-report scales: the case of the individual work performance questionnaire. J Work Organ Psychol 35:195–205
33. Rincy VM, Panchanatham N (2010) Development of a psychometric instrument to measure work life balance. Cont J Soc Sci
34. Sinval J, Marôco J (2020) Short index of job satisfaction: validity evidence from Portugal and Brazil
35. Van Rooij B, De Bruijn AL, Folmer C, Kooistra E, Kuiper ME, Brownlee M, Olthuis E, Fine A (2020) Compliance with COVID-19 mitigation measures in the United States
36. Ramawickrama J, Opatha HHDNP, Kumari MD (2017) A synthesis towards the construct of job performance. Int Bus Res 10(10):66–81
37. Narayanamurthy G, Tortorella G (2021) Impact of COVID-19 outbreak on employee performance—moderating role of industry 4.0 base technologies. Int J Prod Econ 234(3)
38. Tarka P (2017) The comparison of estimation methods on the parameter estimates and fit indices in SEM model under 7-point Likert scale. Arch Data Sci 2:1–16
39. Hair JF, Ringle CM, Sarstedt M (2011) PLS-SEM: indeed a silver bullet. J Mark Theory Practice 19(2):139–152
40. Saunders LJ, Russell RA, Crabb DP (2012) The coefficient of determination: what determines a useful R2 statistic? Invest Opthalmol Vis Sci 53(11):6830
41. Zhang D (2017) A coefficient of determination for generalized linear models. Am Stat 71(4):310–316
42. Ziaul HM, Tehjeeb N (2020) Young people's perceived service quality and environmental performance of hybrid electric bus service. Travel Behav Soc 20:133–143
43. Hayajneh N, Suifan T, Obeidat B, Abuhashesh M, Alshurideh M, Masa'deh R (2021) The relationship between organizational changes and job satisfaction through the mediating role of job stress in the Jordanian telecommunication sector. Manage Sci Lett 315–326
44. Malatjie I (2019) The impact of organizational change on employee job satisfaction at the National School of Government. Afr J Public Sector Dev Gov 2(1):84–99

45. Chen HL, Fang SC (2019) Job stressors and job performance: modeling of moderating mediation effects of stress mindset. Res J Bus Manage 6(1):35–45
46. Jarinto K (2010) Eustress: a key to improving job satisfaction and health among Thai managers comparing US, Japanese, and Thai companies using SEM analysis. SSRN Electron J 50(4):99–129

Sustainability in the Digital Age: Impact of Digital Transformation on Sustainability

Rula Wedaa and Gulin Idil S. Bolatan

Abstract Digital transformation has emerged as a focal point in the organizational landscape in the past few years, signalling a shift in effective operations and competitive advantage. This study dives into the diverse world of digital transformation, with a particular emphasis on the seamless integration of modern digital technology. These technologies include artificial intelligence, big data analytics, cloud computing, the Internet of Things, and other advancements targeted at changing many aspects of corporate operations, strategies, and ethos, as it explores the impact of digital transformation on sustainability, focusing on economic, environmental, and social aspects.

Keywords Digital transformation · Sustainability · Technology management · Digitalization

1 Introduction

The concept of digital transformation has garnered significant attention in the corporate realm in recent years, digital transformation has become a significant focus in corporate environments, revolutionizing operational efficiency and gaining a competitive edge. This research focuses on the integration of digital technologies, such as big data analytics, cybersecurity measures, and enhanced customer experiences, as digital transformation continues to reshape industries and economies, there is a growing interest in understanding its impact on sustainability. Sustainability, often defined as the ability to meet present needs without compromising the ability of

R. Wedaa
Management Engineering Department Kestel, Alanya Alaaddin Keykubat University, Antalya, Türkiye

G. I. S. Bolatan (✉)
Industrial Engineering Department Kestel, Alanya Alaaddin Keykubat University, Antalya, Türkiye
e-mail: gulin.bolatan@alanya.edu.tr

F. Çebi et al. (eds.), *Engineering and Technology Management in Challenging Times*, Lecture Notes in Networks and Systems 1137,
https://doi.org/10.1007/978-3-031-70935-7_19

future generations to meet their own needs, encompasses economic, environmental, and social dimensions [1]. The pursuit of sustainability has become a global priority, with organizations and governments recognizing the importance of addressing environmental challenges, promoting social equity, and ensuring long-term economic viability.

The intersection of digital transformation and sustainability presents both opportunities and challenges. On one hand, digital technologies have the potential to drive sustainability by enabling more efficient resource utilization, reducing waste, and facilitating the transition to renewable energy sources [1]. For example, the use of data analytics and IoT devices can optimize energy consumption in buildings, leading to energy savings and reduced carbon emissions. Additionally, digital platforms and technologies can enhance supply chain transparency and traceability, supporting sustainable sourcing and responsible production practices [2].

2 Literature Review

2.1 Digital Transformation

The term "digital transformation" refers to a comprehensive and diverse procedure that involves the integration of complex digital technologies, such as artificial intelligence, big data analytics, cloud computing, the Internet of Things, and other state-of-the-art tools, into different aspects of an entity's operations, tactics, and ethos. The objective of this transformative initiative is to fundamentally revolutionize the organization's operations and enhance its ability to deliver value to its stakeholders.

One important part of digital transformation is digital orientation, which refers to an organization's strategic focus on exploiting digital technology to promote innovation and create value.

Digital intensity is a critical element of digital transformation, denoting the extent to which an organization embraces and leverages digital technology in its operational and procedural frameworks, digital intensity is determined by several factors, including the degree of investment in digital technologies, the integration of these technologies across various functions and departments, and the extent to which they are incorporated into the organization's products, services, and customer interactions. It is imperative to underscore that mere digital intensity does not guarantee financial success in the context of digital transformation, as posited by Nasiri et al. [3].

Digital transformation also has implications for environmental sustainability. Advances in digital technologies, such as artificial intelligence, big data analytics, cloud computing, and the Internet of Things, have led to unprecedented disruptions in society, industry, and organizations [4]. However, there is a lack of studies mapping the impact of digital transformation on environmental sustainability. This research conducts a systematic literature review to identify the disruptions driven by digital transformation in the environmental sustainability domain [4]. By understanding these disruptions, organizations can develop strategies to leverage digital transformation for sustainable practices.

Technologies are at the core of digital transformation. They enable organizations to leverage digital tools and platforms to streamline processes, enhance efficiency, and drive innovation. The main challenge in the digital transformation process is not technology itself but the human factor, such as employee resistance to change, lack of digital knowledge and experience, and lack of motivation [5]. Therefore, organizations need to focus not only on adopting the right technologies but also on addressing the human aspects of digital transformation.

Digital transformation is powered by disruptive digital technologies, such as Cloud computing, the Internet of Things (IoT), artificial intelligence (AI) and block chain these technologies enable businesses to collect, analyze, Employing data to arrive at informed decisions, automate processes, and improve customer experience, for efficient operations, process optimization, and better decision-making capabilities are some of the benefits that these SMACIT technologies deliver to organizations [6].

Digital transformation and strategic reactions become simple tasks when utilizing these technologies, as identified by Sebastian et al. [7], With the end goal being smart cities, leading markets delve into untried tech like cloud computing, mobile and IoT, social media, robotics, and AI. This "go with the times" approach requires established technologies as well as emerging ones to exist in an organization's operation. Example papers explore this area [8, 9].

The incorporation and implementation of different technological advancements, such as cloud computing, artificial intelligence (AI), the Internet of Things (IoT), and blockchain, have a chance to enhance operational efficacy, optimize processes and workflow, and improve the quality of decision-making within organizations [6]. These tools, known as SMACIT technologies, are critical in triggering strategic reactions and facilitating digital transformation [9].

Market leaders not only use validated technological solutions but also investigate Emerging technologies, including social media, cloud computing, the Internet of Things (IoT), artificial intelligence (AI), robotics, and mobile technologies, have become increasingly prevalent in recent years. Aligning with the concept of smart cities and emphasizing the integration of digital technologies in organizational operations [8, 9]. When considering the impact of digital transformation, it is clear that sustainability is critical. Organizations can use digital technologies to reduce their environmental impact and encourage sustainable practices.

Cloud computing, for example, enables data center consolidation, resulting in energy savings and lower carbon emissions [10].

IoT devices enable monitoring and optimization of resource utilization, resulting in more efficient energy and material use [11].

Organizations. According to [12], the integration of digital technologies results in increased productivity, agility, and competitiveness, which eventually leads to improved financial performance and market position. According to [13], organizations that adopt digital transformation are more likely to adapt to developing customer preferences and market conditions, leading to increased levels of customer satisfaction and loyalty. In addition, digital transformation allows the accumulation and examination of extensive data sets by organizations, which allows for decision-making based on empirical evidence and the identification of novel commercial prospects.

Social media: refers to online platforms and tools that allow users to generate, communicate, and share information, and ideas, Social media applies to digital platforms that enable individuals to generate, disseminate, and interchange data, concepts, and content with other users [14], the pervasiveness of social media in today's communication has fundamentally altered the dynamics of interpersonal engagement. Social media platforms, including Facebook, Twitter, Instagram, and LinkedIn, have attracted an enormous user base globally and have developed as an effective tool for businesses, organizations, and individuals to establish connections with their target audience.

Cloud computing is a cutting-edge technological innovation that facilitates remote access to computing resources such as servers, storage, and applications via the internet, the popularity of cloud computing can be attributed to its attributes of flexibility, scalability, and cost-effectiveness. The development of cloud computing has resulted in a reduction in the expenses incurred by businesses in maintaining their IT infrastructure. In addition, it has improved the accessibility of computing resources for individuals, regardless of their geographical location. The development of cloud computing has allowed the development of new applications and services that were previously impossible [15].

The Internet of Things (IoT): is a system of interrelated physical devices, home appliances, vehicles, and other objects that are equipped with sensors, software, and connectivity features, allowing them to communicate and share data. The Internet of Things (IoT) holds the capacity to revolutionize many different industries including healthcare, transportation, and production. The utilization of Internet of Things (IoT) devices allows the collection and analysis of data in real-time, thus empowering enterprises to make well-informed decisions and enhance their operational effectiveness. The Internet of Things (IoT) has allowed the development of new applications and services that were never before possible [16].

Artificial Intelligence (AI): is the development of computer systems that are capable of carrying out tasks that conventionally require human intelligence, including but not limited to visual perception, speech recognition, decision-making, and language translation. Artificial intelligence (AI) possesses the capability to revolutionize diverse sectors, including healthcare, finance, and transportation. Artificial intelligence has the capability to process vast quantities of data in real-time, thereby facilitating informed decision-making and operational enhancement for businesses. The advent of AI has allowed the development of new applications and services that were previously unobtainable [17].

Robotics is an area of study that involves the creation of machines capable of carrying out tasks that typically necessitate human intelligence, including yet not limited to visual perception, decision-making, and physical movement. The field of robotics believes in guarantee for revolutionizing multiple sectors, including but not limited to manufacturing, healthcare, and transportation. The utilization of robotics technology facilitates the execution of tasks that pose a risk or challenge to human beings, thereby enhancing the efficiency of business operations and minimizing expenses. The field of robotics has facilitated the emergence of novel applications and services that were previously unattainable [18].

Mobile technologies: refer to the advancement of devices like smartphones and tablets that facilitate users in obtaining computing resources and information from any location across the globe. The integration of mobile technologies has become an essential part of modern communication, revolutionizing the way in which individuals engage with one another. The advent of mobile technologies has facilitated the ability of businesses to establish connections with their target audience and offer services while on the move. The emergence of mobile technologies has facilitated the creation of novel applications and services that were previously unattainable [19].

2.2 *Big Data*

The term "big data" refers to a substantial amount of structured and unstructured data that is produced by individuals, institutions, and automated systems. The data is distinguished by its attributes of volume, velocity, variety, and veracity, as stated by [20].

The phenomenon of big data has garnered significant interest from various entities, including private enterprises, public institutions, and academic research centers [21]. The utilization of large-scale data has facilitated enterprises in devising efficacious commercial tactics and sustaining competitiveness, the utilization of big data can also enhance supply chain management, thereby enabling organizations to curtail expenses and enhance operational efficacy.

Younis [22] emphasizes the importance of skilled individuals in organizations who can analyze data and provide valuable insights for informed decision-making. Big data revolutionizes decision-making by enabling real-time data-driven decisions from various sources, such as social media, customer feedback, and sales data.

Galbraith [23] highlights the potential of big data in enhancing organizational decision-making, optimizing product and service quality, and providing a competitive edge, big data can influence an organization's digital transformation, which involves integrating digital technology across all aspects of an enterprise, leading to significant changes in functioning and customer value.

According to [24], big data and digital transformation are closely intertwined, enabling entities to achieve success in the digital era. Big data acquisition facilitates the acquisition of relevant information and perspectives, while digital transformation provides the necessary structure and technological foundation for utilizing large-scale data.

Organizations that employ big data technology have been evaluated to offer a higher level of logistical service, enhanced operational efficiency with their partners, and increased synchronization among specific logistic activities [25], big data is making a significant impact is in supply chain management. The integration of big data analytics with sustainability practices can help mitigate supply chain risks and improve overall sustainability [26]. By leveraging big data, organizations can gain insights into their supply chains, identify areas for improvement, and make data-driven decisions to optimize their operations. However, the application of big data in sustainable operations may face challenges and barriers that need to be addressed for effective implementation.

Big data has the potential to significantly impact sustainability in various domains, including supply chain management, knowledge sharing, and overall organizational and societal sustainability. However, there may be challenges and barriers that need to be addressed for the effective implementation of big data in sustainable practices. Further research and analysis are needed to better understand the trends, developments, and implications of big data for sustainability. By leveraging the power of big data, organizations and societies can make data-driven decisions and optimize their operations to achieve sustainability goals.

The impact of big data on sustainability is not limited to specific industries or sectors. It has broader implications for organizations, regions, and society as a whole [27].

The development of big data has transformative effects on the socio-economic world, and these transformations have important consequences for sustainability. However, there is a lack of studies analyzing the trends and developments in the literature on the nexus between big data and sustainability.

2.3 Cybersecurity

This study includes cybersecurity and privacy concerns pose significant challenges for organizations embracing advanced technologies and digital transformation. Within the context of organizational operations, maintaining system data in the cloud and accessing it through web-based platforms introduces serious security threats. Unlike conventional data systems, industry-specific systems are particularly vulnerable as they can be remotely controlled and rely on inter-object communication [28]. Thus, safeguarding crucial data and establishing a secure infrastructure are critical imperatives to prevent data breaches and ensure the integrity of industrial systems.

The company's ability to innovate is enhanced through digital transformation, which also significantly lowers the costs associated with running the business's administrative functions. However, the success of digitalization initiatives relies on the organization's culture being conducive to change and the proper implementation of digital transformation. An unfavorable atmosphere that is not conducive to change can have negative effects on the company's productivity [29]. Failure to address cybersecurity and privacy concerns can result in tragic consequences, rendering a company irretrievably outmoded. Effective cybersecurity measures are essential to mitigate risks associated with digital transformation initiatives. As organizations embrace digital technologies, they become more interconnected and vulnerable to cyber threats [30]. Cybersecurity, which includes the security of the Internet, the point-to-point protocol, data sovereignty, etc., plays a crucial role in the digital transformation of enterprises [31]. It is related to the fact that businesses and their systems are increasingly being targeted by hackers. Therefore, safeguarding crucial data and establishing a secure infrastructure are critical imperatives to prevent data breaches and ensure the integrity of industrial systems. Within the context of organizational operations, maintaining system data in the cloud and accessing it through web-based platforms introduces serious security threats. Unlike conventional data systems, industry-specific systems are particularly vulnerable as they can be remotely controlled and rely on inter-object communication [27]. As organizations strive to innovate and lower administrative costs through digital transformation, they must also address cybersecurity and privacy concerns that pose significant challenges [30].

Failure to implement effective cybersecurity measures can result in tragic consequences, rendering a company irretrievably outmoded [29] and furthermore, the ability of a company to innovate is enhanced through digital transformation, which can also significantly lower the costs associated with running the business's administrative functions. However, the success of digitalization initiatives relies on the organization's culture being conducive to change and the proper implementation of digital transformation [29].

Cybersecurity is a crucial aspect of digital transformation, as it ensures the protection of digital assets and information [32]. The integration of digital technologies in various sectors has led to an increase in cyber threats, making cybersecurity a critical concern [32]. Several studies have been conducted to analyze the past, present, and future of research on cybersecurity in the context of digital transformation. The significance of customer awareness with respect to cybersecurity has been established as a critical factor. In this context, a range of cybersecurity dimensions, such as cyberattacks, phishing, and hacking, have been subjected to analysis by Johri and Kumar [33].

In digital transformation, the formulation of a cybersecurity threat strategy, management sustainability factors, measures to be done against cybercrime, cyberattack, and cyberterrorism, as well as organizational and corporate culture management, are vital [33]. Network security is another critical component of internet networks.

The incorporation of emerging technologies such as blockchain and the internet of things (IoT) can ensure cybersecurity in various sectors, including healthcare [34]. The impact of digital transformation on data-driven approaches, interoperability, cybersecurity, and data preservation, along with issues of governance and roles in data orientation, have also been analyzed. Cybersecurity is a crucial aspect of digital transformation, and various dimensions need to be considered to ensure the protection of digital assets and information [32]. The incorporation of emerging technologies can also ensure cybersecurity in various sectors [32].

2.4 Costumer Experience

Digital transformation includes not only technological advancements but also necessitates a fundamental shift in the cultural norms and practices that regulate business operations. The adoption of new work techniques necessitates a shift in cognitive perspective and a readiness to embrace alternative approaches. The implementation of digital technologies necessitates a culture that fosters and encourages innovative practices, Experimentation, and also the good continuous learning, customer experience is essentially combination of the customers feelings about the company image and the customer's reaction to it towards the products or services at various stages of the customer's engagement with the brand's product or service. In other words, customer experience is the business image formed by the consumer's view of the specific brand's products or services, it includes the entire process, from customer engagement until the point at which the customer wants and decides to purchase the product.

Customer experience consists of various factors that are critical to the organization's efficiency [35] the initial stage in the customer experience process is to ensure that the customers' engagement with the products is favorable, the second objective is to meet the demands or expectations of the clients and adapt their items to their specifications in the third phase, the firm must determine the weaknesses and dislikes through market research. Implementing digital transformation strategies requires a capability-based framework that focuses on customer experience. Sahu et al. [36] propose such a framework that evaluates strategies and procedures for implementing digital transformation to improve customer experience. The framework also analyzes customer engagement processes to implement digital transformation effectively [36]. Additionally, the study identifies critical factors for reshaping the customer value proposition to enhance customer experience in the digital transformation process by adopting this framework, organizations can align their digital transformation efforts with customer experience goals. Digital transformation has also had a significant impact on the fast-food restaurant industry, altering the way customers interact with restaurants and changing customer value. Daradkeh et al. [37] emphasize the need for fast-food restaurants to enhance their digital presence to maximize customer value in the digital era. The study aims to develop a dynamic and sustainable method for creating customer value in digital transformation by leveraging digital technologies and improving their digital presence, fast-food restaurants can enhance the customer experience and stay competitive in the digital age.

2.5 Culture

Digital transformation is more than simply technology; it also entails a cultural shift in how firms function, It necessitates a shift in thinking as well as a readiness to embrace new ways of functioning, the adoption of digital technologies points out the necessitates an innovative culture, experimentation, and continuous learning, it also requires a focus on customer-centricity, in which organizations focus on their customers' demands and preferences, organizational culture is critical to the successful implementation and longevity of digital transformation efforts. Digital transformation encompasses various aspects of an organization, encompasses everything from business model plans to operational strategies to financial resources to company culture [38].

The significance of corporate culture in integrating big data into organizational operations cannot be overstated [39] understanding the impact of company culture on supporting the integration of big data into organizational processes is critical [39], organizations with an open culture are more inclined to review and adopt big data into their business processes, particularly when their success relies on various stakeholders.

To successfully undertake digital transformation, organizations develop digital transformation strategies, initiate digitalization projects, and adopt IT solutions to adapt to disruption and capitalize on emerging opportunities culture plays a significant role in shaping the attitudes, behaviors, and beliefs of individuals and groups, as organizations embark on digital transformation journeys, they must recognize that cultural norms, values, and practices can either facilitate or hinder the adoption and implementation of new technologies and digital processes and neglecting cultural considerations can lead to resistance, fear, and a lack of engagement from employees, customers, and other stakeholders [40].

2.6 Sustainability

Sustainability refers to having the ability to meet the requirements of the current generation while ensuring that the potential of future generations to satisfy their own needs isn't jeopardized. It is a complex concept that encompasses environmental, social, and economic dimensions. According to [41] the field of sustainability science aims to expand scientific inquiry into areas that exhibit a unique problem-solving focus, intricate physical and social complexities, and intricate ethical and moral terrains.

The relationship between sustainability and digital transformation is complex and multifaceted. On one hand, digital transformation has the potential to contribute to sustainability goals by enabling more efficient resource use, reducing waste, and promoting renewable energy [4]. For example, digital technologies can optimize supply chains, reduce energy consumption, and enable remote work, thereby reducing carbon emissions.

Digitalization can also enhance communication and collaboration, enabling the sharing of knowledge and best practices for sustainable development [42].

Economic sustainability refers to the ability of an economic system to support long-term growth and development while ensuring the efficient allocation of resources. It involves promoting economic prosperity, reducing poverty, and ensuring equitable distribution of wealth. According to [43], sustainable economic practices require a shift towards a circular economy, where resources are used efficiently, waste is minimized, and products are designed for reuse and recycling. This approach, known as "Cradle to Cradle," emphasizes the importance of creating products that have a positive impact on the economy and the environment.

According to [44], sustainability practices involve a management approach in which businesses prioritize collaboration with social, economic, and environmental concerns in their operations, while also engaging with their stakeholders. It is interesting that sustainability has come to be as a crucial concern that has attracted significant attention in modern times. The concept of sustainability includes a range of aspects. It is a complex concept that encompasses environmental, social, and economic dimensions. The concept of sustainable development entails a systematic approach that endeavors to satisfy the requirements of the current generation while safeguarding the potential of future generations to fulfill their own needs. The study conducted by Manuti and Giancaspro [45].

The significance of sustainability is on the rise for corporations due to the mounting pressure to conform to and embrace corporate sustainability practices. Companies that accord priority to sustainability can benefit from multiple benefits. For example, they can reduce their environmental impact, improve their reputation, and increase their competitiveness.

The implementation of sustainable practices can result in financial benefits by optimizing resource utilization and minimizing waste. Moreover, organizations that prioritize sustainability have the potential to attract and retain employees who are driven by a sense of mission and a willingness to work for socially accountable entities [46]. Moreover, the notion of sustainability is multifaceted, encompassing various dimensions such as environmental, social, and economic aspects.

- ***Environmental Sustainability***

The concept of environmental sustainability refers to the responsible utilization and management of natural resources in order to satisfy the requirements of the current population while protecting the potential of coming generations to fulfill their own necessities [47].

The concept of digital transformation refers to the incorporation of digital technologies across all facets of an enterprise, leading to significant modifications in the way organizations function and provide benefits to their clientele [4]. In recent years, the subject of the influence of digital transformation on environmental sustainability has garnered significant attention. Numerous studies have been conducted to investigate the advantages and obstacles of digitalization in advancing sustainability, research has indicated that the implementation of digital transformation can make significant contributions toward encouraging environmental sustainability through digital technologies have the potential to mitigate energy consumption and greenhouse gas emissions through process optimization and waste reduction. The process of digitalization has the potential to facilitate the emergence of novel business models that foster sustainability, such as the sharing economy, which encourages the most efficient use of resources [48] Moreover, digital technologies have the potential to aid in the monitoring and records of environmental performance, thereby empowering entities to monitor their progress toward sustainability objectives.

Nonetheless, the effects of digital transformation on environmental sustainability are not uniformly advantageous. The utilization of digital technologies has the potential to result in an escalation of energy consumption and the production of electronic waste, thereby causing detrimental environmental consequences [49].

Chen and Johansson [50] focus specifically on the environmental sustainability implications of digitalization in manufacturing. The authors note that the rapid development and implementation of digitalization in manufacturing has both positive and negative environmental impacts. They highlight the need for further research to understand the overall implications of digitalization on environmental sustainability in the context of manufacturing, including areas such as product design, production, transportation, and customer service.

Furthermore, the adoption of digital technologies can lead to the displacement of workers and the creation of new forms of inequality, which can have social and environmental consequences [51].

To address these challenges, several studies have proposed strategies for promoting sustainable digital transformation. For instance, organizations can adopt a human-centric approach to digital transformation, which prioritizes the well-being of people and the planet over profit [52] Additionally, organizations can focus on system integration and digital twin technologies, which enable the optimization of processes and the reduction of waste finally, organizations can adopt ethical principles in their digital transformation efforts, such as transparency and accountability, to ensure that their actions align with sustainability goals.

In conclusion, digital transformation has the potential to promote environmental sustainability, but its impact is complex and multifaceted. To ensure that digital transformation contributes to sustainability goals, organizations must adopt a holistic approach that considers the social, economic, and environmental impacts of their actions. Future research should focus on developing strategies for promoting sustainable digital transformation and evaluating the effectiveness of these strategies in achieving sustainability goals [2].

- ***Economic Sustainability***

Economic sustainability is considered one of the fundamental pillars of sustainability. Which is in addition to environmental and social sustainability. It is defined as an economic system's ability to retain its production and viability throughout time without risking the well-being of future generations [53].

The concept of economic sustainability has gained significance in supply chains due to global competition, prompting businesses to adopt sustainable practices [54].

Digitalization plays a crucial role in promoting sustainable growth by supporting renewable energy, infrastructure development, value-added growth, and domestic investment (Integration of digital technologies has shown significant impacts on economic sustainability and social value for both large corporations and small and medium-sized enterprises (SMEs) [55].

One important aspect of digital transformation is its ability to optimize resource utilization, which can lead to economic sustainability. By leveraging digital technologies, businesses can streamline their operations, improve efficiency, and reduce costs. For example, logistics service providers in Thailand have adopted digital transformation practices to enhance their sustainability [56]. This includes implementing digital platforms for managing logistics operations, optimizing routes, and reducing fuel consumption. These digital solutions have not only improved the economic performance of these providers but also contributed to environmental sustainability by reducing carbon emissions.

Moreover, digital transformation has the potential to drive innovation and create new economic opportunities. The use of digital technologies enables businesses to develop new products and services, enter new markets, and enhance customer experiences. This can result in increased revenue and economic growth. In the European Union, digital technologies have been recognized as a tool for business sustainability, with the potential to contribute to economic, environmental, and social sustainability [57]. By embracing digital transformation, businesses can adapt to changing market dynamics, remain competitive, and contribute to the overall economic sustainability of the region. In addition to the direct impact on businesses and the economy, digital transformation also has indirect effects on economic sustainability. For instance, research has examined the influence of digital technologies on economic subjects and their outcomes, including the level of GDP [58].

- ***Social Sustainability***

Social sustainability is considered as one of the third and last dimensions of sustainability, alongside economic and environmental sustainability. The concept of social sustainability pertains to the capacity of a given society to satisfy the requirements of its constituents, while concurrently preserving social unity and parity in the long run. Achieving social sustainability requires addressing various factors, including energy and resources sustainability, economic sustainability, environmental sustainability, and social media. Research has shown that the deployment and use of energy-efficient modern technologies are essential for achieving sustainable development.

The concept of social sustainability maintains an essential place in the field of sustainable development and has garnered increasing significance in the context of the digital revolution. El Hilali et al. [58] conducted a study titled "Reaching sustainability during a digital transformation: a PLS approach" to investigate the correlation between digital transformation and social sustainability. The study mainly looked at the effects of digital transformation on organizations, including small- and medium-sized enterprises (SMEs) as well as other objects [58] contend that digital capabilities have an important effect on improving social sustainability, while the influence of organizational change on social sustainability is considered to be insignificant, the study's results indicate that businesses, including small, medium-sized, and large organizations, can improve their social sustainability performance by implementing advanced technologies, providing employee training programs for developing new skills, and embracing business models that prioritize social responsibility [58]. The findings are consistent with previous studies that have shown how technological advancements can result in positive societal outcomes [59]. Additionally, the research underscores the significance of organizational awareness regarding the probable social consequences linked with digital transformation and adopting preemptive measures to alleviate these consequences.

The digital economy has also played a crucial role in rebuilding and maintaining social governance mechanisms. The digital transformation has affected society at various levels, particularly in the economic and governance domains [60]. The digital economy has not only brought about direct impacts but has also acted as a mediating force, facilitating social reforms and contributing to a sustainable digital economy [60].

3 Research Methods

3.1 Survey Instrument

To explore the impact of digital transformation on organizational performance, we designed a survey instrument based on the Likert-style scale commonly used in social science research. The questionnaire aimed to assess the perceptions of employees regarding various dimensions of digital transformation and their relationship to business outcomes. The survey comprised a set of statements, and respondents were asked to indicate their level of agreement on a five-point scale ranging from 1 (strongly disagree) to 5 (strongly agree). Rigorous validation procedures were employed to ensure the reliability of the questionnaire. The survey instrument underwent thorough evaluation by senior managers and professionals from diverse industry sectors, including organizations that were actively implementing digital transformation initiatives as well as those yet to embark on such endeavors.

3.2 Sample and Data Collection

The research utilized a questionnaire as the means of data collection to explore the impact of digital transformation within organizations. The web-based questionnaire consisted of multiple sections addressing various aspects of digital transformation.

Furthermore, the questionnaire investigated the organization's efforts to reduce its carbon footprint, the generation of positive economic outcomes, the expansion of social impact and communication, and the overall financial performance improvement resulting from the adoption of digital technologies. The participants of the study consisted of managers and engineers and employees working in many different industries, the sample for this study based on the answers we got includes organizations from diverse industries, encompassing metals, machinery, and equipment (40%), food and tobacco (5.5%), construction, wood, furniture, and paper (4.8%), textile and apparel (5.5%), automotive and electronics (15.8%), chemicals and pharmaceuticals (2.4%), energy and mining (3%), and others (23%). The sample consists of both large-sized enterprises (58.8%) and small and medium-sized enterprises (SMEs) (41.2%). The majority of the sample comprises locally owned firms (66.1%), followed by firms with local and foreign ownership (18.8%), and a smaller portion of foreign-owned firms (15.2%). In terms of firm age, the sample consists of young firms (27.3%), middle-aged firms (15.2%), and mature firms (57.6%), with each category representing a relatively equal proportion.

4 Results

In this study, surveys were conducted with 165 firms. The survey results have showed the following findings:

1. Our organization implements Digital Transformation initiatives, such as social media, cloud computing, IOT, artificial intelligence, robotics, and mobile technologies, within our operations.

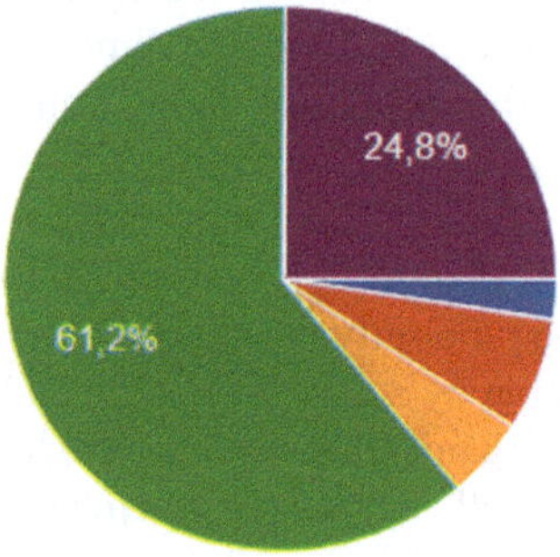

2. Our Organization uses Big Data and Analytics to improve insight customer engagement patterns, data management and informed decision-making to improve outcomes and customer experience.

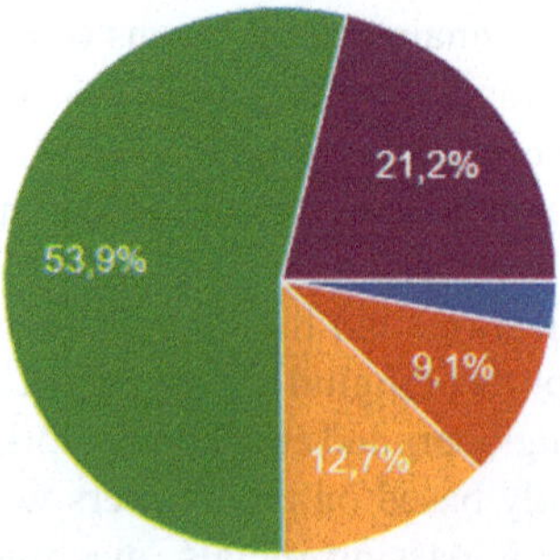

3. Our organization utilizes Digital security platforms to ensure that the data is always safe.

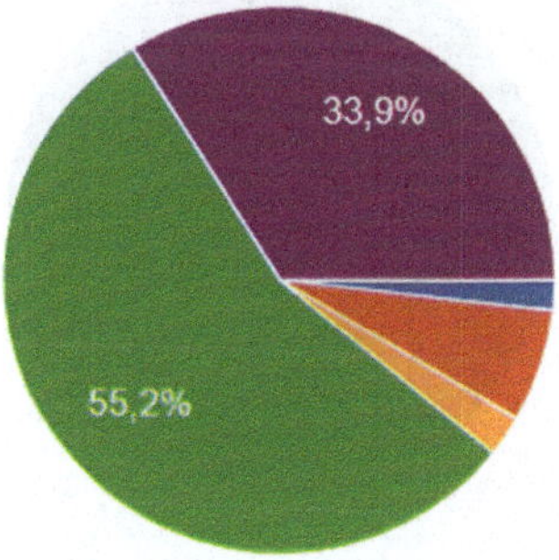

4. Our organization effectively uses online and offline digital platforms to develop customer relations.

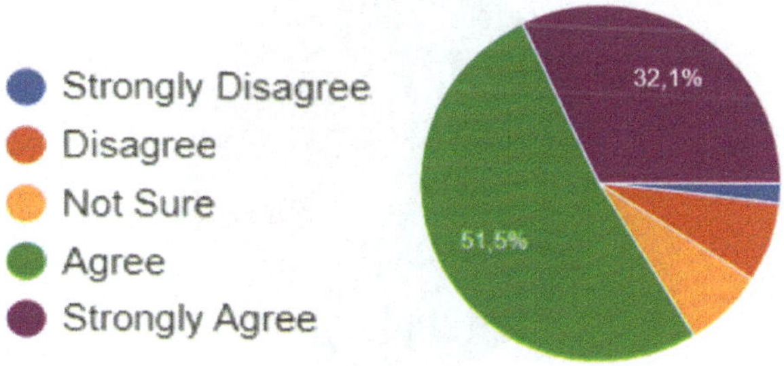

5. With the help of digital transformation and sustainability the customer value increased.

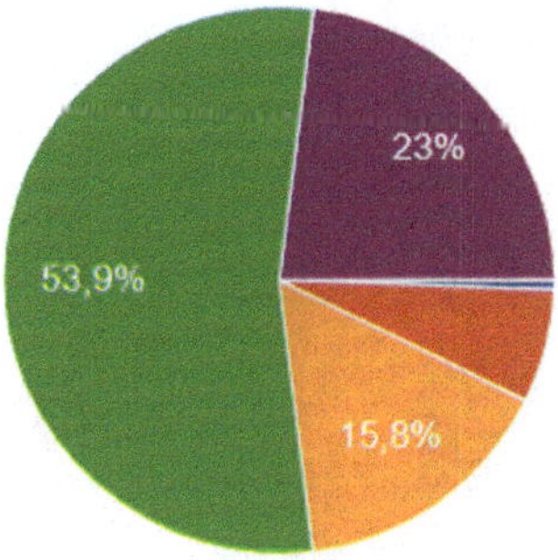

6. Firm performance would not be sustainable without digital transformation.

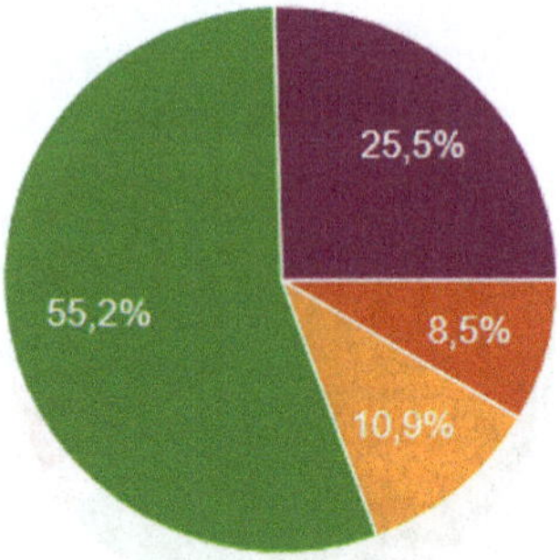

7. Digital transformation and sustainability helped impact our organization's operational performance.

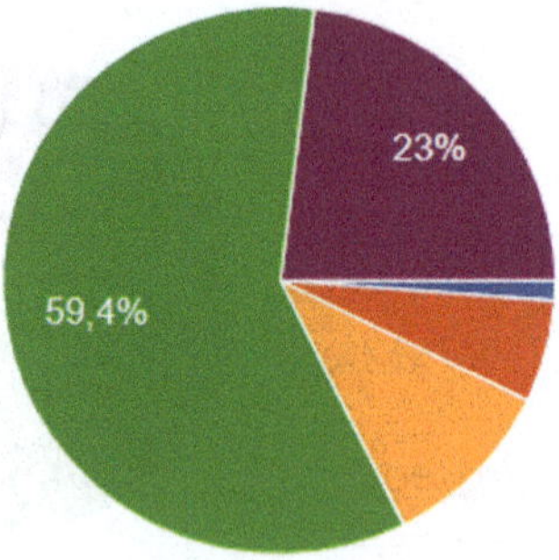

8. There is a significant improvement in my organization's overall financial performance since we adopted digital technologies.

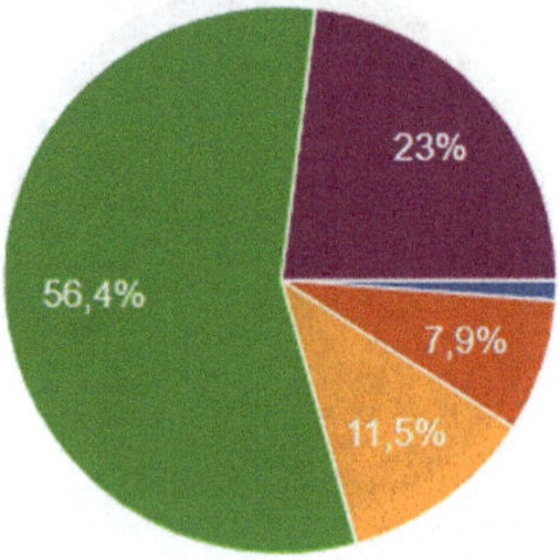

9. Our organization has implemented measures to decrease it's carbon footprint.

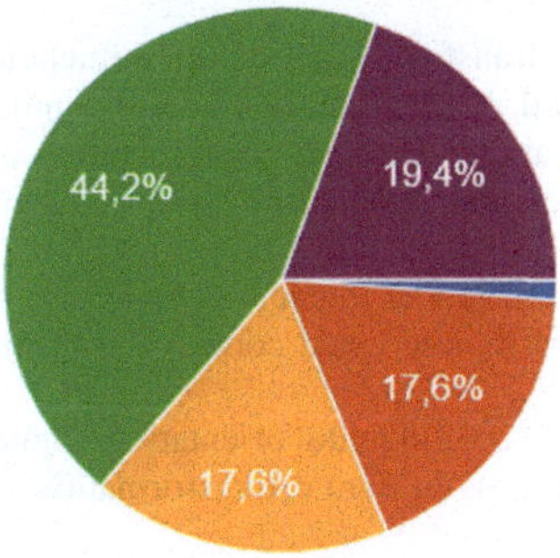

5 Conclusion

Digital transformation has a significant impact on sustainability, affecting economic, environmental, and social aspects. The literature suggests that digital transformation can have a positive impact on sustainability by enabling organizations to reduce their environmental footprint, improve their social responsibility, and enhance their economic performance. Digital technologies can help organizations reduce their carbon footprint by enabling remote work and reducing travel. They can also help organizations improve their social responsibility by enabling them to monitor and report on their environmental and social impact. Finally, digital technologies can help organizations enhance their economic performance by enabling them to improve their operational efficiency, reduce costs, and increase revenue.

However, digital transformation also poses challenges, such as cybersecurity risks, big data management, and organizational culture change. To address these challenges, organizations need to develop a comprehensive strategy that considers the interplay between technology, people, and processes. They also need to adopt a governance framework that ensures responsible and ethical use of digital technologies and embrace a culture of sustainability that aligns their business goals with the broader societal and environmental objectives.

To achieve sustainable digital transformation, organizations need to adopt a governance framework that ensures responsible and ethical use of digital technologies. This framework should include policies and procedures for data privacy and security, as well as guidelines for the ethical use of artificial intelligence and other emerging technologies. Organizations also need to embrace a culture of sustainability that aligns their business goals with broader societal and environmental objectives. This culture should be reflected in the organization's values, mission, and vision, as well as in its day-to-day operations.

References

1. Singhdong P (2011) Digital transformation factors influencing the sustainability of logistics service providers in Thailand. https://doi.org/10.58837/chula.the.2020.287S. Saha S, Beach MC (2011) The impact of patient-centered communication on patients' decision making and evaluations of physicians: a randomized study using video vignettes. Patient Educ Couns 84(3):386–392
2. Zhang C, Chen P, Chen P (2022) The impact of digital transformation on corporate sustainability—new evidence from Chinese listed companies. Front Environ Sci 10.https://doi.org/10.3389/fenvs.2022.1047418
3. Nasiri M, Saunila M, Ukko J (2022) Digital orientation, digital maturity, and digital intensity: determinants of financial success in digital transformation settings. Int J Oper Prod Manag 42(13):274–298
4. Feroz A, Zo H, Chiravuri A (2021) Digital transformation and environmental sustainability: a review and research agenda. Sustainability 13(3):1530. https://doi.org/10.3390/su13031530
5. Kawung G, Elim I, Rompas W, Rogi M (2022) Digital technology transformation of SMEs: Indonesian case study. Am J Multidiscip Res Innov 1(6):56–60. https://doi.org/10.54536/ajmri.v1i6.948
6. Vial G (2021) Understanding digital transformation: a review and a research agenda. Manag Dig Transf 13–66
7. Sebastian I, Ross J, Beath C, Mocker M, Moloney K, Fonstad N (2017) How big old companies navigate digital transformation. MIS Q Executive 16(3):197–213
8. Hess T, Matt C, Benlian A, Wiesböck F (2016) Options for formulating a digital transformation strategy. MIS Q Executive 15(2)
9. Ismail MH, Khater M, Zaki M (2017) Digital business transformation and strategy: what do we know so far. Camb Serv Alliance 10(1):1–35
10. Melville NP (2010) Information systems innovation for environmental sustainability. MIS Q 1–21
11. Nwankpa JK, Merhout JW (2020) Exploring the effect of digital investment on IT innovation. Sustainability 12(18):7374
12. Westerman G, Bonnet D, McAfee A (2014) Leading digital: turning technology into business transformation. Harvard Business Press
13. Berman S, Kesterson L, Townes A, Marshall, Srivathsa R (2012) How cloud computing enables process and business model innovation. Strategy Leadersh 40(4):27–35
14. Weller K (2016) Trying to understand social media users and usage. Online Inf Rev 40(2):256–264. https://doi.org/10.1108/oir-09-2015-0299
15. Nedić V (2019) Gartner's top strategic technology trends. PES 2(1):433–442. https://doi.org/10.24874/pes01.02.041
16. Gill SS, Tuli S, Xu M, Singh I, Singh KV, Lindsay D, Garraghan P et al (2019) Transformative effects of IoT, Blockchain and Artificial Intelligence on cloud computing: evolution, vision, trends and open challenges. Internet Things 8:100118
17. Rahman MM, Khatun F (2023) Challenges and prospective of AI and 5G-enabled technologies in emerging applications during the pandemic. In: Industry 4.0-perspectives and applications. IntechOpen
18. Banjanović-Mehmedović L, Mehmedović F (2020) Intelligent manufacturing systems driven by artificial intelligence in industry 4.0. In: Handbook of research on integrating industry 4.0 in business and manufacturing. IGI Global, pp 31–52
19. Volk M, Sterle J, Sedlar U (2015) Safety and privacy considerations for mobile application design in digital healthcare. Int J Distrib Sens Netw
20. Mishra A (2022) Big data security challenges. Int J Res Public Rev 693–696. https://doi.org/10.55248/gengpi.2022.3.10.31
21. Manogaran G, Thota C, Lopez D (2022) Human-computer interaction with big data analytics. In: Research anthology on big data analytics, architectures, and applications. IGI Global, pp 1578–1596

22. Younis NMM (2020) Big data and the future of the accounting profession. Indian J Sci Technol 13(08):883–892
23. Galbraith J (2014) Organizational design challenges resulting from big data. J Organ Des 1(3):2. https://doi.org/10.7146/jod.8856
24. Al-Sai Z, Abdullah R, Husin H (2020) Critical success factors for big data: a systematic literature review. IEEE Access (8):118940–118956. https://doi.org/10.1109/access.2020.3005461
25. Nagy J, Oláh J, Erdei E, Máté D, Popp J (2018) The role and impact of Industry 4.0 and the internet of things on the business strategy of the value chain—the case of Hungary. Sustainability 10(10):3491
26. Mani V, Delgado C, Hazen B, Patel P (2017) Mitigating supply chain risk via sustainability using big data analytics: evidence from the manufacturing supply chain. Sustainability 9(4):608. https://doi.org/10.3390/su9040608
27. Garrigos-Simon F, Blas S, Narangajavana Y, Buzova D (2021) The nexus between big data and sustainability: an analysis of current trends and developments. Sustainability 13(12):6632. https://doi.org/10.3390/su13126632
28. Khan M, Wu X, Xu X, Dou W (2017) Big data challenges and opportunities in the hype of Industry 4.0. In: 2017 IEEE international conference on communications (ICC). IEEE, pp 1–6
29. van Tonder C, Schachtebeck C, Nieuwenhuizen C, Bossink B (2020) A framework for digital transformation and business model innovation. Manage J Contemp Manage Issues 25(2):111–132
30. Spiekermann S, Acquisti A, Böhme R, Hui KL (2015) The challenges of personal data markets and privacy. Electron Mark 25:161–167
31. Maglaras LA, Kim KH, Janicke H, Ferrag MA, Rallis S, Fragkou P, Cruz TJ (2018) Cyber security of critical infrastructures. Ict Express 4(1):42–45
32. Belkhamza Z (2023). Cybersecurity in digital transformation applications: analysis of past research and future directions. In: ICCWS 2023 18th international conference on cyber warfare and security. Academic Conferences and publishing limited
33. Johri A, Kumar S (2023) Exploring customer awareness towards their cyber security in the Kingdom of Saudi Arabia: a study in the era of banking digital transformation. Hum Behav Emerg Technol
34. Lamba J, Jain E (2023) Advanced cyber security and internet of things for digital transformations of the Indian healthcare sector. In: Research anthology on convergence of blockchain, internet of things, and security. IGI Global, pp 1037–1056
35. Saleem P, Siddik M (2021) Digital marketing: channels and strategies. In: Shah P (ed) Marketing strategies for sustainable growth in a digital era, pp 162–166. ISBN 978-81-946245-1-6
36. Sahu N, Deng H, Molla A (2018) A capability based framework for customer experience focused digital transformation.https://doi.org/10.5130/acis2018.an
37. Daradkeh F, Hassan T, Palei T, Helal M, Mabrouk S, Saleh M, Elshawarbi N et al (2023) Enhancing digital presence for maximizing customer value in fast-food restaurants. Sustainability 15(7):5690. https://doi.org/10.3390/su15075690
38. Hofstede G, Hofstede GJ, Minkov M (2005) Cultures and organizations: software of the mind, vol 2. Mcgraw-hill, New York
39. Katal A, Wazid M, Goudar RH (2013) Big data: issues, challenges, tools and good practices. In: 2013 Sixth international conference on contemporary computing (IC3). IEEE, pp 404–409
40. Schein EH (2010) Organizational culture and leadership, vol 2. Wiley & Sons
41. Nagatsu M, Davis T, DesRoches CT, Koskinen I, MacLeod M, Stojanovic M, Thorén H (2020) Philosophy of science for sustainability science. Sustain Sci 15:1807–1817
42. Linkov I, Trump B, Poinsatte-Jones K, Florin M (2018) Governance strategies for a sustainable digital world. Sustainability 10(2):440. https://doi.org/10.3390/su10020440
43. Peterson M (2004) Cradle to cradle: remaking the way we make things. J Macromark 24(1):78–79. https://doi.org/10.1177/0276146704264148
44. Beji R, Yousfi O, Loukil N, Omri A (2021) Board diversity and corporate social responsibility: empirical evidence from France. J Bus Ethics 173:133–155

45. Manuti A, Giancaspro ML (2019) People make the difference: an explorative study on the relationship between organizational practices, employees' resources, and organizational behavior
46. Kouatli I (2019) The contemporary definition of university social responsibility with quantifiable sustainability. Soc Responsib J 15(7):888–909
47. El Hilali W, El Manouar A, Idrissi MAJ (2021) The mediating role of big data analytics in enhancing firms' commitment to sustainability. Int J Adv Technol Eng Explor 8(80):932
48. Costa I, Riccotta R, Montini P, Stefani E, de Souza Goes R, Gaspar MA, Larieira CLC et al (2022) The degree of contribution of digital transformation technology on company sustainability areas. Sustainability 14(1):462
49. Xue F, Zhao X, Tan Y (2022) Digital transformation of manufacturing enterprises: an empirical study on the relationships between digital transformation, boundary spanning, and sustainable competitive advantage. Discret Dyn Nat Soc 2022:1–16
50. Chen X, Johansson B (2020) Environmental sustainability of digitalization in manufacturing: a review. Sustainability 12(24):10298. https://doi.org/10.3390/su122410298
51. Abugabel A (2023) The impact of digital transformation on sustainable development: the mediating role of development of human resources management practices. An empirical study on private hospitals. الإدارية للعلوم الإسكندرية
52. Al-Qaruty TM, Dabbaghia MN, Shaban EA, Al-Doori J (2022) The digital transformation of human resources and its impact on the organizational strategic sustainability: the mediating role of the conscious leadership. Int J Health Sci 6(S7):4320–4344. https://doi.org/10.53730/ijhs.v6nS7.12800
53. Newport D, Chesnes T, Lindner A (2003) The "environmental sustainability" problem: ensuring that sustainability stands on three legs. Int J Sustain Higher Educ
54. Caradonna JL (2014) Sustainability: a history. Oxford University Press
55. Vrontis D, Chaudhuri R, Chatterjee S (2022) Adoption of digital technologies by SMEs for sustainability and value creation: moderating role of entrepreneurial orientation. Sustainability 14(13):7949
56. Singhdong P, Digital transformation factors influencing the sustainability of logistics service providers in Thailand. https://doi.org/10.58837/chula.the.2020.287
57. Prostova D, Tikhonova A, Sosnina N (2019) The influence of digital technologies on the development of economic subjects. https://doi.org/10.2991/iscde-19.2019.84
58. El Hilali W, El Manouar A, Idrissi MAJ (2020) Reaching sustainability during a digital transformation: a PLS approach. Int J Innov Sci 12(1):52–79
59. Brynjolfsson E, McAfee A (2014) The second machine age: work, progress, and prosperity in a time of brilliant technologies. WW Norton & Company
60. Niu F (2022) The role of the digital economy in rebuilding and maintaining social governance mechanisms. Front Public Health 9. https://doi.org/10.3389/fpubh.2021.819727

The Effect of Green Supply Chain Management Practices on Business Performance: Examining the Mediating Role of Competitive Advantage

Hasan Kürşat Güleş, Emel Gelmez, and Vural Çağlıyan

Abstract In particular, the globalization process, among many other factors, causes the competitive environment in which businesses operate to sharpen both quantitatively and qualitatively. As a result, businesses have started to apply new methods/approaches and techniques that will provide a sustainable competitive advantage over rivals. In addition, the increasing number of environmental problems in recent years has made it necessary for businesses to take environmental aspects into account while continuing their activities. In the light of these explanations, businesses today consider new methods/techniques that will also take environmental factors into account when developing and implementing competitive strategies that will provide sustainable competitive advantage. In this context, one of the approaches/techniques that has come to the fore recently is Green Supply Chain Management (GSCM). GSCM practices, whose effects are evaluated in theory and practice, need to be adopted by businesses in both gaining competitive advantage (CA) and in increasing business performance (BP). In this context, the major objective of this study is to determine the mediation effect of CA on the relationship between GSCM on BP. Within the framework of this main aim, a survey has been applied to the businesses operating in the automotive sub-industry that are registered with the Konya Chamber of Industry (KCI). Four basic hypotheses have been developed within the scope of the study. Before testing the hypotheses, the GSCM practice levels of the business have been determined first. Then, CA and BP of the businesses have been measured. Findings of this research reveal that GSCM practices have a positive and significant effect on CA, and CA has a positive and significant effect on BP. In addition, it is concluded that CA has a mediating role in the effect of GSCM practices on BP.

H. K. Güleş · E. Gelmez (✉) · V. Çağlıyan
Department of Business Administration, Selçuk University, 42075 Konya, Turkey
e-mail: emelgelmez@selcuk.edu.tr

H. K. Güleş
e-mail: kursat.gules@selcuk.edu.tr

V. Çağlıyan
e-mail: vcagliyan@selcuk.edu.tr

F. Çebi et al. (eds.), *Engineering and Technology Management in Challenging Times*,
Lecture Notes in Networks and Systems 1137,
https://doi.org/10.1007/978-3-031-70935-7_20

Keywords Green supply chain management practices · Competitive advantage · Business performance · Automotive sub-industry

1 Introduction

A new approach to supply chain management has emerged with the inclusion of environmental responsibility in production [1]. Adding the term *"green"* to supply chain management aims to include environmentally friendly thinking in all processes of the supply chain, such as green purchasing, green production, green material management, green distribution, green marketing and reverse logistics [2].

GSCM has been defined by Sarkis et al. [3] as "*integrating environmental concerns into the inter-organizational practices of supply chain management including reverse logistics*". The dimensions of GSCM applications have been evaluated in different ways in the literature [4–14]. Within the scope of this study, GSCM practices have been examined in five dimensions that are: *internal environmental management (IEM), green purchasing (GP), cooperation with customers (CWS), eco-design (ED) and investment recovery (IR)* [6].

GSCM is also considered an important and potential issue for businesses to develop, gain CA and improve their performance by achieving sectoral success [15]. Therefore, it is important to determine the relationship between GSCM practices and CA. When the relevant literature is examined, it is seen that the relationship between GSCM practices and CA has been examined in numerous studies [16–20].

GSCM practices have a number of effects on BP. Through reducing energy consumption and material use, improving stakeholder participation, reducing costs and improving product quality green supply chain practices can increase business performance [21]. It should be considered that the level of adoption of GSCM practices has a significant impact on improving the performance of businesses [12]. The relationship between GSCM practices and BP has been evaluated in various studies [12, 19, 21–23].

A business can be said to have a CA, if its activities cannot be imitated by its current or potential competitors or if imitation is too costly [24]. The fact that a business has a CA is usually an indication that it may have one or more of the capabilities of a lower price, higher quality and reliability, as well as shorter delivery time compared to its competitors. Therefore, these capabilities will increase the overall performance of businesses [25]. The relevant literature shows that the interaction between CA and BP has been examined in different studies [25–29].

In addition to the above explanations, there are a number of studies in the literature that focus on the relationship between GSCM practices, CA and BP [19, 30–32]. Based on these explanations, the aim of the study is to ***"determine the mediating role of CA in the effect of GSCM practices on BP in automotive sub-industry businesses"***. In this context, considering the limited number of studies in the literature on determining the mediating role of CA in the relationship between GSCM practices and BP, it is thought that this study will make significant contributions to the literature and businesses.

To achieve the above-mentioned main objective of the study in the second part of this paper firstly the adopted research methodology is explained and information about the sample is presented. Then, the hypotheses and conceptual model of the research are developed. In the third part of the study, the research results are shown. Research results are evaluated in the last section and based on these evaluations contributions, limitations of the research and suggestions for future research are made.

2 Research Methodology

2.1 Research Method and Sample

This is a pilot study of field research conducted on automotive sub-industry businesses registered to KCI. Within the scope of the quantitative research method, the survey method was used to collect data and the convenience sampling method was utilized. Survey data has been collected by meeting participants face to face between February and July, 2023.

The survey instrument gathers information on green supply chain practices, CA, and BP. GSCM practices were measured using the scale developed by Zhu et al. [6]. The relevant scale consists of five dimensions. These dimensions are *IEM* (7 items), *GP* (5 items), *CWS* (3 items), *ED* (3 items) and *IR* (3 items). The scale consisting of five dimensions adapted by Li et al. [25], based on Zhang's study [33], has been used to measure CA. These dimensions are *price/cost* (2 items), *quality* (4 items), *delivery reliability* (3 items), *product innovation* (3 items) and *time to market* (4 items). BP was measured using the scale used by Li et al. [25].

It has been determined that there are 400[1] businesses operating in the automotive sub-industry within the scope of Konya province. Analyses has been made based on data obtained from 102 businesses through convenience sampling. +−0.10 sampling error is p = 0.5 for a sample of 500 businesses, and the number of samples is 81 for q = 0.5 [34]. Considering this information, it can be said that the sample has the ability to represent the businesses operating in the Konya automotive sub-industry.

2.2 Hypotheses and Conceptual Model of the Research

The hypotheses within the framework of the aim of the study and the conceptual model seen in Fig. 1 are as follows:

[1] As of the end of 2022, there are a total of 400 businesses operating in the automotive sub-industry registered with KCI.

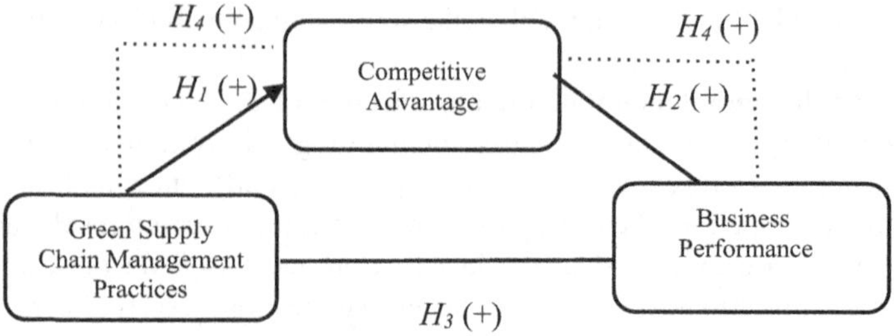

Fig. 1 Hypotheses regarding the conceptual model

H_1: GSCM practices have a positive effect on CA.
H_2: CA has a positive effect on BP.
H_3: GSCM practices have a positive effect on BP.
H_4: CA has a mediating role in the effect of GSCM practices on BP.

3 Research Results

3.1 Characteristics of the Sample

As in each company experts in their field answered the questions of the survey instrument it is evaluated as an indicator that increases the reliability of the answers received. In this context, the questions of the survey have been answered by senior managers and/or managers who are knowledgeable about the subjects within the scope of the survey. The titles of the respondents are respectively assistant general manager (58.8%), general manager (11.8%), business owner (10.8%), technical department manager (7.8%), production manager (6.9%) and marketing manager (3.9%). The operating periods of the businesses vary between 5 and 96 years and the average operating period is 29 years. The number of employees of the businesses changes between 20 and 500 and the average number of employees is 56.

4 Research Findings

Within the scope of the research, firstly the reliability of the scales used was evaluated. Cronbach Alpha (α) values of the scales in Tables 1, 2 and 3 are *0.956, 0.911 and 0.794,* respectively. In this context, the α values of the scales used in this study are reliable because they are above the acceptable lower limit of 0.70 [35, 36]. Findings regarding GSCM practices of businesses are presented in Table 1.

When Table 1 is examined, it is observed that the total mean score (3.17) for GSCM practices application level of the businesses is above the medium/average

Table 1 Findings on green supply chain management practices

GSCM practices	Mean	S.D
Excess stocks/materials are recovered	3.74	1.29
Scrap and second-hand materials are sold	3.63	1.60
Excess capital equipment is sold	3.47	1.55
IR	**3.61**	**1.18**
Products are designed and/or manufactured by avoiding or reducing the use of hazardous materials	3.54	1.29
Products are designed to be reusable and recyclable of materials and/or components	3.46	1.23
Products are designed for less material/energy consumption	3.26	1.19
ED	**3.42**	**1.16**
CWS is made for green packaged products	3.38	1.18
CWS is made for the design of cleaner production processes	3.26	1.17
CWS is made for the production of eco-designed products	2.96	1.17
CWS	**3.20**	**1.06**
Eco-labelling (the label that shows the conformity of the products with environmental standards) is made on the products	3.25	1.24
Suppliers' internal management and environmental audits are carried out	3.22	1.14
We work with suppliers with Environmental Management Certificate (ISO 14000–14001)	3.01	1.16
Environmentally friendly practices of second-tier suppliers are evaluated	2.83	1.21
Care is taken to cooperate with our suppliers on environmental issues	2.70	1.46
GP	**3.00**	**1.06**
Mid-level managers support GSCM	3.43	1.26
Environmental compliance and audit programs are implemented effectively	3.32	1.24
Our business has ISO 14001 certificate	3.30	1.28
There is cross-functional cooperation for environmental improvement activities	3.28	1.23
Total quality environmental management is adopted	3.26	1.17
The business has environmental management systems	2.95	1.82
GSCM is adopted by senior managers	1.26	0.76
IEM	**2.97**	**1.03**
Total	**3.17**	**0.94**

Notes (i) n = 102, (ii) On the scale, 1 = strongly disagree, and 5 = strongly agree

Table 2 Findings on competitive advantage

Competitive advantages	Mean	S.D
We deliver the required types of products	4.27	0.85
We provide dependable delivery	4.23	0.85
We deliver customers' orders on time	4.00	0.95
Delivery dependability	**4.17**	**0.79**
We offer highly reliable products	4.38	0.84
We offer very durable products	4.31	0.84
We provide high quality products to our customers	4.20	0.81
We can compete based on quality	2.89	1.27
Quality	**3.95**	**0.74**
We offer personalized products	4.21	0.85
We respond well to customer demand for "new" features	3.50	1.15
We change our product offering to meet customer needs	3.15	1.44
Product innovation	**3.62**	**0.95**
We deliver the product to the market quickly	3.88	0.95
We are the first in the market in terms of introducing new products	3.72	0.95
We have time-to-market lower than the industry average	3.07	1.05
We are able to develop products quickly	2.97	1.06
Time to market	**3.41**	**0.76**
We offer competitive prices	3.68	1.26
We are able to offer prices as low or lower than our competitors	3.11	1.23
Price/Cost	**3.39**	**0.95**
Total	**3.72**	**0.68**

Notes (i) n = 102, (ii) On the scale, 1 = strongly disagree, and 5 = strongly agree

Table 3 Findings on business performance

Business performance	Mean	S.D
Market share growth	3.51	0.89
Profit margin on sales	3.20	0.70
Growth of sales	3.07	0.80
Growth in the return on investments	3.05	0.64
Market share	3.02	0.70
General competitive position	2.46	0.80
Total	**3.05**	**0.53**

Notes (i) n = 102, (ii) On the scale, 1 = significantly decreased, and 5 = significantly increased value of the measurement scale. When the GSCM practices scale is evaluated in terms of sub-dimensions, the highest value is the *investment recovery* dimension (3.61), while the others are *ED* (3.42), *CWS* (3.20), *GP* (3.00) and *IEM* (2.97). Among these dimensions, the fact that the *IEM* dimension is below the mean value (2.97) is a remarkable issue. When this dimension is evaluated in the context of items, the item "*GSCM is adopted by senior managers*" has a very low value (1.26); It is considered by the senior management as an indication that this issue should be evaluated and focused on more detail. In addition, the fact that the item "*the business has environmental management systems*" is below the mean value (2.95) is evaluated by the businesses as an indicator that the environmental management systems are not given importance as required by the global competition.

The item "*CWS is made for the production of eco-designed products*" in the dimension of cooperation with customers is also below the mean value (2.96). This situation is associated with the fact that the businesses within the scope of the research are sub-industrialists. Since the sub-industry businesses mainly produce products designed by the main industry, it can be said that they cannot act independently in terms of eco-designed products. In the GP dimension, the items "*care is taken to cooperate with our suppliers on environmental issues* (2.70)" and "environmentally friendly practices of second-tier suppliers are evaluated (2.83)" are below the mean value. In this context, it could be said that businesses should cooperate with their suppliers on environmental issues and adopt environmentally friendly practices as a requirement of global competition.

The findings regarding the CA of the businesses are shown in Table 2.

When Table 2 is examined, it is observed that the total mean score (3.72) of the CA scale is above the medium/average value of the CA scale. When the CA scale is evaluated in terms of dimensions, *delivery dependability* has the highest value (4.17), while *price/cost* (3.39) has the lowest values.

In Table 2, the item "*we are able to develop products quickly*" draws attention within the time to market dimension (2.97). This situation can be explained by the fact that the sub-industry businesses produce the products designed by the main industry.

Although the *quality* dimension of the CA scale is above the mean (3.95), the item "*we can compete based on quality*" is below the mean (2.89). At this point, it could be argued that businesses should make more efforts to improve quality.

The findings regarding the BP of the businesses are shown in Table 3.

It is observed that the total performance mean score (3.05) of the businesses is above the medium/average value of the measurement scale. It is seen that the item "*market share growth*" has the highest value on the scale (3.51), while the item "*general competitive position*" has the lowest value (2.46). When Table 3 is evaluated in general terms, the increase in the growth of the market share of the businesses and the profit margin on sales draw attention compared to other items.

The direction and severity of the relationship between green supply chain practices, CA and BP variables was determined by using Pearson correlation. Table 4 shows the results.

Table 4 Correlation matrix of variables

Variables	GSCM Practices	CA	BP
GSCM practices	1		
CA	0.622**	1	
BP	0.446**	0.477**	1

Notes: (i) n = 102, (ii) **p < 0.001

When Table 4 is examined, there is a positive and moderate statistically significant ($p < 0.001$) relationship between GSCM practices and CA (0.622), GSCM practices and BP (0.446), CA and BP (0.477).

After determining the relationships between variables, the systematic proposed by Hayes [37] was used to examine the causal relationships, test the research hypotheses and evaluate the mediating effect of CA in accordance with the conceptual model of the research, and the following model was proposed:

According to the model adapted from Hayes [37], GSCM practices and CA affect BP. It is also assumed that CA has a mediating role in the effect of GSCM practices on BP. In accordance with the systematic developed by Hayes [37] and shown in Fig. 2, the mediation relationship model number 4 has been analysed with the SPSS 22 program using the Process Makro plug-in. In this context, the regression analysis results have been examined and the results are illustrated in Table 5.

Regression analysis based on the bootstrap technique has been conducted to examine causal relationships, test research hypotheses and evaluate the mediating effect of CA. 5000 resampling options have been used in the analysis with the *bootstrap technique.* In order for the research hypotheses to be supported in the mediation effect analysis conducted with the bootstrap technique, the values in the 95% confidence interval (CI) obtained as a result of the analysis should not include the value zero (0) [38]. Table 5 shows that GSCM practices significantly and positively affect CA ($\beta = 0.452$; $p < 0.001$). GSCM practices explain approximately 39% of the variation in CA (R2 = 0.386). With this result, H_1 is supported. Additionally, when Table 5 is examined, it is seen that GSCM practices and CA variables explain BP together within the scope of regression analysis. It is seen that CA affects BP positively and significantly ($\beta = 0.255$; $p < 0.05$). Additionally, GSCM practices have a

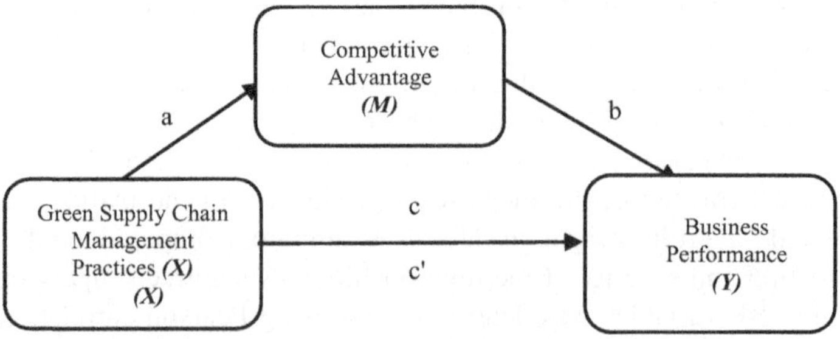

Fig. 2 Mediation model. *Source* Adapted from Hayes [37]

Table 5 Regression analysis

	CA *(M)*			BP *(Y)*		
Predictor variables	B	S.E	95% CI	β	S.E	95% CI
GSCM practices *(X)*	0.452**	0.570	[0.339–0.565]	0.139*	0.063	[0.015–0.264]
CA *(M)*	–	–	–	0.255*	0.086	[0.084–0.427]
Constant	2.290**	0.188	[1.916–2.663]	1.661**	0.258	[1.154–2.169]
	$R = 0.622$, $R^2 = 0.386$, $F = 62.959$*			$R = 0.514$, $R^2 = 0.264$, $F = 17.441$*		

Notes: *$p < 0.05$, **$p < 0.001$, S.E. = Standard Error, β = Unstandardized Beta Coefficient

Table 6 Mediation analysis

Independent variable (X)	Dependent variable (Y)	Mediator variable (M)	β	BootSE	BootLLCI	BootULCI
GSCM	BP	CA	0.115	0.047	0.032	0.220

Note β = Unstandardized beta coefficient

positive and significant effect on BP (β = 0.139; $p < 0.05$). GSCM practices and CA explain 26% of the variation in BP (R2 = 0.264). Based on these results, the H_2 and H_3 hypotheses of the research are supported.

Data regarding the mediation analysis to test the final hypothesis of the study are presented in Table 6.

Whether CA has an indirect effect on BP has been determined according to the confidence intervals obtained with the bootstrap technique. Accordingly, the indirect effect on the effect of CA on BP is significant, therefore CA mediates the relationship between GSCM practices and BP (β = 0.115[2]; 95% BCA CI [0.032; 0.220]). As a result of bootstrap analysis, corrected bias and accelerated confidence interval values (BCA CI) do not include the value 0 (zero). In addition, when the full and partial standardized effect sizes of X on Y are examined, the confidence intervals in both effect values are significant because they do not include the value 0 (zero). Effect sizes can be interpreted as low if they are close to K^2=0.01, medium if they are close to K^2=0.09, and high if they are close to K^2=0.25 [38]. In the analyses, the fully standardized effect size *(K^2)* is ***0.202***, and since this value has an effect size close to the high value, it can be said that the mediation effect in the model is high. In the light of these results, the hypothesis *(H_4)* of the research, which has been put forward as "*CA has a mediating role in the effect of GSCM practices on BP*", is supported.

Based on the explanations, the findings regarding the mediating role of CA in the effect of GSCM practices on BP are summarized in Fig. 3 and Table 8.[3]

[2] The β value is in Table 8 and it has been given in Appendix A.

[3] Table 8 has been given in Appendix A.

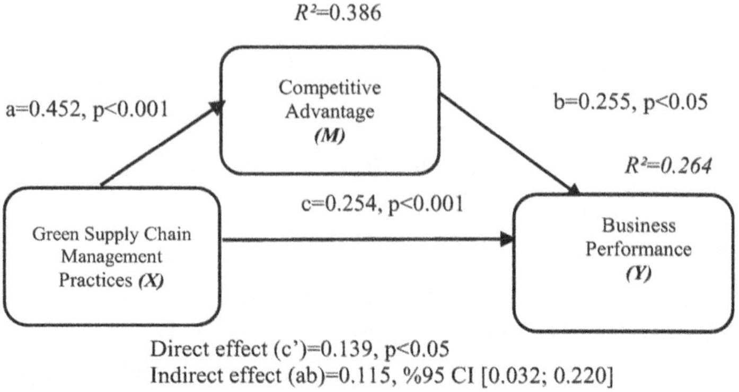

Fig. 3 The mediating role of competitive advantage in the green supply chain management-business relationship

The results obtained from the analyses regarding the research hypotheses and some studies in the literature that support these results are presented in Table 7.

In their study, Al-khawaldah et al. [16] have assessed the GSCM practices as green product, green manufacturing and green marketing. It has been concluded that each dimension affects CA positively. In the study they conducted on PROPER companies in Indonesia; Novitasari and Agustia [19] have examined the mediating role of CA in the effect of GSCM on BP. Based on the research findings, it has been concluded that GSCM has a positive effect on CA and CA effects positively BP; whereas has been found that the effect of GSCM on BP is insignificant. In addition, Novitasari and Agustia [19] have concluded in their study that CA mediates the relationship between GSCM and BP.

In the study conducted by Li et al. [25], it is found that supply chain management has a positive effect on the BP, supply chain management effects positively CA, and CA has a positive effect on BP. A positive relationship between CA on the BP has also been supported in the studies of Abeysekara et al. [26], Agha et al. [27], Hariandi et al. [28] and Koban and İşcioğlu [29].

The effect of GSCM practices on the supply chain performance and the performance of the businesses operating in Azerbaijan has been scrutinized in the study conducted by Safirli and Avunduk [22]. Research findings indicate that the GSCM

Table 7 Summary table of hypotheses

Hypotheses	Results	Similar studies in the literature
H_1	Supported	Al-khawaldah et al. [16]; Novitasari and Agustia [19]
H_2	Supported	Novitasari and Agustia [19]; Li et al. [25]; Abeysekara et al. [26]; Agha et al. [27]; Hariandi et al. [28]; Koban and İşcioğlu [29]
H_3	Supported	Abdallah and Al-Ghwayeen [12]; Safarli and Avunduk [22]
H_4	Supported	Novitasari and Agustia [19]

practices have a positive effect on the BP and the supply chain performance effects positively BP. In the study conducted by Abdallah and Al-Ghwayeen [12], the impact of the GSCM on environmental performance, operational performance and BP has been examined in the context of Jordan. One of the subjects studied within the scope of the study is the mediation role of environmental and operational performance in the relationship between GSCM and the BP. Research findings show that GSCM positively effects BP. Based on these explanations it could be said that the outcomes of this research are coherent with the existing literature.

5 Conclusion and Evaluation

Establishing the mediation effect of CA on the relationship between GSCM on BP in automotive sub-industry businesses was the main focus of this research. To achieve this objective information about GSCM practices, CA and BP of businesses were gathered. Taking into account these variables regression analysis based on the bootstrap technique has been performed to test the hypotheses established within the scope of the research. As a result of the analysis, it is found that GSCM practices and CA have a positive and significant effect on BP. It has also been found that CA plays a full mediating role between GSCM practices and BP.

The IEM dimension of businesses on the scale of GSCM practices has a value below the average. In this context, it is recommended that the businesses within the scope of the research should increase the importance they attach to environmental management systems, both as a requirement of global competition and to comply with legal regulations, and to provide the support of business executives in this regard.

The GP dimension mean score, which is another dimension of the GSCM practices scale, has an average value (3.00). In this context, it is suggested that sub-industry businesses should attach more importance to environmental issues in their relations with their suppliers and work together on environmentally friendly practices. In addition, the fact that both the sub-industry businesses and their suppliers have the internationally recognized "Environmental Management Certificate" and that they work together on environmentally and friendly practices have been evaluated as a positive and supporting element in this process.

It is noteworthy that although the quality dimension of the CA scale of businesses is above the mean value, the level of competitiveness based on quality is below the mean value. Considering the critical role of quality in providing CA, since quality as an order qualifier has become a competitive priority today, it is seen as a necessity for businesses to make more efforts in quality improvement.

In generalizing the findings, like any other study this research also have some limitations. The sample within the scope of the research includes businesses operating in the automotive sub-industry in Konya. This situation is considered as a limitation in terms of generalizing the research results, and although Konya automotive sub-industry has competitive power at national and international levels, the research findings should be evaluated in the context of Konya automotive sub-industry. From

Table 8 The mediating role of competitive advantage in the green supply chain management-business performance relationship

	Effect (β)	S.E	t	p	LLCI	ULCI
Total effect of X on Y (c path)	0.254	0.051	4.984	< 0.001	0.153	0.356
Direct effect of X on Y (c′ path)	0.139	0.063	2.217	< 0.05	0.015	0.264
Effect of X on M (a path)	0.452	0.570	7.935	< 0.001	0.339	0.565
Effect of M on Y (b path)	0.254	0.086	2.953	< 0.05	0.084	0.426
	Effect (β)	BootSE	Boot LLCI	Boot ULCI		
Indirect effect of X on Y (a × b)	0.115	0.047	0.032	0.220		
Partially standardized indirect effect of X on Y	0.216	0.084	0.062	0.398		
Completely standardized indirect effect of X on Y	0.202	0.080	0.055	0.374		

this point on, it is recommended to conduct studies at national and international levels, both with a larger sample in the automotive sub-industry and in different sectors. In this way, the findings obtained in this study regarding the Konya automotive sub-industry can be generalized and inter-sectoral comparisons can be made.

Appendix A

See Table 8.

References

1. Alcaraz JLG, Reza JRD, Soto KCA, Escobedo GH, Happonen A, Vidal RPI, Macías EJ (2022) Effect of green supply chain management practices on environmental performance: case of Mexican manufacturing companies. Mathematics 10(11):1877
2. Masoumik SM, Abdul-Rashid SH, Olugu EU (2014) Gaining competitive advantage through strategic green supply chain management: From a literature review towards a conceptual model. Int J Supply Chain Manage 3(3):49–58

3. Sarkis J, Zhu Q, Lai KH (2011) An organizational theoretic review of green supply chain management literature. Int J Prod Econ 130(1):1–15
4. Green KW, Zelbst PJ, Meacham J, Bhadauria VS (2012) Green supply chain management practices: impact on performance. Supply Chain Manage Int J 17(3):290–305
5. Laosirihongthong T, Adebanjo D, Choon Tan K (2013) Green supply chain management practices and performance. Ind Manag Data Syst 113(8):1088–1109
6. Zhu Q, Sarkis J, Lai KH (2008) Confirmation of a measurement model for green supply chain management practices implementation. Int J Prod Econ 111(2):261–273
7. Chien MK, Shih LH (2007) An empirical study of the implementation of green supply chain management practices in the electrical and electronic industry and their relation to organizational performances. Int J Environ Sci Tech 4(3):383–394
8. Khan SAR, Qianli D (2017) Impact of green supply chain management practices on firms' performance: an empirical study from the perspective of Pakistan. Environ Sci Pollut Res 24:16829–16844
9. Zhu Q, Geng Y, Fujita T, Hashimoto S (2010) Green supply chain management in leading manufacturers case studies in Japanese large companies. Manag Res Rev 33(4):380–392
10. Hsu CW, Hu AH (2008) Green supply chain management in the electronic industry. Int J Environ Sci Technol 5(2):205–216
11. Muma BO, Nyaoga RB, Matwere RB, Nyambega E (2014) Green supply chain management and environmental performance among tea processing firms in Kericho County-Kenya. Int J Econ Finance Manage Sci 2(5):270–276
12. Abdallah AB, Al-Ghwayeen WS (2019) Green supply chain management and business performance. Bus Process Manag J 26(2):489–512
13. Amjad A, Abbass K, Hussain Y, Khan F, Sadiq S (2022) Effects of the green supply chain management practices on firm performance and sustainable development. Environ Sci Pollut Res 29(44):66622–66639
14. AlBrakat NSA, Al-Hawary SIS, Muflih SM (2023) The effect of green supply chain on the export performance of the Jordanian pharmaceutical industry. Uncertain Supply Chain Manage 11(2):613–624
15. Abbas H, Tong S (2023) Green supply chain management practices of firms with competitive strategic alliances—a study of the automobile industry. Sustainability 15(3):2156
16. Al-khawaldah R, Al-zoubi W, Alshaer S, Almarshad M, ALShalabi F, Altahrawi M, Al-Hawary S (2022) Green supply chain management and competitive advantage: the mediating role of organizational ambidexterity. Uncertain Supply Chain Manage 10(3):961–972
17. Liu L, Tang M, Xue F (2012) The impact of manufacturing firms' green supply chain management on competitive advantage. Adv Mater Res 472–475:3349–3354
18. Sharabati AAA (2021) Green supply chain management and competitive advantage of Jordanian pharmaceutical industry. Sustainability 13(23):13315
19. Novitasari M, Agustia D (2022) Assessing the impact of green supply chain management, competitive advantage and firm performance in PROPER companies in Indonesia. Oper Supply Chain Manage 15(3):395–409
20. Hilal F (2022) The effect of green supply chain management practices and competitive advantage on financial performance. Int J Bus 27(1):1–13
21. Kılıçlı Y, Kıpçak E (2023) Yeşil tedarik zinciri yönetimi uygulamaları ile performans arasındaki ilişkinin incelenmesi: bir meta-analiz çalışması. J Transp Logist 7(2):261–287
22. Safarli V, Avunduk H (2023) Yeşil tedarik zinciri uygulamalarının tedarik zincirine etkisi üzerine Azerbaycan işletmelerinde bir araştırma. J Bus Digital Age 6(Special Issue):15–29
23. Novitasari M, Agustia D (2021) Green supply chain management and firm performance: the mediating effect of green innovation. J Ind Eng Manage 14(2):391–403
24. Chen CJ (2019) Developing a model for supply chain agility and innovativeness to enhance firms' competitive advantage. Manag Decis 57(7):1511–1534
25. Li S, Ragu-Nathan B, Ragu-Nathan TS, Rao SS (2006) The impact of supply chain management practices on competitive advantage and organizational performance. Omega 34(2):107–124

26. Abeysekara N, Wang H, Kuruppuarachchi D (2019) Effect of supply-chain resilience on firm performance and competitive advantage: A study of the Sri Lankan apparel industry. Bus Process Manage J 25(7):1673–1695
27. Agha S, Alrubaiee L, Jamhour M (2012) Effect of core competence on competitive advantage and organizational performance. Int J Bus Manage 7(1):192–204
28. Hariandi MSI (2019) E-commerce, competitive advantage and business performance of Banyuwangi small and medium-sized enterprises. Int J Sci Technol Res 8(8):1219–1220
29. Koban E, İşcioğlu TE (2020) Lojistik işletmelerinin pazarlama yeteneklerinin rekabet avantajı ve işletme performansına etkisi. Uluslararası İktisadi ve İdari İncelemeler Dergisi 26:35–52
30. Novitasari M, Alshebami AS, Sudrajat MA (2021) The role of green supply chain management in predicting Indonesian firms' performance: competitive advantage and board size influence. Indonesian J Sustain Account Manage 5(1):137–149
31. Astawa IK, Pirzada K, Budarma IK, Widhari CIS, Suardani AAP (2021) The effect of green supply chain management practices on the competitive advantages and organizational performance. Polish J Manage Stud 24(1):45–60
32. Tran H, Hoang N, Do V, Nguyen T, Nguyen V, Phan T, Doan T (2022) Impact of green supply chain management on competitive advantage and firm performance in Vietnam. Uncertain Supply Chain Manage 10(4):1175–1190
33. Zhang Q (2001) Technology infusion enabled value chain flexibility: a learning and capability-based perspective. Doctoral dissertation, University of Toledo, Toledo, OH
34. Yazıcıoğlu Y, Erdoğan S (2014) SPSS uygulamalı bilimsel araştırma yöntemleri. Detay Yayıncılık, Ankara
35. Altunışık R, Coşkun R, Bayraktaroğlu S, Yıldırım E (2010) Sosyal bilimlerde araştırma yöntemleri SPSS uygulamalı. 6. baskı, Sakarya Kitabevi, Ankara
36. Gürbüz S, Şahin F (2018) Sosyal bilimlerde araştırma yöntemleri. Gözden geçirilmiş güncellenmiş 5. Baskı, Seçkin Yayıncılık, Ankara
37. Hayes AF (2013) Introduction to mediation, moderation, and conditional process analysis. The Guilford Press, New York
38. Gürbüz S (2019) Sosyal bilimlerde aracı, düzenleyici ve durumsal etki analizleri. Seçkin Yayıncılık, Ankara

The Role of STEM Education in the Growth and Sustainability of Smart Cities

Esra Damar

Abstract One of the most important paradigms to improve civic life is considered today as building of smart cities based on innovative industries. This fact requires, naturally, creative, high level skilled and proactive people profile for solving sophisticated challenges. That's why "smart people" pillar of the smart city concept reflects the most considerable aspect of the whole paradigm. It can be said, in this context, smarting a city can be accomplished by smart people not only smart technologies. It is not possible to say that a smart city technologically-capable for the future has also a sustainability capability for prosperity if the citizens are not smart. Smart people are the key triggers to unlock economic and social value in a smart city concept. The main determinants of the smart people are interconnected with STEM (Science, Technology, Engineering, and Mathematics) skills of the people to success the science, mathematics, and engineering and those needed to use technology effectively. This study focuses on proving the close relationship between the smart people component, which is one of the most important components in making these cities smart, and people's STEM skills in line with the findings of the EU (European Union) project titled "A Roadmap for Smart City": carried out in 2019–2020 under the coordination of Çorum Municipality in partnership with Szarvas Municipality from Hungary, Telšiai Municipality from Lithuania, Hitit University from Turkey and Çorum Chamber of Commerce and Industry within the scope of Town Twinning Action Grant Scheme (TTGS) between Turkey and EU.

Keywords Smart city · Smart people · STEM skills

E. Damar (✉)
Hitit University, Çorum 19030, Turkey
e-mail: esradamar@hitit.edu.tr

F. Çebi et al. (eds.), *Engineering and Technology Management in Challenging Times*,
Lecture Notes in Networks and Systems 1137,
https://doi.org/10.1007/978-3-031-70935-7_21

1 Introduction

The construction of smart cities based on innovative industries is currently regarded as one of the most significant paradigms to improve civic life. For a sustainable urban lifestyle, smart cities, which are one of the key components of economic and social development, need to be outfitted with smart solutions from top to bottom. Because of this, it's crucial for the system's operation that the city's components are all carefully designed as a whole, and that everything from production to industry, education to social and cultural life is connected into smart networks and made easier to access. This fact necessitates, naturally, the profile of creative, highly trained, and proactive individuals to address complex problems. Because of this, the "smart people" pillar of the concept of a smart city reflects the most important feature of the entire paradigm. In this sense, it may be claimed that smart people, not only smart technologies, are capable of making a city smarter. If the populace is not intelligent, it is impossible to claim that a smart city that is technologically prepared for the future also has a sustainability capability for prosperity. Indeed, smart people are the key elements that make up a smart city. The smart people component; Meeting the needs of tomorrow's employers, Having access to information and technology, Embracing creativity and innovation, Exploring new ways of doing things, Community leaders include forward-thinking people supported by mentors, services and incentive programmes. The key to unlocking the economic and social potential of a smart city concept is smart people. The primary indicators of smart people are closely related to the STEM (Science, Technology, Engineering, and Mathematics) skills required for success in science, mathematics, engineering, and the successful use of technology. Today's world is facing increasingly complex and challenging problems. To meet these challenges, it is crucial to promote education among the new generations, equipping them with the most useful and powerful tools. To do this, we need to see the world as it is now, a place full of technology, where science is everywhere, from health and infrastructure to energy and the environment, so that helping people develop Science, Technology, Engineering and Mathematics (STEM) skills has become an essential part of basic literacy in today's knowledge. Learning STEM is not just a bare and brutal necessity of life, it can be fun for people and therefore a tool to communicate other important ideas and teach other subjects more effectively [1]. In the literature reviews, there are studies that reveal the importance of people's STEM skills in the construction of smart cities [1–4]. In the study conducted by Lim et al. [2] for young people, a smart city was designed and implemented through STEM-oriented education and training processes. A similar study was conducted by Scrodazzi et al. [1], and they made smart city applications more fun and understandable for young people with the help of STEM modules by providing an example of how robotics technology is used to make young students think about active citizenship and environmental issues. Furthermore, Karwowski et al. [5] addressed sustainable STEM education and workforce development and leadership approach through the smart cities initiative. The aim of this study, in the light of the findings of an EU project implemented in three countries, focuses on revealing the close relationship between the smart people component, which is one of the most important elements in making cities smart, and people's STEM skills.

2 As a Requirement of Technological Development Priority of STEM in the Education

STEM, which has been mentioned as a new paradigm in education, training and learning policies since the early two thousand years, covers the disciplines of science, technology, engineering and mathematics in its paradigmatic integrity, and with the proposal of Georgette Yakman, a graduate student in 2006, it started to be accepted as STE(A)M (Science, Technology, Engineering, Art and Mathematics) to include art [6]. However, Yakman argues that the inclusion of the arts in STEM is not a mere addition of disciplines; fine arts and design thinking will lead to critical reasoning such as "what, how, who and why" in the inherent structure of STEM [7]. Therefore, in macro-scale STEM-based education, it is important to teach students not only the subject matter but also how to learn, how to ask questions, how to experiment and how to create. As a matter of fact, the Ministry of National Education defines STEM as an interdisciplinary educational approach that combines science, technology, engineering, arts and mathematics and aims to develop students' higher-order thinking skills, productivity skills and innovation capacities [8].

The point that draws attention in all definitions that try to reveal what STEM is that STEM is defined as "discipline", "approach", "method" and these definitions do not have a satisfactory equivalent. In our view, STEM is an innovative paradigm developed by the United States of America (USA) for education, teaching and learning policies. The primary justification for defining it as a paradigm is that it contains a hierarchical structure that references specific disciplines in its internal loop. It also lacks an accredited methodology and is often perceived as a vague theory because the interdisciplinary relationship is not firmly embedded in the teaching methodology[9].

Although there is no generally agreed definition of what STEM is, experts generally agree that STEM workers use knowledge of science, technology, engineering and mathematics to understand the world and solve problems [10].

In the literature, it is generally accepted that the acronym STEM came to prominence with the decision of the US National Science Foundation and some funding organizations to support four areas of interest through education, and that the name, originally thought to be SMET, was changed to STEM in 2001 [11]. While this information is not wrong, the intellectual impositions that triggered STEM go back much further. The words of Ronald REAGAN, governor of the state of California in 1967, when he started a revolution in education by changing the scope and purpose of public universities, point to radical changes in educational policies:

> Higher education should prepare students for jobs... Employers are looking for employees who have the ability to answer questions that will arise in the future but that we cannot imagine asking now. If a university is not a place where intellectual curiosity is to be encouraged, and subsidized, then it is nothing [12].

In the same parallel, Barak OBAMA's statements pointed to the pioneering mission of the United States in technological developments.

> Since we cannot compete in a global economy by paying our employees less or producing goods that are of inferior quality, we can never be prosperous. That works against us. The secret to our success, as it always has been, will be to compete by creating new products, new industries, and preserving our position as the global leader in scientific research and technological advancement. Our future depends entirely on it [13].

Indeed, in the same vein, there are serious concerns that educational relationships and roles need to be redefined in their institutional integrity in order to transform knowledge into economic benefits, and that courses are not tailored to market needs [14].

Given the fact that today's knowledge economy needs highly skilled personnel at all levels to cope with rapidly changing industrialization, it has been emphasized that in order to meet the needs, higher education institutions need to contribute to the global economy and redefine curriculum, pedagogy and assessment policies to graduate students with the desired qualities and competencies [15].

STEM has been consistently supported by both the public and business sectors, as it is seen as a major force for technological advancement in the US. In order to increase the number of college graduates with STEM degrees and to promote science and mathematics education, the American Competitiveness Council, established by US President George W. BUSH, called for a significant increase in federal funding for STEM education development in 2006. The Council also recommended doubling the spending budget for this purpose. The STEM Education Coalition is another US-based initiative. The STEM Education Coalition, an alliance of more than 500 business, professional and educational organizations, supports educators and students in organizations that offer STEM-related programs, such as the U.S. Department of Education and the National Science Foundation, and promotes STEM education so that the United States can continue to be an economic and technological leader in the global marketplace in the twenty-first century. On December 1, 2022, leading U.S. semiconductor and computer companies, as well as educational and professional associations from science, technology, engineering and mathematics (STEM) fields, issued a letter of recommendation for funding STEM workforce and education. The letter states that this support for diversifying STEM fields through higher education programs is critical to the application and learning of skills in science, mathematics and technology and to sustaining global leadership in innovation. In the CHIPS bill signed by US President Joe BIDEN on August 9, 2022, among many provisions related to STEM education; under the title of graduate STEM education; facilitating career opportunities for graduate students and postdoctoral researchers, additional financial support to facilitate their personal and professional development, supporting research on the graduate education system, increasing the number of new graduates to meet the workforce demand [16].

Important STEM initiatives are also being made in the private sector in addition to the public sector. For instance, Intel, one of the top innovators in computer technology, portrays itself as a model STEM-centered business and claims that its trainings are intended to support teachers in enhancing STEM classrooms and motivating students to pursue careers as future industry leaders. Intel runs programmes to assist instructors in successfully engaging students in order to produce engaging lessons and bring STEM to the forefront of the classroom. These programmes provide advice on lesson plans, curriculum design, and interactive multimedia tools. In explaining its vision for STEM, it emphasizes the need to maximize course outcomes by putting technology itself at the center of learning activities, rather than focusing on teaching how to use technology to prepare students for the Fourth Industrial Revolution [17]. Similar approaches and initiatives have been seen in other countries. In the UK, the Department for Business Innovation and Skills, which was authorized as the authority responsible for higher education in 2009, states its core mission as follows: to build a dynamic and competitive UK economy by fostering innovation, entrepreneurship and science, and by creating the conditions for business success by providing skills and opportunities for all. To achieve this, we will strengthen world-class universities and support a globally open economy [18]. In Australia, the Department of Education, Employment and Labor Relations, which is responsible for funding investments in higher education, explains its purpose; Shaping Australian higher education and research; building a modern, productive, internationally competitive Australian economy through world-leading, strategically focused infrastructure investment [19].

In 2000, in Lisbon, Portugal, the European Council discussed strengthening employment, economic reforms and social reconciliation as part of a knowledge-based economy and put forward the Lisbon Strategy, which aims to become "the most competitive and dynamic knowledge-based economy in the world, capable of sustainable economic growth" for the next 10 years. In this context, the "European Research Area (ERA)" was established to ensure scientific excellence and competitiveness. For the European Union, universities in particular have a unique position at the crossroads of education, research, innovation and service to society and the economy [20]. Targets 210, 443, 353.1, 443.1, 443.1, 443.4 in the Eleventh Development Plan include targets on training scientific researchers, undergraduate and graduate scholarships, science workshops or national and international scientific activities to increase qualified manpower [21]

3 Smart City Concept as a Paradigm

In this part of our study, definitions in the literature are given by mentioning the concepts contained in the smart city concept. In this context, the importance of the relationship between information society and smart cities is mentioned.

Smart cities refer to projects that leverage ICT-based innovation to better organize urban services and provide new economic opportunities in cities, albeit there is no widely recognized definition of what they are. Smart cities frequently concentrate on projects that leverage digital innovation to improve the efficiency of urban service delivery and so raise a community's overall competitiveness. But there is disagreement on the definition of a "smart city," which has a wider context. One of the most widely used definitions comes from the European Commission, which states that a smart city is one where existing networks and services are improved for the benefit of its citizens and businesses by using digital and communications technology [22]. According to the United Nations, a smart city approach uses opportunities from digitalization, clean energy and technologies, as well as cutting-edge transport technologies, giving residents options to make more environmentally friendly decisions and encouraging sustainable economic growth, as well as allowing cities to improve their service delivery [23]. The Smart Cities Council, a group of major smart city technology players like Cisco, IBM, Intel, and Qualcomm, offers the following definition of a smart city: "A smart city collects data from devices and sensors implanted in its roads, power grids, buildings, and other assets. It distributes the data using a clever communications infrastructure, which often combines wired and wireless technology. Then, it employs intelligent software to produce useful data and digitally enhanced services" [24]. A smart city is one that, in the words of IBM, makes the best use of all the connected information that is currently available to better understand and control its operations and optimise the use of limited resources. Smart cities, in the words of Cisco, "implement scalable solutions that take use of ICT to increase efficiencies, cut costs, and enhance quality of life". According to Japan, a smart city is a sustainable city or region that uses ICT and other cutting-edge technology to address a variety of issues and governs itself (via planning, development, management, and operation) for overall optimisation. Smart cities, according to the UK Department of Business, Innovation, and Skills, are a process rather than a final product that improves a city's living standards, resilience, and capacity to deal with difficulties through increased citizeparticipation, hard infrastructure, social capital, and digital technology. According to their definition, a smart city is one that successfully integrates physical, digital, and human systems into the built environment to provide its residents with a sustainable, wealthy, and inclusive future[25].

Smart city acts with the following orientations by making use of advanced technology to make the city more liveable and sustainable.

- People-oriented,
- Eco friendly,
- Promoting participation and transparency,
- Open to continuous development and learning,
- Data privacy and ethical rules.

Six key indicators of a smart city have been developed by Giffinger and the European Smart Cities research group at the Regional Science Centre of the Vienna University of Technology. Table 1 proposes a classification model for smart city applications, divided into six areas. This model is now widely used and adopted as a reference by both academics and practitioners around the world [26].

- **Smart People** refers to those who are working in a society that encourages creativity and innovation, where they have access to education, training, human resources, and capacity management, and where ICT makes it possible for them to work in an environment that makes it possible for them to do so [27].

Table 1 Functions of smart cities

Smart economy (Competitiveness)	Smart people (Social and human capital)
• Innovative spirit • Entrepreneurship • Economic image and trademarks • Productivity • Flexibility of labour market • International embeddedness • Ability to transform	• Level of qualification • Affinity to life long learning • Social and ethnic plurality • Flexibility • Creativity • Cosmopolitanism/Open mindedness • Participation in public life
Smart governance (Participation)	Smart mobility (Transport and ICT)
• Participation in decision-making • Public and social services • Transparent governance • Political strategies and perspectives	• Local accessibility • (Inter-)national accessibility • Availability of ICT-infrastructure • Sustainable, innovative and safe transport systems
Smart environment (Natural resources)	Smart environment (Natural resources)
• Attractively of natural conditions • Pollution • Environmental protection • Sustainable resources management	• Cultural facilities • Health conditions • Individual safety • Housing quality • Education facilities • Touristic attractively • Social cohesion

- **Smart Governance** refer to governance consisting of e-government services that ensure transparency, participation and collaboration in public administration processes such as analysis, planning, implementation and policy making [27].
- **Smart Living** is a set of practices that aim to increase the quality of life by addressing the basic needs of society with a rational approach, while at the same time aiming to ensure social integrity [27].
- **Smart Environment** with the support of information and communication technologies, it is necessary to control green areas and water resources to ensure sustainability in the environment and nature. Renewable energy, sustainable resource management, smart energy grids, smart meter reading devices, advanced air pollution monitoring systems, nature-friendly green buildings, green city structures, energy-efficient smart street lighting and solid waste management systems can be evaluated in this field [27].
- **Smart Mobility** refer to IT-enabled and integrated transportation systems. It covers sustainable, safe and interconnected transportation systems that include trams, buses, trains, subways, cars, bicycles and pedestrians using one or more modes of transportation [27].
- **Smart Economy** aims to use existing resources efficiently in the face of increasing consumption factors in every field, to develop measures for increasing consumption and to increase the quality of life. It is the handling of a city within the framework of smart industries. Competitiveness, brand value and sharing economy are prominent concepts [27].

Chourabi et al. [28] tried to identify the success factors of smart city applications as eight items. These factors are as follows:

- Management and Organization: Many smart city initiatives are managed and heavily used by governments to provide better services to citizens.
- Technology: Requires the use of next generation integrated hardware, software and networking technologies. ICT enables citizens to make smarter and more accurate decisions using advanced analytical techniques.
- Policy: Policy is critical to understanding the appropriate use of information systems. For this, organizations need to develop policy not only through laws and regulations, but also by assessing what people accept, actions or behaviours.
- Governance: Smart city projects involve multiple stakeholders. Therefore, many cities need to coordinate to develop better practices to manage these projects and initiatives.
- People Communities: Smart city applications enable citizens to participate in the governance of the city and become active users.
- Economy: Economy is the main driver for the development of smart city applications. A city that is economically competitive with other cities is considered to have at least one of the characteristics of a smart city. One of the most important indicators for measuring increased city competitiveness is the extent of a city's economic strength.
- Built Infrastructure: The current state and quality of ICT infrastructure is important for smart cities. The development of ICT infrastructure (its availability and performance) is essential for the development of smart cities.
- Natural Environment: The basic premise of the smart city concept is to use technology to increase sustainability and better manage natural resources. Protection of natural resources and related infrastructure becomes particularly important. Smart cities not only improve the use and efficiency of resources, but also play a role in enriching the lives of the people living there.

On the contrary, there are some threat elements in the process of smart city formation. In this context, especially the imbalance in income distribution affects individuals' access to and effective use of applications. The fact that technology cannot get rid of the reality accessible only with material resources is part of this problem. The fact that information and communication technologies are controlled by a certain group of people and directed by application and content providers actually deepens the problem. It is observed that the data flow on the Internet circulates through developed countries or through developed cities in the country. Communication technologies play a major role in how to bridge the gap between developed countries and others. At the same time, the use of communication technologies is an important need for equality not only for countries but also for individuals. Today, while access to the Internet is possible through mobile or fixed devices, these devices carry ownership costs and usage costs, which can widen the gap in income inequality. In addition, in order to make effective and efficient use of the internet and applications that can be downloaded to smartphones and tablets, the ability to use these hardware and devices is required. In order to benefit from these applications at a high level, it is necessary to be computer literate. In addition, in order to use and benefit from information in what we call the information society, technology, access and data transfer should be recognized as human rights and considered as one of the basic needs in the hierarchy of needs. While the use of technology, access and data transfer is currently based on purchasing power and material means, it should be a more inclusive, collective, homogeneous, human-oriented, democratic and egalitarian model that includes the sharing of real technology and knowledge. Financial capital should not be a prerequisite for the realization and utilization of innovations [29].

3.1 Smart Human Component on the Way to Becoming a Smart City

The core trigger of smart city paradigm is to support innovation at all levels to create sustainable, quality jobs by the way of creating a suitable entrepreneurial eco-system. This kind of eco-system requires stimulating education and skills to increase highly-skilled workforce. In order to keep up with the rapid developments in technology, today's knowledge economy needs highly qualified workers at all levels. Higher education institutions around the world are working to redesign curriculum, pedagogy, and assessment procedures to fit contemporary social needs and make sure that all students have the appropriate skills and abilities by the time they complete a bachelor's degree [30].

In STEM education, which is one of these configurations and has emerged as an innovative paradigm in education and training processes since the early two thousands of years, it is important to teach students not only the subject matter but also how to learn, how to ask questions, how to experiment and how to create. As a matter of fact, existing definitions show STEM as an interdisciplinary educational approach that combines science, technology, engineering, art and mathematics and aims to develop students' higher-order thinking skills, productivity skills and innovation capacities. STEM practitioners agree that they use their knowledge of science, technology, engineering and mathematics to understand the world and solve problems. The idea of smart cities was developed as a result of using ICT infrastructure to achieve a greater, more sustainable level of economic development and at the same time create a better quality of life. Considering the definitions of smart city and STEM, there are basically two factors for the success of these paradigms; the first is that both concepts are based on science, technology, engineering and mathematics, and the other is related to the ability of people to use the technology brought by both paradigms at a high level. In this context, when designing smart cities, people's STEM skills are extremely important for the functioning of the system. The success of projects relating to smart cities depends on collaboration between the various parties involved, including technology developers, service providers, city developers, city administration, inhabitants, and local businesses. The importance of human capital

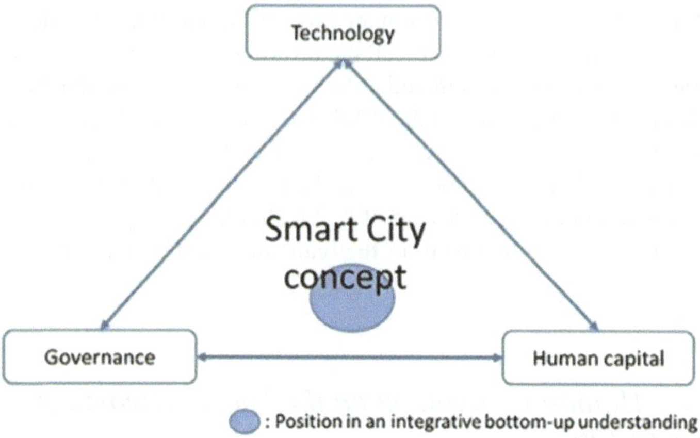

Fig. 1 Integrative bottom-up understanding of smart city in a three-dimension perspective

and education in urban development is highlighted. The creation of new markets and goods necessitates a labour force that is more and more skilled, which defines the relationship between human capital and urban development [31].

According to Giffinger in Fig. 1, social innovation must be emphasised as a second vital feature. Governance and human capital are crucial in a smart city viewpoint based on learning processes incorporating technological breakthroughs into urban development. Cities have the opportunity to enact integrative solutions in the smart city approach thanks to the interaction between technical and social components in a learning process [32].

Smart citizens and intelligent individuals are seen as the foundation of smart cities in terms of human capital [33–37].

Smart people in a city are determined by the following 7 criteria, according to [26];

- Level of qualification,
- Affinity to life-long learning,
- Social and ethnic plurality,
- Flexibility,
- Creativity,
- Cosmopolitanism/open-mindedness,
- Participation in public life.

However, cities become more inclusive, creative, and sustainable when inhabitants are aware, educated, and involved and have access to the opportunities and capabilities of the information society [28, 36, 38]. The cities that grow their creative sectors, establish a creative culture, and draw a creative and competent workforce will be the "smartest" in this knowledge-intensive and globalized economy [39–41]. In order to coordinate the interchange of knowledge and the production of new knowledge,

knowledge management frameworks, human skills, and innovation support institutions are crucial in smart cities [42]. By encouraging additional human capital growth, Glaeser and Berry [31] even showed how a strong human capital base makes cities smarter. They also anticipated that this tendency would continue in the future. On the other hand, networks, trust, and social cohesiveness are the fundamental components of social capital [43]. A number of papers [34, 39, 40, 42, 44] specifically highlight social capital as a crucial success factor in smart city strategies. In order to pool their intelligence in the city's favour and spread knowledge widely, all smart city stakeholders must be involved and work together for a smart city to succeed [45]. In order to promote social inclusion and prevent social polarization, social capital and relational capital should be wisely managed [34, 39, 40]. We consider a city to be smart when it makes investments in its people and social capital as well as in its traditional (transport) and modern (ICT) communication infrastructure, all of which support sustainable economic growth and high quality of life while judiciously managing its natural resources.

4 Key Study

In this part of the study, the findings of the EU project titled "A Roadmap for Smart City" carried out in 2019–2020 within the scope of the Town Twinning Action Grant Scheme (TTGS) between Turkey and EU under the coordination of Çorum Municipality in partnership with Szarvas Municipality of Hungary, Telšiai Municipality of Lithuania, Hitit University from Turkey and Çorum Chamber of Commerce and Industry from Turkey are presented (see Fig. 2) [25].

4.1 Çorum

The city of Çorum is one of the cities called as Anatolian tigers, and is a city that has made a name for itself in the soil industry and machinery-metal industry sector. Çorum has become one of Turkey's major export centres with the increase in exports in recent years have reached 1.5 billion dollars in 2019 and exports has made exports to 138 different countries. Çorum, which ranked 21st in 2018 with exports of 741 million 965 thousand dollars, reached 1 billion 541 million 873 thousand dollars last year and entered the 15th place among 19 provinces that export over 1 billion dollars together with Trabzon. Although Samsun is the natural centre and largest city of the Central Black Sea as a port city, Çorum has the advantage of being the locomotive of regional development for years with its entrepreneurial spirit, industrial potential and historical treasures [25].

As a result of field studies and literature reviews, it has been observed that smart city initiatives differ from city to city. In addition to the needs of citizens, the demographic structure, geographical location, economic infrastructure and social-cultural

Fig. 2 Çorum-Szarvas-Telšiai

infrastructure of a city are important parameters for smart city orientations. Çorum is currently a city that using smart city applications in various fields. Some of these applications are;

- Smart public transportation applications,
- 3D City information system,
- Using renewable energy sources,
- Water management system,
- Smart waste collection system.

4.2 Szarvas

The city of Szarvas, located in the center of Békés County, is considered to be the pearl of the South Hungarian Plain. Among the factors that reinforce this feature are the following:

- Rich in natural beauties,
- Favorable geographic conditions of the town and its surroundings,
- In the border of the city there is the river Körös flowing 30 km length,
- Target area of tourism, popular tourism destination,

- The town is part of the Körös Valley nature reserve,
- Rich geothermal energy, thermal wells.

This natural treasure provides many advantages to Szarvas. Besides these features, the historical background of the town and the features coming from the well-established urbanism tradition will be the reference point for determining the projections of Szarvas in the coming period. These are;

- Old school town, the first agricultural school in the country was founded in the town,
- Centre of agricultural sciences,
- Favorable place for swimming and fishing, water sports,
- Fresh air; air pollution level is low,
- Ethnic groups live in peace in the town: Slovakians, Gypsies (Romas),
- No traffic problems for the crossing traffic.

Szarvas has some applications in the smart environment, smart mobility, smart people and smart living areas from smart city components. Some examples of its work in these areas are;

- City security monitoring centre,
- Waste monitoring centre,
- Bicycle paths and inclusive cultural city activities.

Szarvas should determine smart city orientations in order to be a more liveable and attractive city by improving these applications and adding new applications by increasing the quality of life of its citizens [25].

4.3 Telšiai

There are 39.317 people living in the city of Teliai district in Lithuania. The Samogitia region's capital, Teliai, is situated on the banks of Lake Mastis. The administrative, economic, religious, and educational centres of Samogitia are located in Teliai, a tiny modern European town. Teliai's development is due to its inhabitants' activity, education, and business acumen. One of Lithuania's seven protected old towns is the one in the city of Teliai. EDEN (European Destinations of Excellence), the organisation that designated Teliai as a tourism destination in 2013, also declared Teliai as the official 2016 Lithuanian Capital of Culture.

Value proposition of Telšiai are;

- There are well-developed industrial zones with readymade infrastructure and buildings within Teliai city that are ideal for both greenfield and brownfield developments.
- For decades, the region's growth has been shaped by long-standing significant textile industry, food processing, and wood processing traditions. These industries have a pool of qualified individuals.

- From preschools to colleges and universities, Teliai is home to all kinds of educational institutions. Teliai is also one of Lithuania's most prosperous small towns, undertaking infrastructure projects that were supported by the EU.
- The Teliai area is proud of its cultural centres, theatres, exhibition halls, regional parks, museums, mounds, galleries, public libraries, and other amenities.

Samogitian cultural treasure provides many advantages to Telšiai. Besides these features, the historical background of the town and the features coming from the well-established urbanism tradition will be the reference point for determining the projections of Telšiai in the coming period. Telšiai has some applications in the smart environment, smart mobility, smart people and smart living areas from smart city components. Some examples of its work in these areas are;

- Water management system,
- Interactive tourism monuments,
- Bicycle sharing system and cycle paths,
- Public–private collaborations in project,
- Development and inclusive cultural city activities.

Telšiai should determine smart city orientations in order to be a more attractive and liveable city by improving these applications and adding new applications by increasing the quality of life of its citizens and guests [25].

4.4 Current Situation Analysis

The scale of the research; Students, NGOs, neighbourhood headmen, and managers of public and private institutions and the people living in Çorum, Szarvas and Telšiai constitute the scope of the research. Within the scope of the research, a total of 367 people were researched: 196 people from Çorum, 86 people from Szarvas and 85 people from Telšiai. Survey questions were prepared by participatory method and a set of questions was created to identify the city's priorities, goals, requirements and obstacles in order to become a smart city [25].

4.5 Methodology

The research was carried out via the survey method on the internet. Due to the widespread use of the internet and the increase in the use, faster and more participation was achieved in the surveys conducted over the internet instead of the classical survey method. The questionnaires were made during the activities via the online survey application at https://www.menti.com/ and the results were shared with the participants during the events. At the same time, on preparatory studies such as literature review, determination of process organization, current situation analysis,

Swot analysis, interviews and round table meetings with the stakeholders have done for determination of basic policies and rules, presentation of the projections and proposals [25].

4.6 *Case of Çorum*

After informing about smart cities in the awareness raising meetings, the relevant questions were asked to the participants.

Question-1: The first question was about which one of the six smart city component is more important to make a city smarter?

The first question is about which one of the six smart city component is more important to make a city smarter: Smart Economy, Smart Governance, Smart People, Smart Mobility, Smart Environment and Smart Living.

Considering the results in Table 2, the smart city component that Çorum should give priority has emerged as Smart People with 34.7%. Subject of "Smart People" are expected to give the highest priority with 34.7%; is an individual with high awareness, participation and creativity, lifelong learning, including information technologies in his life, the main element of human and social capital and the focus of city life. By enhancing people's abilities to use and create information and communication technologies, the smart people component seeks to create a diverse society that fosters innovation and creativity. At the same time, smart communities formed by smart people are also an important actor in the development of cities and becoming attraction centres, effectively using applications in smart cities. It should also be remembered that the concept of smart people in smart city applications is the focus and without smart people, smart cities cannot exist and succeed. The empowerment of the smart person is again through education, and the priority of STEM in education is the key trigger of technological progress. The necessities for Çorum regarding this component;

Table 2 Priorities for Çorum

Priorities for Çorum	Municipality Staffs	NGO's	Students	Total Frequency	Percentage (%)
Smart Economy	4	7	21	33	17.1
Smart Governance	6	2	5	7	3.6
Smart People	6	20	26	67	34.7
Smart Mobility	20	2	10	15	7.8
Smart Environment	2	6	4	12	6.2
Smart Living	5	14	40	59	30.6
SUM	36	51	106	193	100.0

- Creating policies for conscious technology use, organizing social campaigns, implementing social protection and responsibility projects,
- To carry out activities in the focus of the smart city in order to create awareness and to maintain the awareness of belonging to the city, ownership of the city, participation in city institutions and dealing with city problems,
- Carrying out studies to reduce the digital gap between different segments of society and to encourage and increase the use of technology,
- Increasing the quality and quantity of services for family, women, children and youth with smart city solutions,
- Services for the poor, homeless with the improvement of smart city applications,
- Carrying out activities in the smart city focus for the disabled and the elderly to participate effectively in social and economic life.
- Developing smart applications in the social, cultural, artistic and sportive areas of the city, which serves the quality, efficient and fun life of people.
- In cities, it is aimed to produce smart city solutions that will enable individuals with different lifestyles and members to harmonize on a wide common ground and live peacefully and productively [25].

Question-2: In your opinion, what should be the primary goal of becoming a smart city?

The second question is about determining the primary purpose for the smart city. Smart city's brand value contribution, resource and time savings, environmental sustainability and life facilitation options were asked.

Considering the results in Table 3, Providing smart technologies and services that will facilitate the lives of individuals and institutions is the basis of the 'Smart Cities' approach. It has gained an important dimension in terms of providing systems and technologies that facilitate human life in all areas from education to health, transportation to energy, security to tourism, economy to mobility, from environment to housing. Cities, which facilitate the life of citizens with the policies it follows, and smart applications, attract more attention from people and increase the level of satisfaction of citizens. In other words, the level of citizen satisfaction increases in direct proportion with the level of intellectuality of the city. As a result, it turns out that the purpose of Smart cities is to increase and facilitate the quality of life of those living in that city [25].

Question-3: Which of the following is the primary requirement for becoming a smart city?

The third question is to reveal the priority need of the city of Çorum under the options of Strong cooperation between institutions, a clear vision, Effective smart city platform, Public–private cooperation, Citizen participation, Technological infrastructure, Project dissemination ability.

Considering the results in Table 4, Citizen participation and Technological infrastructure, with a rate of 22.1%, have been identified as priority requirements to become a smart city. On the way to becoming a smart city, in order to achieve the goals, good

Table 3 Goals for Çorum

Goals for Çorum	Municipality Staffs	NGO's	Students	Total Frequency	Percentage (%)
Increasing the city's brand value	4	11	22	37	18.9
Resource and time saving	6	15	30	51	26.0
Ensuring environmental sustainability	6	17	16	39	19.9
Making the life of the citizen easier	20	8	41	69	35.2
SUM	36	51	109	196	100.0

planning and transformation depends on effective management at every stage of the journey. Looking at the successful and unsuccessful examples in the world, some factors seem critical in terms of success in the smart city transformation journey. In this way, regarding the citizen participation, which emerged as a priority requirement in the survey,

- Supporting citizen participation,
- Strengthening communication between managers and the public,
- Transparent management,
- Producing approaches and solutions to the needs of the city from a citizen perspective.

Attention should be paid to organizing and managing citizen participation in a way that does not remain dysfunctional. In the process of designing smart city

Table 4 Requirements for Çorum

Requirements for Çorum	Municipality Staffs	NGO's	Students	Total Frequency	Percentage (%)
Strong cooperation between institutions	5	16	8	29	14.9
A clear vision	7	2	12	21	10.8
Effective smart city platform	3	9	24	36	18.5
Public-private cooperation	1	5	3	9	4.6
Citizen participation	15	12	16	43	22.1
Technological infrastructure	2	5	36	43	22.1
Project dissemination ability	3	1	10	14	7.2
SUM	36	50	109	195	100.0

solutions, it is very important to use appropriate tools to include and strengthen hard-to-reach population groups such as poverty and/or social exclusion, youth and the elderly, immigrants, and the disabled, and to target gender balance in participation processes. Creative industries and arts can be used more to encourage dialogue and participation of all segments of society. In addition, city officials and urban experts should be trained to carry out meaningful and ethical participation processes of citizens. Regarding the technological infrastructure, which appeared in the survey as a priority requirement on the way to become a smart city, It means that the city should focus more on technological infrastructure, on the way to becoming a smart city;

- Improving the technological infrastructures of institutions,
- Providing technological transformation,
- With the contribution of technology, speed and quality should be increased in services.

Technological infrastructure will be provided by strengthening and technological development for the establishment of a more effective, efficient, competitive, smart city. In addition to all these, the cities that are desired to be smart must have a cloud data centre with the necessary technological infrastructure and have cloud servers in this data centre [25].

Question-4: Which of the following is the most important obstacle for becoming a smart city?

The fourth question is to reveal the most important obstacle under the options of the lack of financial capacity, insufficient cooperation between the institutions, politic issues, the lack of expertise in ICT, the lack of citizen participation, the lack of visionary leadership, Insufficient data.

Considering the results in Table 5, 30.4%, the insufficiency of financial situations has been identified as the biggest obstacle to become a smart city. Due to financial deficiencies, cities have problems in making smart city requirements, applications and investments as much as they want. The fact the technological solutions, such as digitalization and digital transformations, are high cost of investment, the inability to realize the necessary investments, the difficulties of the institutions to solve this problem reveals how important financial opportunities are.

Studies have been carried out to develop existing smart city applications for Çorum and in which are as new applications should be implemented more.

As a result of the current situation and the analysis studies carried out, Çorum can continue its development by focusing on some sectors, creating a goal unity and vision in the city [25].

Under the framework studies, four different aspects has been found to define the image of Çorum in future;

1. Competitive City (based on Smart Economy component)
2. Clean City (based on Smart Environment component)
3. Connected City (based on Smart Mobility component)

Table 5 Obstacles for Çorum

Obstacles for Çorum	Municipality Staffs	NGO's	Students	Total Frequency	Percentage %
The lack of financial capacity	5	15	38	58	30.4
Insufficient cooperation between the institutions	0	8	10	18	9.4
Politic issues	6	3	13	22	11.5
The lack of expertise in ICT	2	3	16	21	11.0
The lack of citizen participation	13	4	8	25	13.1
The lack of visionary leadership	6	8	11	25	13.1
Insufficient date	4	6	12	22	11.5
SUM	36	47	108	191	100.0

4. Creative City (based on Smart Living component and Smart People).

4.7 *Case of Szarvas*

Current Situation Analysis The scale of the research; the scope of the research consists of municipal employees, students and managers of public and private institutions and people living in Szarvas. Within the scope of the research, a total of 86 people were interviewed.

Question-1: In your opinion, which of the following smart city components should be given priority in Szarvas?

The first question is about which one of the six smart city components is more important to make a city smarter: Smart Economy, Smart Mobility, Smart Environment, Smart People, Smart Living, and Smart Governance.

According to the results in Table 6, the smart city component that Szarvas should give priority was Smart Living, with a rate of 44.2%. Prioritizing the following service areas to improve the quality of life will contribute to the realization of the requirements of this component [25].

- Cultural and educational opportunities
- Health conditions
- Personal security
- Housing quality
- Educational equipment
- Touristic activities/facilities
- Social solidarity.

Table 6 Priorities for Szarvas

Priorities for Szarvas	Frequency	Percentage (%)
Smart Economy	4	4.7
Smart Governance	5	5.8
Smart People	9	10.5
Smart Mobility	10	11.6
Smart Environment	20	23.3
Smart Living	38	44.2
SUM	86	100.0

Question-2: In your opinion, what should be the primary goal of becoming a smart city?

The second question is about determining the primary purpose for the smart city. Smart city's brand value contribution, resource and time savings, environmental sustainability and life facilitation options were asked.

According to the results in Table 7, the primary goal of being a smart city is "Providing environmental sustainability" with the highest rate of 43.6%. Sustainable Cities are settlements where human communities can live economically, socially and environmentally healthy and quality and transfer this to future generations, consuming resources as little as possible and showing the ability to renew.

- Turn to the least polluting renewable energy sources,
- Aim to reduce the ecological footprint as much as possible,
- Plan the land use in the most effective way,
- Emphasis on transportation options that take into account the needs and environmental priorities of the residents,
- Reusing wastes as raw materials or energy using composting, recycling and/or other recycling methods should minimize the negative impact of wastes on climate change [25].

Table 7 Goals for Szarvas

Goals for Szarvas	Frequency	Percentage (%)
Increasing the city's brand value	10	12.8
Resource and time saving	23	29.5
Ensuring environmental sustainability	34	43.6
Making the life of the citizen easier	11	14.1
SUM	78	100.0

Question-3: Which of the following is the primary requirement for becoming a smart city?

The third question is to reveal the priority need of the city of Szarvas under the options of Strong cooperation between institutions, a clear vision, Effective smart city platform, Public–private cooperation, Citizen participation, Technological infrastructure, Project dissemination ability.

According to the results in Table 8, "Technological infrastructure" has been identified as a primary requirement for becoming a smart city with a rate of 22.5%. It means that Szarvas City the city should focus more on technological infrastructure.

On the way to becoming a smart city;

- Improving the technological infrastructures of institutions,
- Providing technological transformation,
- With the contribution of technology, speed and quality should be increased in services.
- More effective, efficient, competitive and technological infrastructure will be provided with a smart city for the establishment of strengthening and technological development [25].

Question-4: Which of the following is the most important obstacle for becoming a smart city?

The third question is to reveal the most important obstacle under the options of the lack of financial capacity, insufficient cooperation between the institutions, politic issues, the lack of expertise in ICT, the lack of citizen participation, the lack of visionary leadership, Insufficient data.

According to the results in Table 9, the lack of financial capacity in Szarvas City has been identified as the biggest obstacle to becoming a smart city with a rate of 32.1%. Due to financial deficiencies, cities have problems in making smart city requirements, applications and investments as much as they want. The fact that the smart investment cost is high and the municipalities carry out the smart city transformation with their own limited resources and the unlimited needs of the city are the most important

Table 8 Requirement for Szarvas

Requirement for Szarvas	Frequency	Percentage (%)
Strong cooperation between institutions	16	20
A clear vision	12	15
Effective smart city platform	8	10
Public-private cooperation	6	7.5
Citizen participation	6	7.5
Technological infrastructure	18	22.5
Project dissemination ability	14	17.5
SUM	80	100

Table 9 Obstacles for Szarvas

Obstacles for Szarvas	Frequency	Percentage (%)
The lack of financial capacity	25	32.1
Insufficient cooperation between the institutions	10	12.8
Politic issues	2	2.6
The lack of expertise in ICT	9	11.5
The lack of citizen participation	15	19.2
The lack of visionary leadership	2	2.6
Insufficient date	15	19,2
SUM	78	100,0

problems of the municipalities. Starting in smart city applications that can fund both own cost and subsequent costs in the short term will help reduce the negative impact of the financial situation [25].

Under the framework studies, two different aspects has been found to define the image of Szarvas in future;

1. Slow Szarvas (Based on Smart Environment and Smart Living)
2. Smile Szarvas (Based on Smart Living and Smart People).

4.8 *Case of Telšiai*

Current Situation Analysis The scale of the research; The scope of the research consists of municipal employees, students and managers of public and private institutions and people living in Telšiai. Within the scope of the research, a total of 85 people were interviewed.

Question-1: In your opinion, which of the following smart city components should be given priority in Telšiai?

The first question is about which one of the six smart city component is more important to make a city smarter: Smart Economy, Smart Mobility, Smart Environment, Smart People, Smart Living, and Smart Governance.

According to the results in Table 10, the smart city component that Telšiai should give priority was Smart Living, with a rate of 42.4%. Prioritizing the following service areas to improve the quality of life will contribute to the realization of the requirements of this component [25].

- Cultural and educational opportunities
- Health conditions
- Personal security
- Housing quality
- Educational equipment

Table 10 Priorities for Telšiai

Priorities for Telšiai	Frequency	Percentage (%)
Smart Economy	25	29.4
Smart Governance	7	8.2
Smart People	5	5.9
Smart Mobility	3	3.5
Smart Environment	9	10.6
Smart Living	36	42.4
SUM	85	100.0

- Touristic activities/facilities
- Social solidarity.

Question-2: In your opinion, what should be the primary goal of becoming a smart city?

The second question is about determining the primary purpose for the smart city. Smart city's brand value contribution, resource and time savings, environmental sustainability and life facilitation options were asked.

According to the results in Table 11, the primary goal of being a smart city is "Making the life of the citizen easier" with the highest rate of 36.9% Providing smart technologies and services that will facilitate the lives of individuals and institutions is the basis of the 'Smart Cities' approach. It has gained an important dimension in terms of providing systems and technologies that facilitate human life in all areas from education to health, transportation to energy, security to tourism, economy to mobility, from environment to accommodation. Cities, which facilitate the life of citizens with the policies it follows, and smart applications, attract more attention from people and increase the level of satisfaction of citizens. In other words, the level of citizen satisfaction increases in proportion to the level of intellectuality of the city. As a result, the purpose of Smart cities is to improve and facilitate the quality of life of the inhabitants [25].

Table 11 Goals for Telšiai

Goals for Telšiai	Frequency	Percentage (%)
Increasing the city's brand value	20	23.8
Resource and time saving	16	19.1
Ensuring environmental sustainability	17	20.2
Making the life of the citizen easier	31	36.9
SUM	84	100.0

Question-3: Which of the following is the primary requirement for becoming a smart city?

The third question is to reveal the priority need of the city under the options of Strong cooperation between institutions, A clear vision, Effective smart city platform, Public–private cooperation, Citizen participation, Technological infrastructure, Project dissemination ability.

According to the results in Table 12, "A clear vision" has been identified as a primary requirement for becoming a smart city with a rate of 28.6%. It shows that the citizens of Telšiai give priority to intellectual approach. This means that it is necessary to have an inclusive vision in smart city transformation and the most basic step is to determine the "city-specific vision, target and strategy, refer to the definition of priorities [25].

Question-4: Which of the following is the most important obstacle for becoming a smart city?

The fourth question is to reveal the most important obstacle under the options of the lack of financial capacity, insufficient cooperation between the institutions, politic issues, the lack of expertise in ICT, the lack of citizen participation, the lack of visionary leadership, Insufficient data.

According to the results in Table 13, the lack of financial capacity in Telšiai has been identified as the biggest obstacle to becoming a smart city with a rate of 39.3%. Due to financial deficiencies, cities have problems in making smart city requirements, applications and investments as much as they want. The fact that the smart investment cost is high and the municipalities carry out the smart city transformation with their own limited resources and the unlimited needs of the city are the most important problems of the municipalities. Starting in smart city applications that can fund both own cost and subsequent costs in the short term will help reduce the negative impact of the financial situation.

Table 12 Requirements for Telšiai

Requirement for Telšiai	Frequency	Percentage (%)
Strong cooperation between institutions	12	14.3
A clear vision	24	28.6
Effective smart city platform	3	3.6
Public-private cooperation	13	15.5
Citizen participation	17	20.2
Technological infrastructure	12	14.3
Project dissemination ability	3	3.6
SUM	84	100.0

Table 13 Obstacles for Telšiai

Obstacles for Telšiai	Frequency	Percentage (%)
The lack of financial capacity	33	39.3
Insufficient cooperation between the institutions	8	9.5
Politic issues	12	14.3
The lack of expertise in ICT	5	6.0
The lack of citizen participation	19	22.6
The lack of visionary leadership	4	4.8
Insufficient date	3	3.6
SUM	84	100.0

Under the framework studies, two different aspects has been found to define the image of Telšiai in future;

1. Traditional Telšiai (Based on Smart People and Smart Living)
2. Touristic Telšiai (Based on Smart Living and Smart Economy).

As a result of field studies and literature reviews, it has been observed that smart city initiatives It differs from city to city. In addition to the needs of citizens, a city's demographic structure, geographical location, economic infrastructure and social-cultural infrastructure are important parameters for smart city orientations [25].

5 Discussion

In order to find the answer to the question "What are the prominent factors in the smartening of a city?" three cities that have limited smart city applications but are trying to put forward advanced models in this regard have been selected as examples. As a result of the field studies and literature reviews, it has been observed that smart city initiatives differ from city to city and from country to country. It is natural that each smart city initiative is special for each city; in addition to the needs of citizens, a city's demographic structure, geographical location, economic infrastructure and social-cultural infrastructure are important parameters for smart city orientations. Within the scope of the project; as a result of literature review, determination of process organization, current situation analysis, Swot analysis, interviews with stakeholders, four projections for Çorum and two projections for Szarvas and Telšiai were determined. Although these projections vary from city to city, the only thing common in all definitions and characteristics of smart cities is citizen participation. Therefore, human factors and human development in smart cities is a very important area to be analyzed. The role of education in the development of the smart human component is undoubtedly very important. It is now widely acknowledged that the subjects of science, technology, engineering, and mathematics are essential to the global development. Studies show that young people are not eager to pursue a career in these fields. The performance and interest of young people in the domains of Science, Technology, Engineering, and Mathematics can be boosted via STEM-focused innovative education and training methods, making smart cities that use ICT technology smarter.

Acknowledgements I would like to express my gratitude to the esteemed scientist Mehmet Aydınkal, who made a significant contribution to the intellectual part of the study and passed away suddenly during the preparation phase.

References

1. Scaradozzi D, Screpanti L, Cesaretti L, Mazzieri E, Storti M, Brandoni M, Longhi A (2016) Rethink Loreto: we build our smart city A STEM education experience for introducing smart city concept with the educational robotics. In: 9th annual international conference of education, research and innovation (ICERI). IATED, Seville, Spain, pp 750–758
2. Lim TC, Wilson B, Grohs JR, Pingel TJ (2022) Community-engaged heat resilience planning: lessons from a youth smart city STEM program. Landsc Urban Plan 226:1–13
3. Bonora L, Martelli F, Marchi V (2019) An amazing way to learn STEM concepts developing sustainable cities idea in the citizens of the future: the methodology of Erasmus+ project DIGITgame (digital improvement by game in smart city projecting). In: Proceedings of the 10th international conference on e-education, e-business, e-management and e-learning. Association for Computing Machinery, Tokyo, Japan, pp 18–22
4. Maranda AP, Szukalska A, Wilczyński L (2021) Learner-centered pedagogical approach based on smart city concept to enhance the development of STEM skills of children complex, intelligent and software intensive systems. In: Barolli L, Poniszewska-Maranda A, Enokido T (eds) Complex, intelligent and software intensive systems, vol 1194. Springer, Cham, pp 210–221
5. Karwowski W, Amaba T (2014) A system-of-systems engineering approach to leadership and innovation: sustainable STEM education and workforce development through the Smart Cities initiative. QScience Proc 12(3)
6. https://steameducation.eu/space.html. Accessed 11 Apr 2024
7. https://www.k12digest.com/steam-an-educational-framework-to-relate-things-to-each-other-and-reality/. Accessed 11 Apr 2024
8. MEB (2016) Milli Eğitim Bakanlığı STEM Eğitim Raporu. Yenilik ve Eğitim Teknolojileri Genel Müdürlüğü (YEĞİTEK). https://yegitek.meb.gov.tr/STEM_Egitimi_Raporu.pdf. Accessed 11 Apr 2024
9. Damar E (2023) Yükseköğretimde STE(A) M Algisinin Cinsiyet Odakli İncelenmesi ve Kariyer Planlamasina Etkisi. Kadın ve Yaşam, p 131
10. Vilorio D (2014) STEM 101: intro to tomorrow's jobs, occupational outlook quarterly. https://www.bls.gov/careeroutlook/2014/spring/art01.pdf. Accessed 11 Apr 2024
11. Sanders M (2009) STEM, STEM education, STEM mania. The Technology Teacher; Reston 68(4):20–26
12. https://www.wpr.org/reagans-1967-speech-changed-purpose-college-forever-says-journalist. Accessed 11 Apr 2024
13. https://obamawhitehouse.archives.gov/innovation/strategy/introduction. Accessed 11 Apr 2024
14. Bok D (2005) Universities in the marketplace: the commercialization of higher education. Oxford, NJ; Princeton University Press, Princeton

15. Kirst M, Stevens ML (2015) Remaking college: the changing ecology of higher education. Stanford University Press, Redwood City
16. http://www.stemedcoalition.org/2022/08/08/stem-education-in-the-chips-science-and-competitiveness-bill/. Accessed 11 Apr 2024
17. https://www.intel.com/content/www/us/en/education/k12/teachers/stem-education.html. Accessed 11 Apr 2024
18. DBIS (2009) Department for business innovation & skills. http://www.bis.gov.uk/about. Accessed 11 Apr 2024
19. DEEWR (2009) Department of education, employment and workplace relations. http://www.deewr.gov.au/HigherEducation/Programs/EIF/Pages/default.aspx. Accessed 29 Jan 2023
20. TÜSİAD (2000) Lizbon Avrupa Konseyi AB Başkanlar Konseyi Belgesi. https://tusiad.org/tr/yayinlar/raporlar/item/download/8086_bcff752d69010c8bfbacce24f9545cca. Accessed 11 Apr 2024
21. Türkiye Cumhuriyeti Cumhurbaşkanlığı Strateji ve Bütçe Başkanlığı On Birinci Kalkınma Planı, https://www.sbb.gov.tr/wp-content/uploads/2022/07/On_Birinci_Kalkinma_Plani-2019-2023.pdf. Accessed 11 Apr 2024
22. European Commission (2014) EC digital agenda for Europe: smart cities. http://eige.europa.eu/resources/digital_agenda_en.pdf. Accessed 11 Apr 2024
23. United Nations (2016). http://nua.unhabitat.org/pillars.asp?PillarId=7&ln=1. Accessed 29 Jan 2023
24. http://smartcitiescouncil.com/article/our-vision. Accessed 1 July 2023
25. https://www.smarttwinning.org/. Accessed 1 July 2023
26. Giffinger R, Fertner C, Kramar H, Kalasek R, Pichler-Milanovid N, Meijers E, Smart cities; ranking of European medium-sized cities, Vienna, UT, p 11. https://www.smart-cities.eu/download/smart_cities_final_report.pdf. Accessed 11 Apr 2024
27. Tübitak Bilgem (2017) Akıllı şehir stratejisi ve eylem planı. http://yte.bilgem.tubitak.gov.tr/sites/images/bilgem/tubitak_bilgem_yte_akillisehirlersunumu_301117.pdf. Accessed 11 Apr 2024
28. Chourabi H, Nam T, Walker S, Gil-Garcia JR, Mellouli S, Nahon K, Pardo T, Scholl HJ (2012) Understanding smart cities: an integrative framework. In: Paper presented at the proceedings of the 45th international conference on system sciences, Hawaii
29. Armağan MV (2018) Bilgi Toplumunda Akıllı Şehirler ve Katılımcı Yurttaşlık, Doktora Tezi, Gazi Üniversitesi, Ankara
30. Chan RY (2016) Understanding the purpose of higher education: an analysis of the economic and social benefits for completing a college degree. J Educ Policy Plan Adm (JEPPA) 6(5):1–40
31. Glaeser EL, Berry CR (2006) Why are smart places getting smarter?. Taubman Cente Policy Brief 2006-2. Taubman Centre, Cambridge, MA
32. Giffinger R, Lu H (2015) The smart city perspective: a necessary change from technical to urban innovations
33. Hollands RG (2008) Will the real smart city please stand up? City 12(3):303–320
34. Lombardi P, Giordano S, Farouh H, Yousef W (2012) Modelling the smart city performance. Eur J Soc Sci Res 25(2):137–149
35. Neves BB (2009) Are digital cities intelligent? The Portuguese case. Int J Innov Reg Dev 1(4):443–463
36. Paskaleva KA (2011) The smart city: a nexus for open innovation? Intell Build Int 3(3):153–171
37. Pelton JN, Singh IB (2009) Future cities. Designing better, smarter, more sustainable and secure cities. Intelligent Community Forum, New York
38. Aurigi A (2006) New technologies, same dilemmas: policy and design issues for the augmented city. J Urban Technol 13(3):5–28
39. Caragliu A, Del Bo C, Nijkamp P (2009) Smart cities in Europe Serie research memoranda 0048. VU University Amsterdam, Faculty of Economics, Business Administration and Econometrics
40. Caragliu A, Del Bo C, Nijkamp P (2011) Smart cities in Europe. J Urban Technol 18(2):65–82
41. Florida R (2005) The flight of the creative class. The new global competition for talent. Harper Business, HarperCollins

42. Komninos N (2009) Smart cities: towards interactive and global innovation environments. Int J Innov Reg Dev 1(4)
43. Falk I (2000) Human capital and social capital: what's the difference? Adult Learn Commentary (28)
44. Nam T, Pardo TA (2011) Conceptualizing smart city with dimensions of technology, people, and institutions. In: Proceedings of the 12th annual international digital government research conference: digital government innovation in challenging times, pp 282–291
45. INTA (2012) Report on the results of the "Smart City Workshop", Paris. Retrieved 3–4 Sept 2012

Transition to Smart Agriculture: Case of Kasaplar Village

Mehmet Çam and Ömür Yaşar Saatçıoğlu

Abstract The developments in information and communication technologies (ICT) have brought about radical changes for society and industry. Especially the growth and globalization of digital technologies like the Internet of Things, Artificial Intelligence (AI), and robotics towards the end of the twentieth century initiated a transformation process in social and economic life. Sustainability is another trend that should be considered which also caused an inevitable transformation process due to the requirements. Digital transformation has also affected all sectors. Agriculture sector was one of the sectors that changed considerably due to the smart technologies and sustainability. Smart agriculture practices were deemed to be important to achieve United Nation's Sustainable Development Goals. Turkey is one of the countries that developed smart agriculture practices. Hopefully, good examples of Smart Agriculture practices were developed in Kasaplar Village, Aydın, Turkey. This research aims to investigate the reasons behind smart agriculture practices. Furthermore, a mixed research method is used to learn the effects of smart agriculture applications in Kasaplar Village.

Keywords Smart agriculture · Kasaplar village · Tabit

1 Introduction

According to the World Economic Forum (WEF), "Food plays a central role in human societies and is vital to the well-being of people and the planet. But a fundamental transformation is required to meet the goals of an inclusive, efficient, sustainable and healthy food system" [1].

M. Çam
Dokuz Eylul University, İzmir, Turkey

Ö. Y. Saatçıoğlu (✉)
Logistics Department, Dokuz Eylul University, Maritime Faculty, İzmir, Turkey
e-mail: yasar.saatci@deu.edu.tr

F. Çebi et al. (eds.), *Engineering and Technology Management in Challenging Times*, Lecture Notes in Networks and Systems 1137,
https://doi.org/10.1007/978-3-031-70935-7_22

Poverty and malnutrition continue to be one of the biggest problems in many parts of the world. According to the information on the website of the UN Food and Agriculture Organization (FAO), the Millennium Development Goals were implemented in the early 2000s, but complete success was not achieved. For this reason, with the decisions taken within the framework of the Sustainable Development Goals in 2015, new targets were set to completely eliminate poverty and hunger by 2030 [2]. To this end; first of all, supporting social protection programs that aim to ensure that households increase and diversify their consumption by increasing their own production has come to the fore. In developing countries, such programs cover approximately 1.5 billion poor people, one-third of whom are extremely poor [3]. In this context, food value chain, which includes data production, processing technologies, logistics, sales and exports, needs to be supported with technologies to solve climate change and food problems. WEF [1] argues that modern information technology is a critical enabler for the development of food supply chain, especially regarding food traceability. It has been determined that technologies will help achieve the result of minimizing losses and environmental impacts by coordinating storage, delivery times, volumes and transportation routes. It is expected that new values will be created through the application of artificial intelligence and big data in matters such as meteorological data, product growth data, market conditions, food trends and needs. It is envisaged that the use of technology will produce the following outputs [4].

- To determine crop yields appropriate to needs, develop business plans along with weather forecasts, share experience and knowledge, and create a plan for agriculture by expanding the customer base.
- To provide consumers with the agricultural products they need at the desired time.
- Delivering agricultural products to consumers with driverless vehicles when needed.

Additionally, for society at large, these solutions will help increase food production, stabilize supply, address labor shortages in agricultural areas, reduce food waste, and encourage consumption. The development of technological trends in the agricultural sector requires a more efficient and sustainable solution integrated with smart systems, considering the worldwide food need. Therefore, it becomes necessary to focus on developing technologies such as the internet of things (IoT), artificial intelligence, machine learning and the use of large volume data in agriculture [5]. In this context, technology is more than a new trend in the agricultural sector. As a result, trends in technology have affected agriculture and food. It is predicted that the decreasing food and production level with climate change will change our perception of future agriculture or create a new field by creating a different future or paradigm breaks with the support of human mind and agricultural technology. Baierle et al. [6] stated that digital transformation and innovation processes are also making their way into the food industry; giving rise to smart agriculture.

1.1 Smart Agriculture

With the increasing world population, agriculture is becoming increasingly important. Nowadays, agriculture aims to make nutrition more sustainable, as well as providing nutrition with existing resources. Agriculture was one of humanity's first endeavors. It has changed throughout time in order to fulfill the needs of an expanding global population, adjust to shifting climate circumstances, and solve sustainability-related issues [7]. Agriculture has faced major changes in the last half century, and today, with the development of technology, an agricultural system known as smart agriculture has emerged. Smart agriculture means strengthening agricultural activities with automation and technology playing a role in replacing or complementing agricultural activities. In this way, the efficiency of agricultural activities will increase, and a sustainable agricultural practice will be created. Encouraging the use of technology in the agricultural sector and making agricultural activities smarter and more efficient will also lead to the realization of the Sustainable Development Goals designed by the United Nations. In this way, the agricultural sector will help solve the social problems of the whole world and support economic development. The farming system is driven by smart agriculture technologies, a data-driven agricultural operating system from planting to harvest to optimize agricultural management and reduce negative impacts on the environment. The smart farming system prioritizes the concept of ecology in agricultural ecosystems, while crop growth patterns are built on the basis of extracted and interpreted data about production and the environment (weather, soil, water). Smart agriculture is a trend that seems to be crucial for the transition towards a sustainable future, fundamental for the long-term prosperity of the planet [8]. However, as stated in [9]; there are technological, economic, political, social and environmental barriers for the effective implementation of smart agriculture practices.

Despite all the advantages of technology, smart agriculture requires actors to be critical, innovative, and adaptable. It has been demonstrated that especially agricultural actors or farmers who have mastery of technology, problem solving, social networking, negotiation skills and multidisciplinary cooperation skills are important for the future production increase and production needs in response to the increase in food demand.

The best way to move towards a modern and sustainable agriculture that can demonstrate the full power of data-driven management to face the challenges of food production in the twenty-first century is to implement a data-driven decision system by combining ancient knowledge from the past with modern agriculture. Especially when we look at it on an agricultural basis, it is expected that many paradigms will evolve in the next 10 years and make robot and human collaboration necessary in terms of efficiency and production [10].

In the last century, agriculture has transformed from a labor-intensive sector into a sector where mechanization and power-intensive production systems are used. With this transformation, there was a continuous decrease in labor force. For example, in 1900, 41% of the workforce in the USA was employed in agriculture, but by 2000, this rate dropped to only 2% [11]. The rate of labor force employed in agriculture in European Union countries increased from 7.32 to 2.98%, from 29.76 to 18.11% in Turkey, from 1.91 to 1.36% in the USA and in the Eurozone, it decreased from 7.32 to 2.98%. [12]. These decline rates reflect changes in the agricultural sector due to technological advances, economic development and increase in productivity. Furthermore, the smart agriculture market has grown considerably over time. A compound annual growth rate of 10.2% between 2022 and 2030 should fuel this increase [7]. Agricultural activities based on manual labor were replaced by mechanized agricultural activities, productivity in agriculture increased and the share of the agricultural sector in economic development decreased. However, the agricultural sector is still considered an important sector in many countries and a large proportion of the world's population is still employed in the agricultural sector. Therefore, issues such as sustainability of the agricultural sector, productivity increase in agriculture, technological advances and the share of the agricultural sector in economic development need to be examined further.

1.2 Agriculture Sector and Smart Agriculture in Turkey

Turkey ranks second after the USA in terms of agricultural areas and arable lands [13]. However, export performance is not high due to the ineffective use of resources in the agricultural sector. It is seen that successful countries, especially the Netherlands and Israel, base agricultural efficiency due to the technology. In recent years, Turkey has been developing its research and development activities in the field of agricultural technology with the support of state policies, universities and the private sector. In this context, digital transformation projects in agriculture are being carried out by various universities, especially Ege, Boğaziçi, Ankara universities and Konya Food and Agriculture University.

Smart Agriculture 4.0 and Tabit Smart village project, carried out in partnership with Izmir Commodity Exchange, are among the studies carried out in this field. Producers use agricultural technologies to monitor weather conditions and pests with satellite and warning systems, reduce costs in workers and inputs, and increase productivity. Technology companies offer solutions that will increase automation and efficiency in the field-to-table process through the use of digital technologies in agriculture. Unions, cooperatives, public institutions and universities carry out studies on big data analysis, traceability, sustainability and effective use of resources. Thus, the digital transformation of the agricultural sector is supported, and efforts are made to increase efficiency and sustainability in agricultural production.

TAK (Tabit Smart Village) was chosen as the case study because it is a good example of human and machine cooperation in rural areas in Turkey in the smart agriculture sector. In the study, data were collected from face-to-face interviews and observation.

Tabit Smart Agriculture technologies is a social enterprise founded in 2004. Based on the distance of farmers living in rural areas to information and technology, it has implemented Turkey's first communication, e-learning network and e-commerce media in the field of agriculture. By modeling farmer credit cards together with banks, it enabled farmers to find financial resources without being harmed and visited approximately 12,000 villages to explain the work done. Thanks to the Vodafone Farmers Club project, it has initiated early warning information in agriculture by providing information and services to 1 million 500 thousand farmers and their families in Turkey with special applications for farmers. This project has become an example for African and Middle Eastern countries and has delivered quality information to 7.5 million farmers in the MENA region. A special farmer's computer software, supported by Intel, has been produced that can be used by anyone who can read and write and allows asking questions to experts on information, communication, marketing, and other topics they need. It has also carried out various social responsibility projects in order to improve water awareness in rural areas and prevent tractor accidents. Currently, development of Smart Village Project is still in progress by working on technologies that will reduce expenses while increasing productivity with easy-to-use, accessible prices and providing applied agricultural technologies training for farmers living in rural areas. The Smart Village project was created as a new rural life model that will combine family farming with technological agriculture, aiming to increase the efficiency and profitability of small producers which brings together the opportunities provided by advanced technology to large investors with the rural agricultural sector. The project, which was implemented for the first time in the world and in Turkey, aimed to increase productivity, profitability and living standards in rural areas by synthesizing knowledge from academic, public and private sectors, farmers' experiences and ancient experiences with high technology. It aims to prevent migration from villages to cities, which has increased all over the world, especially in recent years, due to economic and social reasons. In Smart Village, which works to shape the future of agriculture, the devices used at every point of agricultural and livestock activities are made smart in order to use agricultural lands efficiently. While measuring the soil surface with smart systems, it is possible to monitor the fields with the data transferred to smartphones through sensors, irrigation and automation systems. Measurements and all data can be controlled from farmers' smartphones and tablets. Along with socialization, education and sports activities for children, youth and women, smart irrigation systems, smart greenhouse and fruit growing, humidity sensors, weather station, early warning systems. There are smart agriculture projects in Smart Village such as: smart pasture, mass milking and animal tracking systems, beekeeping and tracking automation, agricultural data monitoring, farmer decision support center, renewable energy solution center, soil leaf analysis laboratories and experience workshops [14].

1.3 Emergence of TAK Project

The emergence of TAK is actually based on the "agricultural marketing" project of TABİT (Agricultural Information and Communication Technologies). With this project, problems and solutions in rural areas in Turkey were tried to be revealed through this website, but the fact that these solutions were not adopted by those living in rural areas gave rise to the idea of developing the issues described in this project on a sample village. However, after a certain period of time, it was observed that these solutions were not implemented by the people living in rural areas. For this reason, instead of visiting every village, the founders of the TAK Project were laid with the idea of creating a center, an exemplary village, from which everyone can be inspired and that can adapt to anywhere. The project was started with the idea of providing an area where farmers who want to be informed about the practices explained by visiting the villages or farmers who want to apply the projects in their own fields can come, receive training and practice. The land, where the application is made, is important in the project because it is a special location. The biggest financing of the project belongs to TABİT. Although TABİT is the founder of this project, the company also transfers its own sustainable income to TAK. Additionally, Vodafone Türkiye also transferred financing to the project in the establishment of the smart village; it provides support in developing technology in the smart village and disseminating it by creating a global network. Using its own dealership networks, Vodafone ensures that a technology developed at TAK is sold in Turkey and worldwide.

The business model implemented in the smart village is to ensure that the villagers benefit by increasing their welfare and quality of life with a completely social benefit-oriented approach. As shown in Fig. 1, it carries out R&D studies to achieve optimal efficiency with the support of farmers' ancient knowledge and technology, and pursues a sustainability-oriented goal by sharing this information with the public sphere. In line with the goal of smart society, the fact that development in rural areas is carried out with the power of technology focusing on people shows the complementary role of human consciousness and technology.

1.4 Mathematical Model

The provinces of Izmir, Aydın, Manisa, Muğla and Mersin have been determined as the provinces where 96% of the products with high productivity potential in Turkey can be grown. These 5 provinces are in a more advantageous position than other provinces due to factors such as location, crop cultivation and climate. Especially Aydın Province has many products that can be grown with high efficiency, apart from products such as tea, hazelnuts, kiwi and banana, which can only be grown economically in the Black Sea region. In other provinces, the crop cultivation potential varies between 80 and 90%. 5 provinces were evaluated according to certain parameters:

- Izmir has the ability to adapt to advanced and technological innovations.
- Muğla does not have an international airport and the current airport is used mostly for tourism purposes.
- Mersin, Adana Airport does not have sufficient capacity,
- The distance between Manisa and international airport was also high.

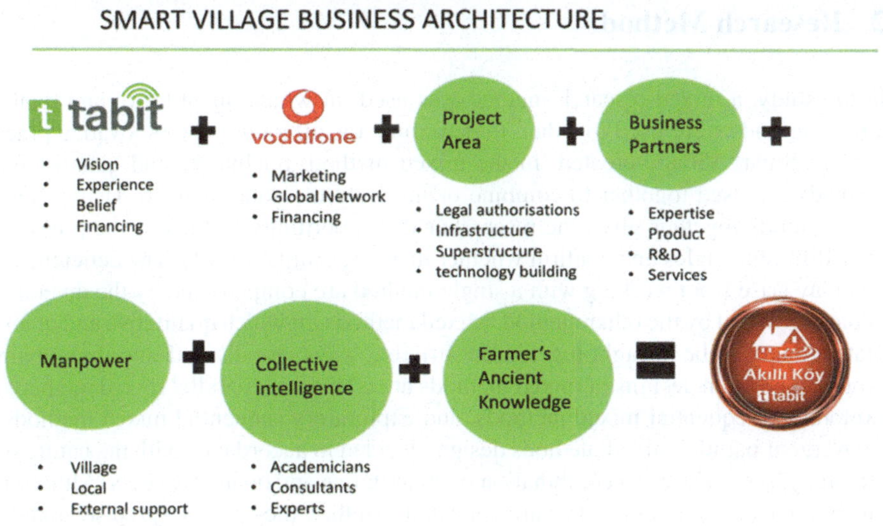

Fig. 1 Smart village business model [15]

In addition, TAK provides the basic criteria used in selecting areas among candidate villages suitable for establishing a smart village.

It should be a location where agricultural production is as efficient as possible.

- It should be away from industrial and tourism areas,
- It must be a village with similar characteristics (economic, social, demographic) of the people living in the countryside,
- It must be very close to the international airport,
- It must be a village in the West with a development level close to Türkiye's average,
- The education level of the village people is listed as being incompatible or difficult in the use of technology.

After Aydın province was determined for the project, the location selection process suitable for the TAK project started and as a result of the examinations, Kasaplar District in Koçarlı locality was determined as the most suitable place. This decision was taken taking into account the proximity of plant and animal production areas and the fact that the existing area would be productive due to the fact that it is a swamp. The project team dried the swamp area with the drainage method and turned it into a productive agricultural land. The founders of the project has thus implemented the TAK project as a rural development model as well as an exemplary project on a national and international scale.

2 Research Method

In the study, a mixed research method was used, in which quantitative and qualitative methods were used together, to determine the effect that the innovative practices in Smart Village created. In the mixed method, qualitative and quantitative methods are used together to combine or integrate the data obtained. Using these two methods together gives the researcher the opportunity to look at the problem from different angles and confirm whether there are similar results. The deficiencies that may arise from working with a single method are compensated by the diversity of data provided by the other method. Mixed methods, in which qualitative and quantitative data can be brought together in different ways, are divided into three main groups. The main designs of mixed methods are convergent parallel mixed methods, exploratory sequential mixed methods, and exploratory sequential mixed methods. Convergent parallel mixed methods design, decided in accordance with the nature of this study, is considered a combination of quantitative and qualitative data. Although quantitative and qualitative data are collected together, they are analyzed separately to check whether the findings confirm each other [16]. The purpose of this method is to collect different but complementary data on the same subject in order to best understand the research problem [17]. If a significant difference is obtained from the application as a result of the findings obtained from the quantitative data of the research, the reasons for this difference can be revealed with qualitative findings; if a significant difference is not obtained, the reasons for this situation can be revealed. Using quantitative and qualitative data together to answer the same research questions is called method triangulation [18]. It is possible to obtain richer data with triangulation method. Research process procedures are provided in accordance with the format recommended by Creswell and Plano [16].

In the qualitative part of the research; it was aimed to examine farmers' opinions about smart village in depth and to make sense of the data obtained in the quantitative part of the research. Content analysis technique was used to analyze these data obtained by qualitative research method. The aim of content analysis is to reach concepts and relationships that can explain the collected data. The collected data is primarily conceptualized and then the resulting concepts are arranged in a logical manner. The basic process is to bring together similar data within the framework of certain themes and organize and interpret them in a way that the reader can understand [19]. In this study, quantitative data examines farmers' use and adaptation of innovative practices, while qualitative data provides the opportunity to present these practices from the farmers' perspective.

2.1 Method Followed in In-depth Interviews

In-depth interviews that contributed to the method of the thesis were held face to face on 26.10.2022 with a total of 26 people, including the founding partners of the

TAK project, the R&D Manager and the Business Manager, and the villagers living in the smart village neighborhood. Interviews and evaluations were discussed under the following 2 headings in line with the TAK Project process:

- Project Emergence Process.
- Project Implementation Process.

2.2 *Research Question*

It is aimed to examine the definition of Human-Centered Society (Society 5.0) (Smart Agriculture) in terms of its impact on agricultural production, sustainability and efficiency by considering its conceptual framework and an application example in Turkey.

The main research question and sub-questions that this study considers have been determined. These main research questions were determined as.

1. How effective are smart agricultural practices in terms of efficiency?
2. How efficient are smart agricultural practices in terms of sustainability?

2.3 *Method and Reliability of the Research*

Some studies have been carried out to increase the validity and reliability of the research. A conceptual framework was created as a result of scanning the literature on the subject in order to increase the internal validity of the research while conducting interviews. In the content analysis, themes and sub-themes were determined in accordance with the purpose of the research, and inappropriate concepts were excluded. Regarding the interview questions, expert opinion was sought regarding the data obtained from the participants and the interpretation of the data. In order to increase the external validity of the research, an attempt was made to explain in detail what was done during the research process. At the same time, the research was supported both quantitatively and qualitatively by analyzing the data in the smart village system. Based on this, the study group, data collection process, analysis and interpretation of data are explained in detail. Research process is explained in Fig. 2 in detail.

2.4 *Data Collection and Analysis*

The data of the research were collected by the "face to face interview" method. In accordance with the purpose of the research, the relevant literature was scanned and open-ended questions were collected. The questions asked in the research include 13 open- ended questions, including managers and project beneficiaries, to get their opinions on the subject regarding the Smart Village located in Kasaplar Village.

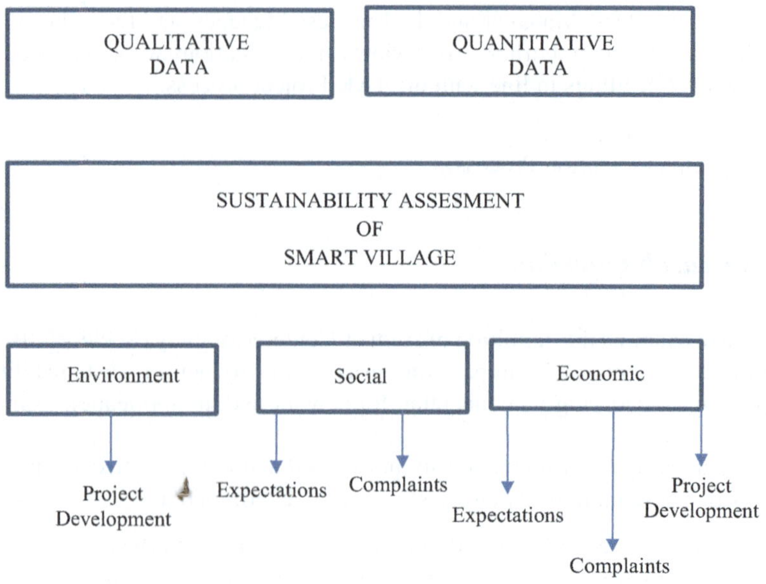

Fig. 2 Research flow

These questions are related to the activities carried out in the Smart village. Data were collected through observation and face-to-face interviews.

Data regarding the research were collected through one-on-one interviews between 25 and 27 September 2022, at a time deemed appropriate by the people in the Smart Village who participated in the research. Before starting the interviews, the villagers were given explanatory information about the purpose of the research. First, users' experiences were observed by asking questions about Smart Village. In addition, they were informed that they could behave comfortably with the villagers during the interview, as in a chat environment, and that they could leave the study at any stage of the research during the interview.

2.5 *Coding of Data*

The following steps were followed in the analysis of the data. First, the significant answers given to the question in the interviews conducted with the villagers were discussed one by one and relevant codes were created; Superordinate categories (themes) were reached by considering the codes that showed similar characteristics together. The data and the comments made based on the codes and themes created from the data were expressed with a descriptive approach. The following coding technique was used in the analysis of data obtained through face-to-face interviews:

P1, P2, P3: Created the codes of the villagers who participated in the interview.

M1, M2, M3: The codes of the managers who participated in the interview were created and participant names were not used, coding was done.

Themes were created based on these concepts. Apart from the headings that fall within the same answer, different themes were created for the answers given. Finally, the analysis of the answers falling into the same category was completed and converted into text.

2.6 Sampling Area

The sample area of the study consists of Aydın Province, Koçarlı District, Kasaplar District and TAK located in this rural settlement. In selecting the sample area; TAK is located in Kasaplar District, the use of smart agricultural tools in this region and the villagers gaining income and efficiency by using technologies. It was especially influential in its selection because it was the first smart village established in rural areas and it was a model for small family farming in rural areas, thanks to technological applications in the agricultural field.

2.7 Findings

In order to find answers to the basic questions of the thesis, to analyze the project, smart village applications and changes, which is the first smart village in Turkey and uses smart agriculture technology and applications in rural areas:

- Interviews with managers in the smart village.
- Interviews with villagers living in the smart village.

It was tried to be examined in terms of technological sustainability, innovation and efficiency.

2.8 Case Study

Nowadays, digitalization and the use of technology in the agricultural sector are becoming increasingly important to ensure sustainability. In this context, many countries are implementing digitalization projects in the agricultural sector and developing smart agricultural applications for farmers. The effectiveness of these practices and their contribution to sustainability is a frequently researched topic. In this context, the research was examined under six headings: permanent income, reservations, expectations, complaints, project development and area of impact. It is also shown in the field impact analysis table.

Project Development. While the smart village, which continues its activities in Kasaplar village, was being established, the founders and managers were evaluated with questions regarding the selection of business partners, the idea of establishment, and the vision of the future.

Social Impact. Defining itself as a social enterprise company, managers of Tabit's answers to questions about its regional and global impact and future work after the establishment of Smart Village are as follows:

M1: *While I was continuing my university education in the Department of Agricultural Marketing in 2003, when I saw the desperation of farmers and how difficult conditions they were trying to produce, I thought that I should definitely work on this subject. My aim was to provide farmers with access to information with existing technologies and reduce their losses by sending them early warning notifications. In 2004, I founded an e-commerce site called agricultural marketing.com and later it turned into a corporate structure as TABİT Company. Together with my teammates within this company, we started to reach farmers through different channels such as SMS and website, and to provide training through village visits. In the same years, smart technologies and Internet of Things systems began to enter our lives. Innovative solutions, including unmanned agriculture, have begun to be used in developed countries. Internet of things has made the biggest change in the agricultural sector.*

Soil, humidity and temperature measurements, monitoring of meteorological data and use in decision support systems have greatly increased efficiency. A new ray of hope has emerged in front of the helplessness of family farmers in the face of nature and increasing input costs. For this purpose, we wanted to understand the current conditions of family farmers and show them exemplary practices.

Based on the saying "The farmer's mind is in his eyes", we imagined a village to introduce farmers to these technologies with live examples. Tabit Smart Village, which includes different agricultural practices from poultry farming to beekeeping, from fruit growing to vegetable cultivation areas, is the only smart village in the world where nearly thirty companies come together, join forces and offer the latest technologies to farmers. The farmers who saw our work left saying, "I can do this too". Change has already begun in many villages. Our future goal is to spread these practices and increase the number of happy farmers who obtain higher quality and more efficient harvests by using less fertilizer, pesticides and natural water resources.

Considering the location and social resistance of Kasaplar village, many rural people, especially villagers, do not change their traditional production methods, are closed to innovation, and most importantly, think that they have more knowledge than experts and act accordingly, making it difficult to use technology and negatively affecting the development process of the smart village. When asked about the villagers' persuasion regarding their participation in the project, their obstacles, and their views on the future, they made the following statements:

M2: *This project emerged as a model project in which farmers combined traditional methods with technology support. In particular, we created a pattern by visiting 12,000 villages in Turkey and examining both the social and cultural textures of the villages. One of the problems that challenged us the most was the problems of adaptation to technology and not giving up the knowledge they knew. It took us about*

2 years to convince the villagers to bring their cows to the smart milking system. The biggest obstacle to progress was the remoteness of the roads of some farmers and the lack of faith in the project for others. However, the truth is that when you are patient enough for your dreams and do not give up, you see someone knocking on your door. Obstacles are a steppingstone to growth. By improving our perspective, we managed to overcome obstacles by thinking like a villager and empathizing. We have carried out various projects together with our farmers in many areas for 18 years.

In this context, we have put forward a business model that solves the financing problem of farmers, and we have implemented many projects such as farmers' special computers and farmers' clubs. During this process, we cooperated with companies such as Vodafone, Şekerbank, Intel and Lassa. Our Vodafone Farmers Club project currently provides information to hundreds of thousands of farmers. It was exported as a business model to 6 countries and became a project that was cited as an example in the world and studied as a case study. Our Smart Village project, which we started 2 years ago, is still continuing. One of our most important goals for the coming period is to spread social benefit to large masses by carrying out visionary projects in the international arena and to raise awareness by touching more farmers.

When agriculture meets technology, it becomes both a profitable and enjoyable business. We want to direct young people to this field and develop technologies that we can export to the world. In line with these ideals, we are establishing a digital farm in a North African country in desertified lands that have never been cultivated, and we are working to create an exemplary model for underdeveloped and developing countries, where climate change and food crisis are increasingly increasing.

Economic Impact. Considering that the smart village reaches 1.5 million farmers in Turkey and 7.5 million in the world, the Mena region has made a huge impact. Based on this, they expressed their future plans and their views on the impact of the smart village within the scope of global goals as follows:

M3: *By bringing together international companies, agricultural companies, universities, local governments, NGOs and farmers for global goals, Smart Village enabled each of them to contribute and share information. He presented the information he produced in a usable way to all his supporters, especially academically, who supported him in line with his goals. Each stakeholder has contributed to the empowerment of small family farmers, to change in the field in which it was established by adding strong elements on agriculture, animal husbandry and nature to the project, to raise awareness in Turkey and to inspire similar studies in other countries. This smart agriculture model, developed by Aydın, created value in terms of the promotion of the city and the branding of the city. It became the first smart village established in Turkey and in the world and created social benefit and economic value in 9 countries with a project such as Vodafone Farmers' Club. When we look at it in general terms, we can say that it has an impact not only on Turkey but also worldwide.*

Environmental Impact. In the smart village, especially the changes made on the land and the transformation of the land structure from a swampy area to fertile land have contributed to the organization of the environment as a living space and the creation of a sustainable area. The smart village manager answered the question as follows:

M3: *All the work carried out in Tabit Smart Village includes transferring information to farmers about the correct methods that protect soil and water in production, and we act with this awareness. It contributes to the reduction of carbon emissions by saving water through technology-based methods. Since we act for sustainable purposes, we use good agricultural methods without the use of chemical pesticides, encourage healthy products for the villagers, and reduce the harm caused using pesticides with technological methods.*

2.9 Evaluation of the Productivity Effect of Technological Applications

Intensive R&D studies are being developed for the ideas developed in the smart village. After the R&D studies are completed, the data is interpreted and made available to farmers and organizations operating in the agricultural field. Providing farmers with prosperity by increasing product yield and quality with technology is in perfect harmony with smart agricultural outputs. IoT applications in smart agriculture is classified as (7): (1) Irrigiation monitoring system, (2) Fertilizer administration, (3) Crop diseases detection, (4) Monitoring, harvesting and detection, (5) Climate conditions monitoring (6) Fire detection.

Smart farming is greatly impacted by weather. Changes in weather is an important factor for the quality and quantity of the agricultural product. Hence, sensors are technologies for getting environmental data such as water level, humidity, and temperature sensors. These sensors are essential for capturing real-time data, giving farmers insightful information, and assisting them in making decisions regarding their agricultural methods.

By using, above mentioned technologies particularly IoT sensors, High productivity increase with less labor cost is one of the methods intended to be demonstrated in the smart village. In our research on this subject, the R&D trials and the opinions of the villagers are as follows:

Production Areas and One-Year Trials. The installation cost of modern production greenhouses makes it difficult for small family farmers to access this method. However, with this production method, the quality of the product grown per unit area has increased and the yield per decare has increased. In addition, unnecessary costs were prevented by analyzing soil and water and providing appropriate nutrients to the plant. Thanks to mulch technology, the soil temperature is maintained, and the healthy development of plants is ensured. Since weed control is achieved through mulch, labor costs are reduced. A high efficiency rate was achieved in the experiments.

Sufficient production could not be achieved by the villagers trying traditional methods instead of these techniques. However, with the application of innovative agricultural methods, lettuce cultivation has resulted in more positive results in terms of yield and profit. As the effect of innovative methods was seen, many villagers continued to grow lettuce after planting and achieved high productivity and income. With the Mulch technology method applied in the trial area, the soil temperature was maintained, and the healthy development of the plant was ensured. Since weed control is provided by mulches, labor costs are reduced.

Technology Supported Melon Production. In traditional melon production in the region, an average of 3–5 tons of harvest is obtained per acre. Due to high temperature risks, applications have been made to increase efficiency. The water needs of the roots were managed using IoT technologies, and thanks to 45 days early production, melons were offered for sale during the peak market period. As a result, profitability reached its maximum level and melon production was carried out when market prices were at their highest (20). In addition, mulch was applied to the roots to prevent melon seedlings from being affected by temporary frost periods, and plant nutrition, humidity and temperature values were controlled using sensor technologies. In this way, a more efficient and profitable model has been created in melon production.

Efficiency and income increase in melon production has been achieved with technological solutions developed in line with the United Nations' Zero Hunger (SDG 2) target. A 150% increase in productivity and an 85% increase in income were achieved with the early production and plant nutrition method against high temperature risks. At the end of the project, it was successfully reported and qualified information was produced. Thus, an increase in welfare has been achieved thanks to technology and human cooperation.

Perennial Trials. The R&D study, carried out in cooperation with Aydın Fig Research Institute, was implemented within the scope of the Smart Village project. By using the wire finishing method, 10 times more efficiency per unit area was achieved. In addition, the branches are directed for hard pruning after the products exceed the height of 40 cm. In this way, a more efficient fig production was achieved.

The plant sprouted by determining its direction according to the sun. Hard pruning is done during the sprouting process to ensure that the tree grows sideways. While it takes 7 years to harvest in classical fig gardens, it can be harvested in 3 years in gardens implemented with the wire training system. In this way, 3–4 tons of product can be harvested per decare. This method is still being tested. The change it creates in the long term is observed through sensors, and the technological impact is measured by analyzing yield and harvest for the coming years.

Efficiency in fig production has been increased with the wire dressing method developed in line with the United Nations' Zero Hunger (SDG 2) target. While productivity in agricultural production can be increased with innovative solutions, qualified employment and socio-economic development have also been supported, contributing to increasing the welfare levels of individuals.

2.10 Evaluation of Technological Applications in Terms of Sustainability

As a result of R&D studies and crop cultivation in the village, in order to evaluate the sustainability and impact of technological applications, questions were asked to the villagers who use these applications, their opinions and evaluations were investigated.

Reservations. They had difficulty changing their habits in the smart village and adapting to technology. Adaptation to technology has been seen as a significant obstacle, especially for farmers living in rural areas. On this issue, farmers were asked questions to evaluate smart village from a social and cultural perspective.

Social Impact. It is observed that the newly developed technological applications in the village have both positive and negative impacts. Farmers answered questions regarding this issue as follows.

K1: *They taught us new techniques for planting new crops, and we had the opportunity to learn more through literacy courses. Thanks to the Vodafone Farmers Club, they taught us how to use computers. I bought furniture for my house with the money I earned, and it gave me a feeling of freedom. We have a belief that women can do anything individually. The children in the village broadened their horizons by seeing new things. We were able to spend time and have fun in social facilities and social living areas.*

K2: *After the lettuce greenhouses were established, they gave us training and showed us how to make the product productive and how to harvest faster by planting it as practices in the smart village campus. We didn't go at first because we didn't know it very well, but when we realized it was useful, we started going. In this way, we, as women, earned economic income. We contributed to our home.*

K3: *I live alone in the village. My children live in the city. They set up greenhouses and taught us how to grow crops in greenhouses. We learn useful information through mobile phone messages. Thus, I save time with technology. I can spend more time chatting with my friends.*

Economic Impact. Below are the answers given to questions about the evaluation of the economic impact of newly developed smart agriculture applications:

K4: *Before the application in the smart village, I did not have the courage to do this. Because we always grew melon and lettuce using the method we knew from our ancestors. After seeing this method, I decided that it was an efficient system and I planted melon and lettuce in my own field by mulching. My expenses decreased and my income increased because my diesel cost decreased because I cultivated the land less. I was able to irrigate the crop by watering the soil under the mulch. I got good yields with dripping. I decided to try again after 1 year, but the mulch was damaged and weeds started to grow everywhere, and then I did not get good results. The weed took away all the yield and choked the melon. Such a thing did not happen with lettuce, but when the practice in the smart village ended, there were no job opportunities for the villagers again and they did not continue this work.*

Farmers' opinions about smart agricultural practices and the practices they implemented with the advice of the managers in the smart village:

K5: *They said to try peach and plum production by dripping, but we could not do it due to our lack of situation, high fertilizer and diesel costs, and time. Secondly, products such as plums and peaches grow in 3 or 4 years, so I must have ready money so that I can plant it and then harvest it. I can pay the expenses with money, but I don't have the finances to do anything new anyway.*

K6: *We were growing figs the way we know, then I went to the smart village and saw the new technique on site. Then I applied it in my own garden. We are now starting to receive the product. Such production was a risk. We thought, 'What if it doesn't work, what will we do?' We took risks but the result was perfect. Those who saw it were very surprised. He says I will do my garden like this.*

Another farmer shared the following opinion, stating that even if there is technology, there will be no production without sufficient credit or incentives unless the customer and supply chain are properly organized:

K7: *Smart Village showed many new products here and even showed us how to generate income by applying the products. When we sew the product, when the costs increase, we lose money on that product and cannot sustain it. Maybe if we get incentives and support, we can produce better. Otherwise, we are afraid to try these new things because of the costs. If the state made an incentive in this regard, producers and villagers could produce more cheaply.*

K8: *These things are very nice, but how can we do it? When fertilizer and medicine are expensive, no one wants to grow crops and deal with them. Because after growing the product, it doesn't even cost me, we keep it in our hands. The state needs to find a solution to this. If it reassures us by providing a purchase guarantee, we will plant our product and try new applications. It becomes very difficult otherwise.*

Expectations. In the meeting we held at the village coffee house in Kasaplar village, farmers were asked to list their expectations from the smart village.

Social Impact. Questions were asked about farmers' expectations about the smart village and its change over time.

K9: *Smart Village has contributed a lot to the promotion of this place. People from foreign states come to see this place. We are very happy. It has promoted this place a lot. I am proud that it is here. The milking system has benefited us greatly. However, there were trainings in the smart village before, now it would be better if the trainings start again. Women learned to read and write, they worked in greenhouses and provided employment opportunities. Having such jobs again would contribute financially to the villagers.*

K10: *When it was first established, there was a lot of demand, but later on, the number of visitors started to decrease. First, events were held, but after the pandemic, they are almost never held. It doesn't contribute much to the villagers economically right now, but they allow us to use the garden for things like weddings and entertainment.*

Economic Impact. The villagers, who had economic difficulties due to the increase in feed costs, made the following statements regarding the issue.

K11: *We are having difficulty in feeding the animals due to the constant increase in feed prices. If a smart village grows forage plants and distributes them to the villagers or gives them to farmers at a low cost, the villagers will be protected against market prices. The agricultural credit cooperative also sells feed to the villagers at constantly increased prices. The cost of fertilizers and pesticides for the product we plant is beyond our limits, and the growing of forage crops and the constant increase in their prices puts a great strain on us economically. After it was transferred to the municipality, it became an untouchable area.*

K12: *Well, feed has become a big problem in the village. We cannot grow forage plants due to costs or we do not have the space to grow them. Caring for animals has become more costly. When not fed, milk yield decreases. It would be great if the smart village gave it to the villagers.*

Complaints. One of the most important problems is that village residents and farmers resist the use of technology and do not want to use applications. Although 5 years have passed since the establishment of the smart village, technology resistance still continues. The most important reason for complaints is that the farmers' perspective is not compatible with the goals of the smart village.

Social Impact. The villagers answered our questions about how the complaints have changed over the years with the smart village in the village and what the complaints that still exist are as follows;

K13: *Smart village is a very useful project, but they did not do what they promised before. At the previous meeting, they said that they would build a road, but even though all these years have passed, there has been no change in any way. An agricultural engineer was supposed to come before, but an agricultural engineer never came to the village. When we removed the pesticide tanks next to the cemetery, we could not benefit from them.*

K14: *There was no one coming from the smart village to sit at the coffee house and chat. In the past, a manager would occasionally stop by and chat with the villagers. There was a communication gap between the villager and the smart village, so the villager was disturbed. Previously, the flowing water in the smart village was cut off and irrigation could not be done due to this water. Such situations occurred because they did not communicate much in the village. When a guest came to the village, I took my guest around. There is no such opportunity in other villages, but it would be possible if they communicate with the villagers a little more or come and go.*

Economic Impact. Among the complaints of the villagers in the smart village, the most prominent are the transfer of the smart pasture to the municipality and the village not taking on more responsibility. The villagers demand that the forage plants be purchased at a cheap price and offered for sale by the smart village, and they claim that the areas where forage plants were previously produced are occupied. The villagers responded to the questions asked on this subject as follows:

K15: *The smart pasture system was previously constantly checked and maintained by the smart village. When the pasture law changed, the management of this place passed to the municipality and everywhere became grass, and since it was not taken care of, it became difficult for animals to enter the pasture or for the villagers to benefit from it. If it can be done, feeding the animals there would be very advantageous in terms of cost to the villagers. If the feed arrives once a month, I load the feed myself. If the smart village can do this, it can be good.*

K16: *We were able to learn what the disease of the cows was in the pasture of Smart Village and take precautions. It has benefited us a lot from a technological perspective. Our costs have decreased. Later, when the management changed, these were no longer done. I feed my cows there, but it's not the same as before.*

As a result, the Smart Village established in Kasaplar village has provided social benefits as a rural development model and raised the living standards of villagers. However, the failure to solve the problems accumulated over the years and the failure to rebuild the broken ties with the public created a negative perception. It has been observed that when farmers encounter innovative technologies and see their benefits, they embrace these technologies. However, the importance of traditional knowledge and experience for the agricultural sector should not be forgotten. Even today's artificial intelligence is not sufficient in the agricultural sector, and farmers' knowledge is the cornerstone of agricultural systems (Table 1).

Therefore, rural development models such as the Smart Village project need to be designed and implemented in accordance with the needs of farmers and centering them. Additionally, the participation and support of local people is extremely important for the success of the projects. For the adoption of innovations, the existence of a

Table 1 Impact analysis of smart village

Case study	Impact	View of villager
Reservations		
Economic	Inability to break away from tradition and economic insecurity	Before the application in the smart village, I did not have the courage to do this job
Expectations		
Social	The need for education and socialization in the village	There were trainings in the smart village before, now it would be better if the trainings start again
Economic	Increasing revenue by reducing costs	We are having a hard time feeding the animals due to the constant increase in feed prices. In this regard, it would be better if a smart village grows forage plants and distributes them to the villagers
Complaints		
Social	Trust	Smart village is a very useful project, but they did not deliver what they promised before
Project development		
Economic	Value creation	Vodafone created social benefit and economic value in 9 countries with the project together with the farmers' club
Environment	Achieving the sustainable development goals	It contributes to the reduction of carbon emissions by saving water through technology-based methods

group of leaders who believe in the advantages of innovation and benefits will accelerate the adoption of innovation. It is important to take knowledge and experience into account when adopting innovations. When farmers meet with innovative technologies and their benefits are seen, they embrace that technology. Rural development models such as Smart Village are an important tool to raise the living standards of farmers. However, the success of the projects can be achieved with the participation and support of the local people.

3 Conclusion

The agricultural sector plays an important role in fighting poverty and hunger around the world. However, it has faced problems such as the decline in agricultural production, global population growth and urbanization, as well as the migration of the workforce working in agriculture to other sectors. It is of critical importance to use technological developments in agricultural research and development studies to solve these problems. Smart agricultural practices contribute to the development of the agricultural sector in many countries. Some countries aim to increase the use of agricultural technologies by developing a national e-agriculture strategy vision.

Smart food supply chains and ecosystems are developed by leveraging technological advances to mitigate risks and meet the challenges facing today's food supply chains, including a rapidly growing global population, rapid urbanization, an aging society, and the uncertainties associated with climate change. In preparing food supply chains of the future, stakeholders (e.g., policy makers, food industry, technology providers, non-governmental organizations or NGOs) should consider the development of a human-centered society that balances the economic dimensions of the food ecosystem with future needs for human health, sustainability, nutrition and food security must have.

In the scenario developed for the smart food supply chain; a smallholder farmer will have a digital device with information feeds and alerts from multiple on-farm technologies and cloud-based service providers (e.g., weather forecast, market pricing, alerts to harvest before rain or frost). In this context, it is aimed to support small farmers technologically in a future scenario and enable them to use an affordable and usable decision support system. In this way, technological adaptation in the agricultural sector will accelerate and the productivity of farmers will increase, and transparency will be created by ensuring the safety of food on the path from table to fork, which is one of the strategic policies of the European Union.

Considering that the innovative practices in Tabit Smart village are suitable research areas for this research. A mixed method was followed, and an analysis was made by observing the behavioral situation, economic situation and other conditions of the farmers. Considering the criteria such as the need for a specific program and resources within the project, especially in order to increase productivity in agricultural production with information and communication technologies, it has been observed that there is an increase in the lives of the villagers in Kasaplar village in terms of income and social welfare when they apply the innovative technologies and methods in the village. However, due to the increase in migration from village to city, the young population not staying in the village and technological barriers, it has been revealed that the smart village has not been able to achieve its intended goals since its establishment in 2014, and it has been observed that the villagers continue to apply old methodologies without giving up ancient knowledge. As a result, considering today's technological advances and population growth, its negative impact on production has become inevitable.

During the interviews, it is seen that agricultural production continues in its traditional dimensions and that efforts are made to increase productivity in production and reduce the use of pesticides, but it is difficult for the villagers to present this as a solution in the short term. The practices carried out in the Smart Village, which is seen as a new generation rural development model, continue despite the social resistance of the villagers or farmers living in the village, and the villagers' use of information and communication technologies-based tools and their inclusion in the education curriculum starting from primary school, together with the state-supported policies that need to be made regarding their dissemination, improve agricultural awareness. It was concluded that it is necessary to increase awareness.

The inclusion of stakeholders and local actors in the negotiation process is crucial, as evidenced by the outcomes of the discussions held inside the village. Problems such as roads, water and internet infrastructure slow down the transition to technology of the people living in the village. The fact that they look for the solution in the smart village instead of local bodies such as the municipality is actually an indication of the trust created by the smart village. Although it has a positive effect, in the long term, unless structural problems are solved, the transformation in rural areas will not be fully realized, even if an innovative model such as smart village is found. One of the most important problems in particular is that the connection between the village people and the smart village has been broken. Although such innovative projects are intended to spread to large masses and create a social benefit in the villages where they are implemented, they have not had the necessary impact due to the disconnection in recent years.

In this context, where food production and agricultural technology are becoming more important day by day, behavioral changes after the pandemic and a decrease in agricultural incentives due to inflation, external reasons, or a shift away from the products harvested from the agricultural area due to not covering the production costs are possible effects. Smart village is a great value for the villagers and the region. Spreading such models to other regions, both in the public and regional areas, and setting an example will gain prestige both nationally and internationally.

In light of the evolving global dynamics, and the changing world order, the agricultural sector has become open to the development and implementation of innovations. These innovations also affect competition globally, and stakeholders who aim to produce sustainably in agricultural markets need to improve themselves. Considering the contribution of technology to increasing productivity and surviving in the aforementioned competitive markets, farmers' adoption of these innovations becomes an element that will make a difference in competition. As in the case of Tabit

Smart Village, policies should be created by considering the structures and unique dynamics of different regions within countries and local factors. When looking at rural development road maps or models, one of the main duties of the public is to prepare the infrastructure that will ensure the effectiveness of such agricultural technologies that provide both social benefit and economic gain. The public also; It must create the relevant financial and legal basis, and especially the institutional structure, to encourage innovations.

Detailed research is also needed on the agricultural sector, innovation capacity and practices in Turkey. It is necessary to determine under what conditions farmers turn to innovation, especially as in the smart village example developed in Kasaplar Village. Policies created by taking into account farmers' level of adaptation to innovations are expected to be successful. Emphasis should be placed on training activities to raise farmers' awareness and their continuity should be ensured by applying the smart agriculture model.

Increasing farmer income, reducing poverty and ensuring rural development by increasing production in agriculture are not unattainable goals. Detailed and comprehensive analyzes and plans are needed to achieve the goals of increasing agricultural production and reducing poverty through rural development.

In conclusion, agriculture continues to serve as the fundamental pillar of the rural economy inside our nation. As the level of understanding regarding the socio-economic structure in rural regions is heightened, there will be a corresponding amplification in the significance and capacity of agricultural output. In the context of the shift towards smart agriculture, it is imperative to ensure the provision of requisite technological infrastructure to farmers, alongside the organization of training sessions and seminars oriented towards this objective. The integration of data-driven Smart Agriculture techniques, in conjunction with both human and autonomous systems, is poised to serve as a pivotal moment in the progression towards a smart society within the realm of Turkish agriculture and for Turkish farmers.

The practical contribution of this research is to determine the impact of the use of digital technologies in the agricultural sector and to use real data to measure the success of the Smart Village project. These data reveal the potential of smart agricultural applications.

Regarding theory, this research shows that technology-based applications are human-centered approaches and that digital technologies can increase not only economic efficiency but also the quality of human life. Therefore, these concepts emphasize that a human-centered approach should be adopted in the digital transformation process.

References

1. WEF (2019) Innovation with a purpose: improving traceability in food value chains through technology innovations. World Economic Forum. http://www3.weforum.org/docs/WEF_Traceability_in_food_value_chains_Digital.pdf. Accessed 2 Nov 2020

2. FAO (2019) The state of food security and nutrition in the world. http://www.fao.org/3/ca5162en/ca5162en.pdf
3. FAO (2015) Decent work indicators for agriculture and rural areas: conceptual issues, data collection challenges and possible areas for improvement. Technical Report. www.fao.org/3/a-i5060e.pdf
4. Alhefetiti F (2018) Society 5.0: a human-centered society that balances economic advancement with the resolution of social problems by a system that highly integrates cyberspace and physical space. Unpublished Postgraduate Thesis The British University in Dubai
5. Spanaki K, Karafili E, Despudi S (2021) AI applications of data sharing in agriculture 4.0 a framework for role-based data access control. Int J Inf Manage 59:102350
6. Baierle IC da Silva FT, Correra RGdF, Schaefer JL, Da Costa MB, Benitez GB, Nara EOB (2022) Competitiveness of food industry in the era of digital transformation towards agriculture 4.0. Sustainability 14:11779
7. Morchid A, Alami RE, Raezah AA, Sabbar Y (2024) Applications of Internet of Things(IoT) and sensors technology to increase food security and agricultural sustainability: benefits and challenges. Ain Shams Eng J 15:102509
8. Cesco S, Sambo P, Borin M, Basso B, Orzes G, Mazzetto F (2023) Smart agriculture and digital twins: applications and challenges in a vision of sustainability. Eur J Agron 146:126890
9. da Silveria F, da Silva SL, Machado FM, Garcia J, Barbedo A, Amaral FG (2023) Farmers' perception of the barriers that hinder the implementation of agriculture 4.0. Agric Syst 208:103656
10. Accenture digital agriculture: improving profitability (2019). http://wise.co.th/wise/References/Digital_Economy/Digital_Agriculture_Accenture.pdf
11. Autor DH (2014) Skills, education, and the rise of earnings inequality among the "other 99 percent." Science 344(6186):843–851
12. Roser M (2013) Employment in agriculture. Our world in data
13. Saygılı F, Kaya AA, Çalışkan ET, Kozal ÖE (2019) Türk Tarımının Global Entegrasyonu ve Tarım 4.0, Izmir Ticaret Borsası
14. Tabit Durum Raporu (2019). http://www.tr.tabit.com.tr
15. Tabit.: 2016–2021 yılı Tabit Akıllı Köy Durumu Raporu (TAKDR). Aydın. pp. 1–45 (2022)
16. Creswell JW, Plano CVL (2015) Karma Yöntem Araştırmaları Tasarımı ve Yürütülmesi. Ankara, Anı Yayıncılık
17. Morse JM (2019) Approaches to qualitative and quantitative methodological triangulation. Nurs Res 40:120–123
18. Türnüklü A (2001) Eğitim Bilim Alanında Aynı Araştırma Sorusunu Yanıtlamak İçin Farklı Araştırma Tekniklerinin Birlikte Kullanılması, Eğitim ve Bilim 26(120):8–13
19. Yıldırım A, Şimşek H (2013) Sosyal Bilimlerde Araştırma Yöntemleri. Seçkin Yayınları, Ankara

Uncertain Times: Caused and Solved by AI

Jason T. Owen

Abstract Artificial Intelligence (AI) has created exciting and uncertain times and for many reasons. Long held norms in life have become unstable and chaotic especially when it comes to the role of a technology manager. One particularly scary factor is the increase in unemployability for long term employees who face jobs being replaced by artificial intelligence engines. Job security has been a strength for the tech industry for years but with AI many have begun to worry that their skills are less marketable on future development teams. AI and machine learning tools to train AI are here to stay; however, this work looks at an unconventional approach to solving the very problem that AI seems to have created. By using the A* search agent this research explores connecting older skills developed over years of participation in the industry with new skills that are more valuable in the new job place by replacing job skills and soft skills with locations on a map and then using the A* algorithm to determine proximity and thus similarity.

Keywords A* agent · Artificial intelligence · Technology management · AI search agent · Technology reskilling · AI uncertainty

1 Introduction

Technological environment uncertainty is just a matter of when, where, and how often calamity will strike. The logistical conditions that arise often require complex solutions. These are ideal problems for artificial intelligence to solve. In a recent article published in Soft Computing [6] an AI deployment for optimal redistribution planning in the aftermath and disruption of resources in disaster management was explored. The research plan presented was to minimize total cost of a relief operations while maximizing allocation of various resources. Clearly a prime example and a

J. T. Owen (✉)
Department of Computer Science and Information Systems, Murray State University, Murray, KY 42071, USA
e-mail: jowen22@murraystate.edu

F. Çebi et al. (eds.), *Engineering and Technology Management in Challenging Times*, Lecture Notes in Networks and Systems 1137,
https://doi.org/10.1007/978-3-031-70935-7_23

powerful opportunity for AI deployment. Unfortunately, AI is also causing some disruption of resources in its own way.

Job security within the technology field has been shaken due to the rapid adoption of AI tools. In 2018 research was presented in the peer reviewed Journal of Service Research relating to the need for AI replacement for service-based jobs. The theory presented argued for AI job replacement and provided a roadmap for methodology helping AI to overtake the need for human interaction where different intelligences were required. Additionally, the research found that AI should be used for service-based tasks and argued that workers should shift their skills to achieve a win–win between machines and human workers. [3] This is not entirely uncommon to technology and not uncommon to technology managers but the fear for employees due to the rise of AI is still quite strong. An article in the peer reviewed Journal of Service Research recently showed five separate studies that revealed how AI was increasing negative outcomes for employees. This was magnified when the employees were engaged in thinking vs feeling based tasks due to the adverse effect on the employee's perceived ability. [7] Regardless, the goal to stay ahead of the competition has led many technology managers to seek out any advantage they can. In a recent article titled "Artificial Intelligence in Higher Education: A Practical Approach", published in the Journal of Higher Education Policy and Management, assertions are made that AI is being used heavily in academia and in various other sectors such as industry, marketing, social media, health, entertainment, astronomy and many more.

Many of the task's AI is being used for are appearing to be clerical in nature; however, that is not the only adaptation of AI in academia or in businesses. For example, risk management is a highly valuable field impacting almost every industry. In his dissertation released in 2022, Alawad [1] explores the application of AI within the field of risk management relating to the railroad industry and asserts that it is both wise and necessary to use all available technologies to mitigate the risk and consequences that might arise. Applying this same rational, it makes sense to apply the use of AI to mitigate the growing risks of reduced job security for future technology managers and their workforces. In this work it is proposed to use an unconventional version of the A* search algorithm to determine optimal paths for potential career shifts for existing and new technology experts by considering soft skills attached to low and high value skills as potential destinations on a virtual map. The target for this strategy is to possibly offer some guidance and aid for technology managers and their employees to consider on their career paths and increase job security while reducing the stresses of technology management in such uncertain times by aiding in repurposing of existing skill sets.

2 Theory

The rapid adoption of AI tools has shaken the very foundation of the security enjoyed by many lifelong technology experts. Technology managers are all now facing uncertain times due to the rapid rise of AI technologies. This theory is that the problem AI created may be solved by AI as well, namely the problem of career reskilling needs and concern in technology managers. The hope is that technology experts such as programmers could potentially have their concerns eased through an unconventional deployment of the A* AI search problem with the end goal of converting current low value skills into high value skills in both current and trending technology markets. This unique application of the A* AI search agent problem may offer a programmer or technology manager a tool to deploy in various decision support systems or possibly within one's own human resource software.

2.1 AI Agent Search Problem

In the book, Artificial Intelligence: A Modern Approach, Russell and Norvig [5] assert that any quality problem for an AI search agent deploys six fundamental stages that include states, initial state, actions, transition model, goal state, and action cost. The search problem being recommended in this research will follow the same model while also expanding upon it. With any AI search agent, the problem must be clearly defined and filtered through the lens of abstraction. The more specific problem to be analyzed for this agent will be compiled considering it being a single agent that this nondeterministic, episodic, dynamic, continuous, and known. The AI search agent in this case will leverage an informed (heuristic) search strategy with a success being achieved with the cheapest path from the initial state to the goal state. This should be applied to the following problem description:

2.1.1 Problem Statement

"Technology managers are facing uncertain times due to the rapid adaptation of AI tools in business technologies. Past programmers are becoming more and more obsolete due to various AI tools being able to replace their current skill sets. This drastic reduction in skill value is painfully obvious to many career programmers and technology managers are facing decisions to release long term employees from roles that are no longer in demand. Many programmers and technology managers seek to find optimal strategies and paths to repurpose the current skills of programmers so that they can quickly gain new more practically applicable high value skills. This will reduce the stresses on technology managers and regain the psychological capital held by long term programmers in relation to the operation of various businesses."

2.2 Initial State

The initial state for this problem could potentially include multiple conditions. It could be a programmer at risk, a recently terminated programmer, or many other variations but for the purpose of increased abstraction for the development of this AI search agent adaptation this research will use an initial state as the current skill set of a programmer who needs to convert existing skills or gain new skills that are higher value.

2.3 Goal State

The goal state for this problem could be a set of new high-value skills required for continued employment or new opportunity advancements. Often an existing employee supervised by technology management has high value skills already; however, based on past expertise and experiences those skills are convertible to a different task thus making the employee more employable than might be realized just in a different context and ultimately makes them more valuable to technology managers.

2.4 Actions/Operators

Potential actions for transitioning from one state to the next will involve various states of iteration taking current skills and if relative to high value skill converting them to new high value skills unless the high value skill is listed in the current skills of the participant. Additionally, any combination of taking an online course, attending workshops, attending boot camps, reading various books all could be a valuable heuristic and thus be a factor to upgrading various skill sets.

2.5 State Space

Programmers, especially long-term programmers have a plethora of skills that could be deemed as high value. The states of a possible career for programmers could include but are not limited to advanced problem solving, complex solution development, technology deployment, leadership, and much more. The abstract states used for this search agent are, high employability, low employability, currently employed, at risk, and unemployed. All of these in conjunction with any potential combination of current and potential new skills that might be acquired. The state spaces could include the current skill of the programmer themselves or the technology manager who is employing programmers. Historical aspects of high-level employable skills must also be tracked by the AI search agent to provide potential similarity analysis with trending high value skill sets.

2.6 Transition Model

To better define how the state changes as actions are taken, a consideration for how the skills currently held are updated to new skill sets. The formula being adapted within this agent is fundamentally evolving,

$$\text{RESULT(Low Value Skill, To High Value Skill)} = \text{Increased Employability}$$

therefore, as each new skill is mapped an additional heuristic becomes available and will be adapted into the problem itself. This would mean the new skill modifies the initial set of heuristics being considered within the problem itself. This creates an ever-evolving model providing the most optimal solutions for potential employable high value skill sets.

2.7 Action Cost Function

To assign a cost to each action or transition state this problem will count $+ 1$ for every new skill gained. For example, taking into consideration the effort required to assimilate new skills, time investments, various resources or costs associated with investing the skill gain.

2.8 Heuristic Function (h)

Estimating the cost to reach the goal from any given state (g) with the heuristic estimate of the given cost to get to it will give an evaluation of the given state. This aides in the efforts to bridge the gap between existing skills and new ones. Once a list of skills that are highly valued are being considered for a given position or role this model will be able to determine not only the convertible skills but also the potential investment needed to gain gaps in skills that are not currently held by the employee.

2.9 Evaluation Function (f)

For the evaluation function in this model the cost of reaching state (g) considering the additional heuristic data and an estimation of the cost to the goal (h) will provide an evaluation of any given state. The nature of the A* search agent algorithm defaults to lower f-values to be explored first. This will be a valuable piece to this algorithm considering an existing skill is fundamentally a heuristic consideration for the calculations of the AI tool.

2.10 Pseudo Code

This application of the A* search algorithm is being called the Racer-One algorithm because it seeks to race new technology experts into great careers but also put existing technology experts back in the race thus helping technology managers adapt their current workforce into newly leveraged positions in such uncertain times.

Racer-One A* AI Search Agent Model

```
function A*(initialState, goalState):
   openList = PriorityQueue() // Priority queue sorted by f-values
   openList.push(initialState)

   cameFrom = {} // Stores the parent state for each state
   gScore = {} // Stores the investmentCost from initial state to a state
   fScore = {} // Stores the estimated total investmentCost from initial state to goal
   gScore[initialState] = 0
   fScore[initialState] = existingHVS(initialState, goalState) // Heuristics = existing
   or new high value skills

   while openList is not empty:
      currentState = openList.pop()

   if currentState == goalState:
      return reconstructPath(cameFrom, currentState)

   for each action in getNewSkillsAdded(currentState):
      newState = applyNewSkill(currentState, action)
      tentativeGScore = gScore[currentState] + investmentCost(currentState,
      action)
      if newState not in gScore or tentativeGScore < gScore[newState]:
         cameFrom[newState] = currentState
         gScore[newState] = tentativeGScore
         fScore[newState] = gScore[newState] + existingHVS (newState,
         goalState)
         if newState not in openList:
            openList.push(newState)
   return failure
function reconstructPath(cameFrom, currentState):
   path = [currentState]
   while currentState in cameFrom:
      currentState = cameFrom[currentState]
      path.insert(0, currentState)
   return path
```

Of course, with the A* search agent model there should be critical considerations made for the heuristic function and the heuristics themselves should guide greater search efficiency as the model moves towards the goal state. It is important to also bring to light the impact specific details of the problem itself may have. Various state.

representations will likely have a great influence on an implementation of the Racer-One A* AI search agent model.

3 Conclusions

3.1 Technology Managerial Implications

The goal for this adaptation of A* is to address a gap in existing literature by providing evidence for the use of AI to solve a problem AI seems to have created. Adding the Racer-One search agent model to various decision support systems dedicated to talent acquisition, retention, and retooling is a potential solution that may bring greater peace to technology managers facing such uncertain times. The traditional mindset has been that robots would take over human work and that's not too far from a reality. In a recent peer reviewed journal article in Electronic Component News a leading robotics company reported that they were deploying AI to replace much of the work humans were doing but robots are not the only thing taking over jobs in the technology industry. [4]

As a former technology manager, the privilege to oversee software developers on various teams was common. Managed team members lived in areas like the central United States, India, Mexico, and even Brazil. Each had a uniquely common skill set, they were expert programmers. Regardless of their ability to code in front-end or back- end technologies they all had a common passion for writing code. While working on a project with one of the world's largest farming and agricultural suppliers there were entire teams of engineers and lifelong programmers experiencing the pain of being let go from their positions due to the belief that they were not needed, and AI could do the work instead. The fear of the uncertainty that comes from the adoption of AI is very real for many programmers. It was quite painful to witness firsthand.

It was this pain that inspired the idea of using AI to solve an AI induced problem. As a technology manager and former programmer, it made sense to want to prevent the loss of the intellectual capital for so many highly skilled employees. For a technology manager, it makes sense that the uncertain times created by the explosion of AI use could benefit a company. Lower cost for AI tools to complete certain jobs makes the adoption of AI appealing. The cost of eliminating long term dedicated employees may be far greater than just salary though. Considering the psychological capital of the team or company and how mass layoffs can impact it is wise. A tool build with the Racer-One algorithm could provide technology managers with a resource to help reallocate current skills that the company has at its disposal while also retaining difficult to grow intellectual capital.

The future of technology management will likely involve the replacement of middle management with regard to decision-making and thus an algorithm such as Racer-One could provide additional value to the current work force but also to the company relying on AI to optimize efforts. This only goes to support the introduction of advanced artificial intelligence into technology stacks and adding an AI search agent such as Racer-One could have a high impact on the understanding of future technology management practice relating to staff optimization. Put more simply, this type of AI search agent could have profound strategic and operational implications for technology mangers.

3.2 Technology Professional Implications

The Racer-One A* search algorithm could potentially secure long-term employability for technology professions in the programming field. Technology is always growing, changing, and adapting to new environments and discoveries. It is inevitable the technical skills must evolve but that doesn't have to mean that previous skills are no longer needed. Many soft skills, along with technical skills can be repurposed to revive themselves as new higher value skills. Racer-One seeks to provide a way to search out the relation of varying aspects of a person's current skills and how they could apply to newer more current skills deemed as high value by technology managers and industry professionals. Rodrigues, Ana Luísa et al. [6] wisely notes that technology skill development is critical to a country's education systems and that a balance of personal, social, soft, and professional skill is crucial for the future careers of technology professionals. It seems straight forward to change the skill sets being taught to new students; however, the concern here is that there are greater challenges for retooling a lifelong career in programming. The Racer-One algorithm aims to assist technology professionals in that challenge.

3.3 Theoretical Implications

It is plausible to see the implications of the Racer-One search agent algorithm going into use in fields outside of technology. To do so will require extensive data from high value skill sets and the various other soft skills the employ based on varying fields of interest; however, the use of such a tool to explore how existing skills could be repurposed due to relativity to more commonly desired high value skills is quite exciting as well as practical. With the growth of uncertainty in the global economy, social contract, and geopolitical conditions there is a great need for tools that leverage AI to streamline and improve existing processes. The use of the Racer-One agent could provide a valuable new asset for retooling workforces in a plethora of fields of interest. For example, far more optimal skill transition planning can be conducted reducing the loss of valuable experience and resources when the frustrations of job displacement arise. Increasing the efficiency of path selection for skills modification and additional skills adoption means less investments of company time and employee efforts. Technology managers would likely find a partner in the effort to sustain psychological capital when they unite the use of the Racer-One agent with their human resource officers. Once Racer-One is embedded into the practice of a technology managers the direct impact could likely be a reduction in the uncertainty of the technology employment markets while simultaneously increasing the company's psychological capital investments.

3.4 *Limitations and Future Research*

Even though the A* search algorithm offers a plethora of advantages, limitations to this model are that it will rely heavily on a data structure of skills that are related to a variety of soft skill sets. An initial data set can be manually created to work from as a start; however, additional machine learning processes will need to be deployed to leverage the incoming data with every use so that the agent can learn as it matures. Similar to most AI tools, the more it is used the smarter it will become, and the more various contexts can be used when processing a given search.

Like GPS mapping, A* algorithm-based systems that have leveraged precalculated routes to speed processing for the AI agent the Racer-One model can increase speed and efficiency. For example, if one were to try to map a path on a GPS from Karaburun Turkey to Kirikkale the AI model will likely default to using a previously known calculation of Istanbul to Ankara first. Then go about calculating Ankara to Kirikkale and Karaburun to Istanbul. This will drastically speed the processing time while simultaneously researching the processing power requirements. As more paths are mapped from low skills to high skills those nodes can be stored and added to varying heuristics making this a far more efficient model for skill conversions. AI is a powerful tool and its benefits are innumerable; however, it is important to consider that it is still growing with each deployment. Future research should be conducted to explore the skills themselves and their soft or sub-skills related to them. This could help with streamlining the algorithm. Additionally, it may be useful for the next generation of developers to begin to track skills they find highly valuable for future searches on the agent as well as providing updated high value skills entering the marketplace.

References

1. Alawad H (2022). Artificial intelligence applications to enhance risk management of railway station operations
2. Aswin A, Ariati C, Kurniawan S (2022) Artificial intelligence in higher education: a practical approach. J High Educ Policy Manage 1–4. https://doi.org/10.1080/1360080x.2022.2156088
3. Huang M-H, Rust RT (2018) Artificial intelligence in service. J Serv Res 21(2):155–172. https://doi.org/10.1177/1094670517752459
4. Rodrigues AL et al (2021) Technological skills in higher education—different needs and different uses. Educ Sci 11(7):326; Intelligent robots take on demanding industry challenges (2019). ECN
5. Russell P (2021) Artificial intelligence: a modern approach, global edition. Pearson Education Limited

6. Sarma D, Das A, Bera UK (2020) An optimal redistribution plan considering aftermath disruption in disaster management. Soft Comput (Berlin, Germany) 24(1):65–82. https://doi.org/10.1007/s00500-019-04287-7
7. Vorobeva D, El Fassi Y, Costa Pinto D, Hildebrand D, Herter MM, Mattila AS (2022) Thinking skills don't protect service workers from replacement by artificial intelligence. J Serv Res JSR 25(4):601–613. https://doi.org/10.1177/10946705221104312

The manufacturer's authorised representative in the EU is Springer Nature Customer Service Centre GmbH, Europaplatz 3, 69115 Heidelberg, Germany. If you have any concerns regarding our products, please contact ProductSafety@springernature.com

Printed and bound by CPI Group (UK) Ltd, Croydon, CR0 4YY
30/09/2026
02225573-0009